**Co- and Post-Translational Modifications of Therapeutic Antibodies and Proteins**

# Co- and Post-Translational Modifications of Therapeutic Antibodies and Proteins

*T. Shantha Raju M.Sc., Ph.D.*

This edition first published 2019


*Registered Office*
John Wiley & Sons, Inc., 111 River Street, Hoboken, NJ 07030, USA

*Editorial Office*
111 River Street, Hoboken, NJ 07030, USA

For details of our global editorial offices, customer services, and more information about Wiley products visit us at www.wiley.com.

Wiley also publishes its books in a variety of electronic formats and by print-on-demand. Some content that appears in standard print versions of this book may not be available in other formats.



*Library of Congress Cataloging-in-Publication Data applied for*

ISBN: 9781119053316

Cover design by Wiley
Cover image: Polygonal space © tj-rabbit/Shutterstock. All other images courtesy of T. Shantha Raju

Set in 10/12pt WarnockPro by SPi Global, Chennai, India

Printed in the United States of America

V10008327_022619

*Dedicated to my*

*Dad (late), Mom, Wife and Son*

# Contents

# Preface

Biotechnology industry is in the forefront of delivering life-saving biopharmaceutical drugs to patients to treat and also often to cure many human diseases. In this context, the biotechnology industry has developed and continue to develop many mega block buster drugs to treat both chronic and life-threatening human diseases. The list of biological drugs biotechnology industry has developed so far are too many and that list now includes proteins, peptides, peptidomimetics, monoclonal antibodies, polyclonal antibodies, antibody fragments, antibody drug conjugates, bispecific antibodies, multispecific antibodies, oligonucleotides, gene therapy, cell therapy, vaccines including oncolytic viral vaccines, etc. Several of these biotherapeutics are for disease maintenance, but some are also to cure many different and life-threatening human diseases. Currently, the biotechnology industry is also focusing on developing strategies for disease prevention and precision medicine. Further, the industry is learning from translational science approaches to identify biomarkers to target and cure human diseases with targeted approach for a very high precision treatment.

All in all, the biotechnology industry has been involved in researching and developing many breakthrough therapies to treat numerous human diseases. The list of human diseases the biotechnology industry is currently focusing includes life-threatening diseases such as cancer and AIDs along with some rare diseases, chronic diseases, metabolic diseases, autoimmune diseases, etc. Although biotherapeutics are very expensive, the biosimilar industry is making significant progress to reduce the cost of biotherapeutics. However, the cost of innovative biotherapeutics as well as biosimilar biological drugs are still very expensive compared to the generic pharmaceutical drugs that are mostly small chemical entities or molecules. Reasons for the high price of biotherapeutic drugs are due to the complexity of their manufacturing processes, purification, analytical characterization, formulation, transport, storage, etc. Accordingly, the cost of goods and services of biotherapeutics is very high compared to the generic pharmaceutical drugs. In addition, biotherapeutics as such are very complex and highly heterogeneous molecules compared to the generic drugs that are mainly small molecule drugs. For example, compared to aspirin that is a small organic molecule drug, recombinantly produced human insulin (rh-insulin) is a much more complex protein molecule. Further, compared to rh-insulin, recombinantly produced human erythropoietin (rh-EPO) is a much more complex and a highly heterogeneous molecule. In addition, monoclonal antibodies are larger in molecular size and are highly complex molecules than erythropoietin.

Accordingly, as the molecular weight of biomolecules increases, their complexity may increase. With the increase in their molecular weight and complexity, the heterogeneity of biomolecules also increases. As the complexity and heterogeneity increases, the analytical characterization of such biomolecules becomes much more difficult. In addition, biotherapeutics may also undergo many different modifications during production, purification, formulation, transportation, storage, etc. For example, biotherapeutics such as proteins undergo co- and post-translational modifications. Such modifications may affect the drugs' safety and efficacy. For example, rh-EPO is a glycoprotein that is very rich in co- and post-translationally modified glycans. The glycans present in rh-EPO are highly complex and highly heterogeneous. Because of the complexity and heterogeneity of glycans, production, purification, formulation, and analytical characterization of rh-EPO is very tedious and difficult which involves the use of many expensive and often state-of-the art instruments. Hence, cost of goods and services (COGS) of producing rh-EPO is very high compared to small molecule-based pharmaceuticals. Additionally, biopharmaceutical drugs may require complex purification schemes to remove impurities from the host cells such as host cell proteins, host cell DNA, etc. In addition, the biopharmaceuticals may be purified to remove residual impurities such as insulin, antibiotics, etc. that are used during their production. This is because such impurities may also impact the drug safety and efficacy. For the biotechnology industry and for the regulators, patient's safety is very important. In addition, the efficacy of the drug is also very important to make sure that the patients get the right drug to treat the right disease/s. Otherwise, it is a waste of time and resources for both the industry as well as to the patients. Many of the biopharmaceutical drugs are life-saving drugs and if the industry does not develop efficacious drugs, patients may lose their life. In certain diseases, patients are not only waiting for the right drug to treat their disease, but they may also be dying because of the lack of efficacious drugs to treat the disease/s.

In this context, it is very important for both the biotechnology industry and the regulators that approve the drugs to understand the structure and function of the biopharmaceutical drugs. The structure–function studies (SFS) of biopharmaceuticals are similar to structure–activity relationship (SAR) studies of small molecule pharmaceutical drugs. The SFS may help to understand the mechanism of action (MOA), mechanism of toxicity (MOT), immunogenicity, pharmacokinetics, pharmacodynamics, etc., of biopharmaceutical drugs. The co- and post-translational modifications (CTMs and PTMs) may significantly impact the SFS of biopharmaceuticals. Often, the CTMs and/or PTMs may become the product-related impurities that may impact MOA, MOT, etc. of therapeutics. Hence, it is also very important to understand the SFS of product-related impurities derived from CTMs and/or PTMs. The product-related impurities derived from CTMs and/or PTMs often impact the biological functions of the biomolecules and hence may become critical quality attributes (CQAs).

Understanding the SFS of biopharmaceuticals including their structural variants involves a detailed study of their primary, secondary, tertiary, quaternary, etc. structure along with their biological functions. Such studies often involve isolating the individual structural variants, characterizing them and comparing their biological functions with the parent molecule. Hence, structure–function studies are very tedious and often involve the use of many complex analytical methods, strategies, and state-of-the-art analytical instruments. Further, SFS should be designed to elucidate the MOA, MOT, immunogenicity, and pharmacokinetic properties of the biopharmaceutical drugs. All these work needs to be carried out by highly skilled scientists.

Vast majority of the biopharmaceutical drugs that are currently marketed and in the clinic are proteins and antibody-based products. These proteins and antibody-based biopharmaceuticals are currently generating more than $100 billion in product revenue, annually. Majority of these drugs are produced using recombinant DNA technology. Some of them are being produced using bacteria such as *Escherichia coli* (*E. coli*), but many of them are being produced using mammalian cells such as Chinese hamster ovary (CHO) cells. The proteins and antibodies produced using recombinant DNA technology also undergo CTMs and PTMs similar to their endogenous counterparts. Hence, understanding the SFS of these recombinantly produced protein and antibody-based biopharmaceutical drugs along with their structural variants is very important to elucidate the MOA, MOT, immunogenicity, pharmacokinetic, and pharmacodynamic properties.

In order to design and understand the SFS of biopharmaceutical drugs and their structural variants, it is very important to have a basic knowledge of the CTMs and PTMs. Hence, this book is written in such a way that the reader will be able to understand the basics of CTMs and PTMs of therapeutic antibodies and proteins and extrapolate such an understanding to other biopharmaceutical drugs. Although it is very difficult to address each and every CTMs and PTMs in one book, attempts were made to address the main aspects of CTMs and PTMs that are relevant to understand the SFS which helps to elucidate MOA, MOT, immunogenicity, pharmacokinetic, and pharmacodynamics properties of antibody- and protein-based biotherapeutics. Individual chapters are dedicated to individual CTMs and/or PTMs in which attempts were made to describe each modification/s in terms of chemistry, biochemistry, biology, and analytical strategy so that the readers not only understand the modification, they may be able to design experiments to elucidate the SFS of their molecules. Upon reading this book, it is my sincere hope that the reader would understand the impact of CTMs and/or PTMs on the SFS of biopharmaceutical drugs. I highly appreciate your time and efforts to read and understand this book. I would also appreciate any constructive suggestions/feedback to improve this book during the possible future editions.

September 28, 2018

*T. Shantha Raju*
West Chester, PA 19382, USA

# About the Author

T. Shantha Raju (Raju) obtained his B.Sc. (in Physics, Chemistry, and Mathematics), M.Sc. (in Organic Chemistry) and Ph.D. (in Bio-Organic Chemistry) degrees from the University of Mysore, Mysore, India. Raju did his postdoctoral studies at York University, Toronto, Canada (on lipopolysaccharides from Helicobacter Pylori), Georgetown University Medical Centre, Washington DC, USA (on mucin glycoproteins and snake venom glycoproteins), and Albert Einstein College of Medicine, Bronx, NY, USA (on glycosylation mutants from CHO cells). Raju has more than 20 years of experience working in the biotechnology industry with Genentech, FibroGen, ProZyme, Carboworld, Johnson & Johnson (Centocor and Janssen Pharmaceuticals), Medimmune, Celgene, and Venn Therapeutics at various increasing capacities ranging from Scientist to Consultant to Senior Director to Head of R&D. He is an expert in Glycobiology and protein structure–function studies to understand the mechanism of action, mechanism of toxicity, immunogenicity, and pharmacokinetics of biotherapeutics. Raju has contributed to the research and development of many biopharmaceuticals including Rituxan, Herceptin, Avastin, Xolair, Raptiva, Remicade, Stelara, Simponi, Darzalex, etc. Raju has published more than 40 research papers, review articles, and book chapters. He has served/serving in the editorial board of many journals including *Analytical Biochemistry, BioProcess International, Drug Discovery and Development, Current Biotechnology, Virology and Immunology Journal*, etc.

# Abbreviations

| | |
|---|---|
| 2-AA | 2-aminobenzoic acid |
| AA | amino acid |
| ADCC | antibody-dependent cellular cytotoxicity |
| ADCP | antibody-dependent cellular phagocytosis |
| ADCR | antibody-dependent cytokine release |
| AdoMet | *S*-adenosyl methionine |
| AGEs | advanced glycation end products |
| Ala | alanine |
| Amu | atomic mass unit |
| APF | animal protein free |
| APFS | accessorized single-dose prefilled syringe |
| API | active pharmaceutical ingredient |
| AQL | acceptable quality limit |
| Arg | arginine |
| Asn | asparagine |
| Asp | aspartic acid |
| AT | antithrombin |
| ATP | adenosine triphosphate |
| AUC | analytical ultracentrifugation |
| BLOQ | below the limit of quantitation |
| CAS | chemical abstract service |
| CCI | container closure integrity |
| CDC | complement dependent cytotoxicity |
| CDG | congenital disorder of glycosylation |
| CDLC | chemical defined lipid concentrates |
| CDR | complementarity-determining region |
| CE | capillary electrophoresis |
| CEX | cation exchange |
| CHO | Chinese hamster ovary |
| CID-MS | collision induced dissociation-mass spectrometry |
| cIEF | capillary iso-electric focusing |

| | |
|---|---|
| CFPS | cell free protein synthesis |
| CMO | contract manufacturing organization |
| CoA | certificate of analysis |
| CPB | carboxypeptidase B |
| CPP | critical process parameter |
| CQA | critical quality attribute |
| CQAs | critical quality attributes |
| CSDs | control strategy documents |
| CTD | common technical document |
| CTM | co-translational modification |
| CV | coefficient of variance |
| Cys | cysteine |
| CZE | capillary zone electrophoresis |
| Da | daltons |
| DAM | diacetyl morphine |
| DNA | deoxyribonucleic acid |
| DO | dissolved oxygen |
| DOPA | 3,4-dihydroxy phenylalanine |
| DS | drug substance |
| DSC | differential scanning calorimetry |
| DTNB | 5,5′-dithiobis-(2-nitrobenzoic acid) |
| EGF | epidermal growth factor |
| ELISA | enzyme linked immunosorbent assay |
| ENDO F | endo-glycosidase F |
| ENDO H | endo-glycosidase H |
| EOPCB | end-of-production cell banks |
| EPO | epogen |
| ER | endoplasmic reticulum |
| FB | final bulk |
| FBS | fetal bovine serum |
| FC | fluorocarbon-coated |
| FcγR | Fcγ receptor |
| FMEA | failure mode effects analysis |
| fMet | formylmethionine |
| FTIR | Fourier transform infrared spectroscopy |
| Fuc | fucose |
| FucT | fucosyltransferase |
| FUV CD | far UV circular dichroism |
| Gal | Galactose |
| GalNAc | *N*-acetylgalactosamine |
| GalNH2 | galactosamine |
| GC | gas chromatography |
| GC–MS | gas chromatography–mass spectrometry |

| | |
|---|---|
| GCP | good clinical practices |
| Gla | glutamic acid |
| Glc | glucose |
| GlcNAc | *N*-acetylglucosamine |
| GlcNH2 | glucosamine |
| GLP | good laboratory practices |
| Glu | glutamine |
| Gly | glycine |
| GM | geometric mean |
| GMP | good manufacturing practices |
| GnT | *N*-acetylglucosaminyltransferase |
| GPI | glycosylphosphatidylinositol |
| GSD | geometric standard deviation |
| GSH | glutathione peroxidase |
| GT | glycosyltransferase |
| HATs | histone acetyltransferases |
| HC | heavy chain |
| HCP | host cell protein |
| HDACs | histone deacetylases |
| HETP | height equivalent to a theoretical plate |
| HIC | hydrophobic interaction chromatography |
| HILIC | hydrophilic interaction liquid chromatography |
| His | histidine |
| HMW | higher-molecular weight |
| HPLC | high performance liquid chromatography |
| HPSEC | high performance size exclusion chromatography |
| HSA | human serum albumin |
| ICMT | isoprenylcysteine carboxyl methyltransferase |
| IdeS | IgG degrading enzyme from *Streptococcus pyogenes* |
| IEC | ion-exchange chromatography |
| IEF | iso-electric focusing |
| Ig | immunoglobulins |
| IgA | immunoglobulin A |
| IgD | immunoglobulin D |
| IgE | immunoglobulin E |
| IgG | immunoglobulin G |
| IgM | immunoglobulin M |
| INN | international non-proprietary name |
| IPCs | in-process controls |
| Iso | isoleucine |
| IVIG | intravenous immunoglobulin |
| JP | Japanese Pharmacopoeia |
| KATs | lysine acetyltransferases |

| | |
|---|---|
| $K_D$ | equilibrium dissociation constant |
| kDa | kilodalton |
| KDACs | lysine deacetylases |
| KS | keratan sulfate |
| LAL | limulus amoebocyte lysate |
| LARGE | like acetylglucosaminyltransferase |
| LC | liquid chromatography |
| LC/MS | liquid chromatography/mass spectrometry |
| LC–MS | liquid chromatography–mass spectrometry |
| Leu | leucine |
| LIMS | laboratory information management system |
| LIVCA | limit of in vitro cell age |
| LLOQ | lower limit of quantitation |
| Ln | natural log |
| LOQ | limit of quantitation |
| LoQ | list of questions |
| LRVs | $\log_{10}$ reduction value |
| Lys | lysine |
| mAb | monoclonal antibody/recombinant monoclonal antibody |
| MALD-TOF-MS | matrix assisted laser desorption time-of-flight mass spectrometry |
| Man | mannose |
| ManAc | *N*-acetylmannosamine |
| ManNH2 | mannosamine |
| MC | microbial controls |
| MCB | master cell bank |
| MeC | 5-methylcytosine |
| Met | methionine |
| MetO | methionine sulfoxide |
| MMSRs | microbial monitoring strategy requirements |
| MOA | mechanism of actions/mode of action |
| MOT | mechanism of toxicity |
| MS | mass spectrometry |
| MTX | methotrexate |
| MVM | minute virus of mice |
| N | mumber of observation |
| NA | not applicable |
| NANA | *N*-acetylneuraminic acid |
| NCPPs | non-critical process parameters |
| ND | not detected |
| NDST | *N*-deacetylase/*N*-sulfotransferase |
| NeuAc | *N*-acetylneuraminic acid |
| NeuGc | *N*-glycolylneuraminic acid |

| | |
|---|---|
| NFAT | nuclear factor of activated T-cells |
| NGNA | *N*-glycolylneuraminic acid |
| NK | natural killer |
| NKPPs | non-key process parameters |
| NP | not performed |
| NUV CD | near UV circular dichroism |
| NWP | normalized water permeability |
| O | occurrence |
| PAPS | 3′-phosphoadenosine-5′phosphosulfate |
| Pas | performance attributes |
| PC | paper chromatography |
| PD | pharmacodynamics |
| PDE | permissible daily exposure |
| PDL | population doubling level |
| PFS-SA | pre-filled syringe sub-assembly |
| Ph.Eur. | European Pharmacopoeia |
| Phe | phenylalanine |
| pI | isoelectric point |
| PIMT | protein isoaspartyl *O*-methyltransferase |
| PK | pharmacokinetics |
| PNGase A | peptide-*N*-glycosidase A |
| PNGase F | peptide-*N*-glycosidase F |
| PP2A | protein phosphatase 2A |
| PPQ | process performance qualification |
| PPME | protein phosphatase methylesterase |
| PRMTs | protein arginine methyltransferases |
| Pro | proline |
| PRS | primary reference standard |
| PSGL-1 | P-selectin glycoprotein ligand-1 |
| PTM | post-translational modification |
| PTMs | post-translational modifications |
| Pup | prokaryotic ubiquitin like protein |
| PV | process validation |
| QA | quality assurance |
| QC | quality control |
| QP | qualified person |
| Q-TOF | quadrupole time-of-flight |
| RAGE | receptor of advanced glycation end product |
| rEPO | recombinant epogen |
| RER | rough endoplasmic reticulum |
| rhDNase | recombinant human deoxyribonuclease |
| RLPs | retrovirus-like particles |
| RNA | ribonucleic acid |

| | |
|---|---|
| RNS | rigid needle shield |
| RP | reversed phase |
| RPC | reversed phase chromatography |
| rPA | recombinant protective antigen |
| RP-HPLC | reversed-phase high performance liquid chromatography |
| RP-HPLC/MS | reversed-phase high performance liquid chromatography with mass spectrometry |
| RPN | risk prioritization number |
| RT | retention time |
| RTPs | recombinant therapeutic proteins |
| rtPA | recombinant tissue plasminogen activator |
| S | severity |
| SAM | *S*-adenosyl methionine |
| SAR | structure–activity relationship |
| SD | standard deviation |
| SEC | size-exclusion chromatography |
| Ser | serine |
| SFS | structure–function studies |
| Sia | sialic acid |
| SOD1 | superoxide dismutase 1 |
| SPC | signal peptidase complex |
| SPR | surface plasmon resonance |
| SRP | signal recognition particle |
| SR | SRP receptor |
| SR-α | SRP receptor subunit-α |
| SR-β | SRP receptor subunit-β |
| ST | sialyltransferase |
| SulT | sulfotransferase |
| TEM | transmission electron microscopy |
| Thr | threonine |
| TLC | thin layer chromatography |
| tPA | tissue plasminogen activator |
| TPST | tyrosylprotein sulfotransferase |
| Try | tryptophan |
| TSE | transmissible spongiform encephalopathy |
| TSPs | thrombospondins |
| TTC | threshold of toxicological concern |
| Tyr | tyrosine |
| UDS | unformulated drug substance |
| UF/DF | ultrafiltration/diafiltration |
| UPB | unprocessed bulk |
| UPLC | ultra-performance liquid chromatography |
| USP | United States pharmacopeia |

| | |
|---|---|
| UV | ultraviolet |
| Val | valine |
| VH | heavy variable region |
| VL | light variable region |
| VMP | Validation Master Plan |
| WCB | Working Cell Bank |
| WRS | working reference standard |

# 1

# Introduction to Co- and Post-translational Modifications of Proteins

Proteins are mostly biologically derived macromolecules containing amino acid residues as their constituent monomers (Marrack and Hoch 1949; Zamecnik 1950). Amino acids are small organic molecules containing amine and carboxylic acid as their functional groups (Sakami and Harrington 1963). The amine group of one amino acid can interact with the carboxylic acid group of another amino acid or another molecule of the same amino acid to form a stable covalent bond (Hertweck 2011). Such covalent bonds between amine groups and carboxylic acid groups of amino acids are often referred to as amide bonds. These amide bonds are also called as peptide bonds. Peptides are mostly linear chain molecules containing two or more amino acid residues that are linked by peptide bonds (Work 1961). The peptide bonds can form extended polypeptide chains resulting in very high molecular weight peptides. Such very high molecular weight peptides are also referred to as proteins. Proteins, unlike most peptides, mostly contain highly ordered structures and are most commonly synthesized by biological processes within a given biological system or organism. However, recently cell-free in vitro synthesis systems are also established using enzymes and/or chemicals to produce proteins (Perez et al. 2016). Such systems are usually referred to as cell-free protein synthesis (CFPS) systems (Semenchenko et al. 2016; Jia et al. 2017).

Biosynthesis of proteins involves the ribosomes in which mRNA translate the nascent polypeptide chains. During the biosynthesis of polypeptide chains amino acids, available within a given cell and/or organism, are used to construct amide/peptide bonds. There are atleast 20 different amino acid residues that exist in nature and several more of their derivatives are also present in living organisms. The chemical structure of 20 different amino acids are shown in Figure 1.1. Physicochemical properties along with the three and one letter abbreviations of these amino acids and some of their derivatives are shown in Table 1.1. In living organisms, about 20 different amino acids are directly translated by RNA that are represented by the triple letter codons in the genetic code during the biosynthesis of the nascent polypeptide chains

*Co- and Post-Translational Modifications of Therapeutic Antibodies and Proteins*, First Edition. T. Shantha Raju.

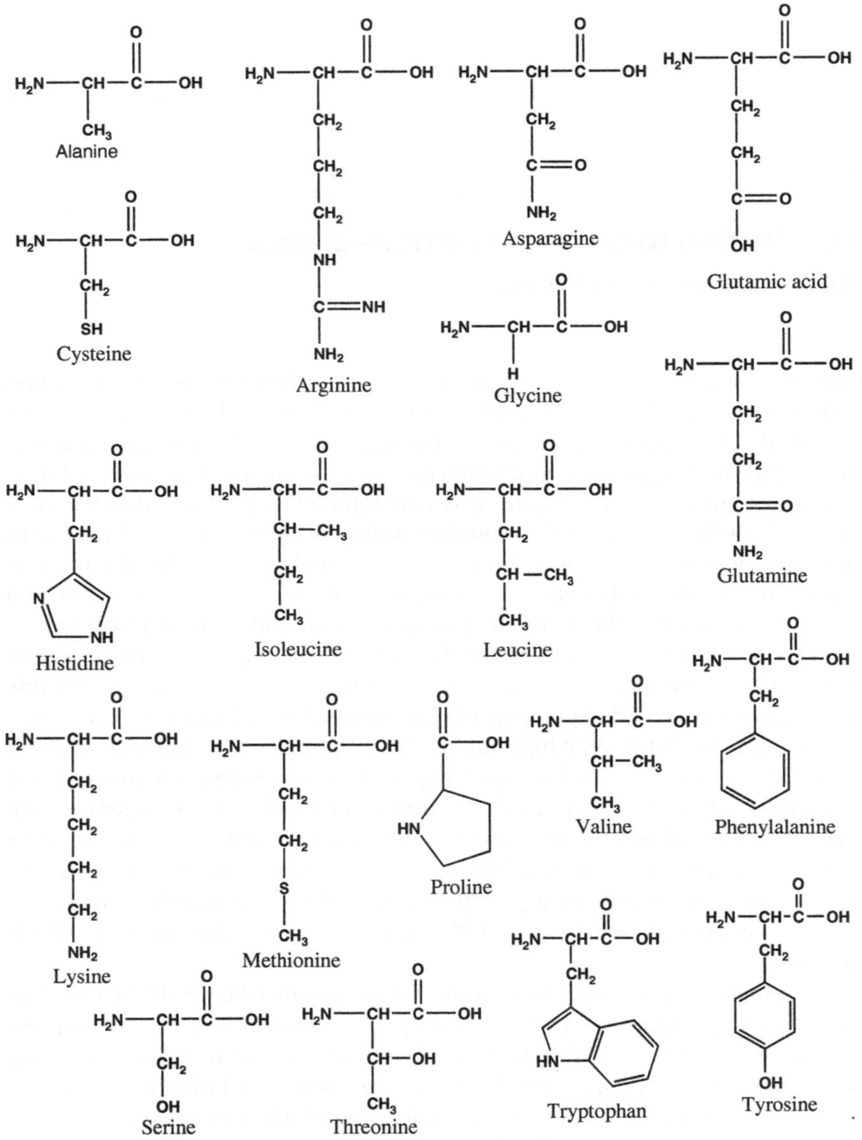

Alanine

Cysteine

Arginine

Asparagine

Glycine

Glutamic acid

Glutamine

Histidine

Isoleucine

Leucine

Lysine

Methionine

Proline

Valine

Phenylalanine

Serine

Threonine

Tryptophan

Tyrosine

**Figure 1.1** Chemical structure of 20 different amino acid residues commonly found in human proteins. (*See insert for color representation of this figure.*)

**Table 1.1** Amino acid residues and their functionality.

| Names (in alphabetic order) | Three letter abbrevia-tions | One letter abbrevia-tions | Nature of functionality | Functional group | Comments |
|---|---|---|---|---|---|
| Alanine | Ala | A | Neutral | Aliphatic | Nonessential |
| Arginine | Arg | R | Basic | Amine | Semiessential |
| Asparagine | Asn | N | Polar | Amide | Nonessential |
| Aspartic acid | Asp | D | Acidic | Carboxylic acid | Nonessential |
| Aspartate or asparagine | Asx | B | Acidic or polar | Carboxylic acid or amide | Nonessential |
| Cysteine | Cys | C | Nucleophilic | Thiol | Nonessential |
| Glutamate | Glu | E | Polar | Amide | Nonessential |
| Glutamine | Gln | Q | Polar | Amide | Nonessential |
| Glutamine or glutamate | Glx | Z | Polar | Amide | Nonessential |
| Glycine | Gly | G | Neutral | Proteogenic | Nonessential |
| Histidine | His | H | Basic | Reactive heterocyclic ring | Nonessential |
| Isoleucine | Ile | I | Hydrophobic | Aliphatic | Essential |
| Leucine | Leu | L | Hydrophobic | Aliphatic | Essential |
| Lysine | Lys | K | Basic | Amine | Essential |
| Methionine | Met | M | Hydrophobic | Aliphatic | Essential |
| Phenylalanine | Phe | F | Hydrophobic | Aromatic | Essential |
| Proline | Pro | P | Hydrophobic | Ring structure | Nonessential |
| Pyrrolysine | Pyl | O | Basic | Pyrroline | Nonessential |
| Selenocysteine | Sec | U | Neutral | Selenol | Nonessential |
| Serine | Ser | S | Polar | Hydroxyl | Nonessential |
| Threonine | Thr | T | Polar | Hydroxyl | Essential |
| Tryptophan | Trp | W | Hydrophobic | Aromatic | Essential |
| Tyrosine | Tyr | Y | Hydrophobic | Aromatic | Nonessential |
| Valine | Val | V | Hydrophobic | Aliphatic | Essential |

(Wu et al. 2013). Upon biosynthesis, the polypeptide chain undergoes different processes to form the active protein, which ultimately might become structural proteins, functional proteins, enzymes, receptors, transporters, translocators, proteases, binding proteins, antibodies, cytokines, etc.

The monomeric components of proteins, i.e. amino acids, are zwitterionic organic molecules at neutral pH (Palacín 1994; Avila-Chávez et al. 1997). The zwitterionic property of amino acids is derived from amine and carboxylic acid functionalities present in them (Bordes and Holmberg 2015). The amine group act as a basic or cationic functionality, whereas carboxylic acid group act as an acidic or anionic functionality. Some amino acids may also contain an additional carboxylic acid group, and such amino acids are referred to as acidic amino acids. Examples of acidic amino acids are aspartic acid (Asp, D) and glutamic acid (Glu, E). Similarly, some amino acids may contain an additional amine group, and such amino acids are often referred as basic amino acids. Examples of basic amino acids are arginine (Arg, R), histidine (His, H), and lysine (Lys, K). Apart from the acidic and basic amino acids, the amino acids that contain only one carboxylic acid group and one amine group are called as neutral amino acids. Some examples of neutral amino acids are alanine (Ala, A), valine (Val, V), and leucine (Leu, L).

The polypeptide chain of peptides and proteins usually contains two terminal ends. One end of the polypeptide chain contains an amino acid in which the amine functional group is free and hence is called as amino termini or N-termini. The other end of the polypeptide chain contains an amino acid with carboxylic acid functional group that is free and hence, is called as carboxyl termini or C-termini. As discussed earlier, in the protein polypeptide chain, the carboxylic acid functional groups of amino acids are linked to amine functional groups of other amino acids to form amide bonds, i.e. polypeptide bonds. Hence, in the polypeptide chain of proteins both the amine and carboxyl functional groups of neutral amino acid residues are involved in the peptide bond formation unless they are in the either C- or N-termini. However, in the acidic amino acid residues the additional carboxylic acid groups are not involved in the peptide bond formation. Similarly, in the basic amino acid residues, the additional amine functional groups are not involved in the peptide bond formation. In addition, in the C-termini, the carboxyl functional group of terminal amino acid residues in the N-termini and the amine functional group of terminal amino acid residues is not involved in the peptide bond formation. These functional groups are also reactive, and hence they may undergo additional modifications.

Some modifications of the available reactive functional groups of amino acids of proteins may occur during the translation of proteins. Such modifications of amino acids during translation of proteins are termed as co-translational modifications (CTMs). In addition, some modifications of functional groups of amino acids may also occur on the nascent polypeptide chains of proteins following the translation. Those modifications that occur following translation of proteins are collectively referred as post-translational modifications (PTMs). Further, PTMs can also occur at the N- and C-termini of proteins and peptides. In the N-termini, the primary amine ($NH_2-$) group can undergo modifications and, in the C-termini, the carboxyl (–COOH) group can undergo modifications (Voet and Pratt Donald Voet 2006).

In addition to acidic, basic, and neutral amino acids, some amino acids may contain additional functional groups such as hydroxyl group (–OH), sulfhydryl group (–SH), aliphatic chain, aromatic ring, etc. For example, Ser and Thr contain a hydroxyl group, Cys contain a thiol functional group, Trp contains an aromatic ring, etc. Such functional groups are not involved in the peptide bond formation but may be involved in the hydrogen bond formation, disulfide bond formation, hydrophobic interactions, etc. These functional groups may also be involved in the formation of the higher-order structure of proteins (Wu and Fuxreiter 2016). Further, many of these functional groups may also undergo additional modifications such as glycosylation, phosphorylation, sulfation, oxidation, etc. Such modifications of these additional functional groups are also collectively referred as CTMs or PTMs. Such modifications may depend on the surface accessibility of these functional groups, participation of groups surrounding them, neighboring group participation, solvent exposure, influence of metal ions, etc.

Hence, proteins may possess many different functional groups that may undergo CTMs or PTMs during and after translation by mRNA. Many of the common CTMs or PTMs of proteins are shown in Table 1.2 in the alphabetical order with no particular emphasis on the modifications. Many CTMs and/or PTMs of proteins may be due to simple chemical or enzymatic reactions. Also, many CTMs and/or PTMs of proteins are due to highly complex chemical and/or enzymatic reactions.

The CTMs or PTMs of proteins may be very simple modifications, or they may be highly complex modifications (Walsh 2010). Regardless of their simplicity or complexity, the CTMs or PTMs may affect the physicochemical, biochemical, biological, pharmacokinetic, immunological, toxicological, etc., properties of proteins (Kessler and Edelmann 2011). Hence, understanding the modifications of proteins is very important not only to understand their structure and functions (SFS) but also to identify, define, and understand their influence on mechanism of actions (MOAs), mechanism of toxicity (MOT), immunogenicity, pharmacokinetic properties, pharmacodynamic properties, etc. (Kurepa and Smalle 2008; Tanaka 2009). Many endogenous proteins have been purified to their maximum homogeneity to study either their structure or functions or a combination of both structure and functions (Dolinnaya et al. 2016; Ishchenko et al. 2017; Kumar et al. 2017; Nishimura 2017). Such studies are also important to identify the proteins to be used as human therapeutics (Galeotti et al. 2017; Kinoshita and Saeki 2017). Some of the examples of endogenous proteins currently used as therapeutics are intravenous immunoglobulin (IVIG), thrombin, Factor VII, Factor VIII, etc. (Parayath and Amiji 2017). Since purifying the endogenous proteins is a very tedious process and often their overall yield is very low, recombinant

**Table 1.2** Common CTMs or PTMs of proteins.

| PTMs | Modified functional groups | Amino acid residues involved in the modification | Comments |
|---|---|---|---|
| Acetylation | –OH, –$NH_2$ | Ser, Thr, Lys, Met, N-terminal AA, etc. | Very common CTM or PTM |
| C-terminal lysine or arginine clipping | | Lys, Arg, N-terminal AA | Proteolysis by carboxypeptidases |
| Cysteinylation | –SH | Cys | Rare CTM or PTM |
| Deamidation | Amide | Asn | Common in RTPs |
| Glycation | –$NH_2$ | Lys | Common PTM |
| Glycosylation | Amide, –OH | Asn, Ser, Thr | Very common CTM or PTM |
| N-glycosylation | Amide | Asn | Very common PTM |
| O-glycosylation | –OH | Ser, Thr | Very common CTM or PTM |
| Hydroxylation | –CH | Pro, Lys, Phe, Trp | Commonly found in Collagen, Gelatin, etc. |
| Methylation | –$NH_2$, –OH | Arg, Lys | |
| Oxidation | –SR, –OH | Met, Cys, Trp, etc. | Common PTM |
| Phosphorylation | –OH | Ser, Thr | Dynamic PTM |
| Prenylation | –OH, –$NH_2$ | Ser, Thr, Lys | Rare PTM |
| Proteolysis | Peptide bonds | All amino acids | Caused by proteases |
| Selenylation | –SH | Met, Cys | Rare PTM |
| Signal peptides | –$NH_2$ | Met in N-termini | Common PTM |
| Sulfation | –OH | Tyr, glycans | Common PTM |
| SUMOylation | –$NH_2$ | Lys | Rare PTM |
| Ubiquitination | N-termini | Lys, Cys, Ser, Thr | Common PTM |

DNA technology was developed to produce therapeutic proteins in large quantities (Good 1972). Recently, advanced technologies have been used to improve the yield of recombinant therapeutic proteins (RTPs) (Fischer et al. 2017). A growing list of RTPs now includes proteins, glycoproteins, peptides, monoclonal antibodies, polyclonal antibodies, antibody mixtures, antibody fusion proteins, antibody fragments, antibody drug conjugates, bispecific antibodies, multispecific antibodies, pegylated proteins, conjugated vaccines, etc. (Patil and Walther 2017).

Almost all the endogenous proteins isolated from human, animal, microbial, etc., sources for various purposes including the study of structure and functions contains CTMs and/or PTMs. Similarly, proteins isolated from endogenous sources to use as human therapeutics also contains CTMs and/or PTMs. Since many of the endogenous proteins are purified from a large pool of human serum, the CTMs and/or PTMs of such proteins are often highly complex and very heterogeneous. Such complexity and heterogeneity may affect the protein functions including bioactivity, pharmacokinetics, immunogenicity, toxicity, etc. Hence, for endogenous protein therapeutics it is very important to characterize the physicochemical and biochemical nature of CTMs or PTMs to understand their MOA, MOT, immunogenicity, etc. It is also necessary to set proper control strategies and/or specifications to monitor the lot-to-lot variations in CTMs and/or PTMs of endogenously derived protein therapeutics. This is because many of the CTMs or PTMs might be critical quality attributes (CQAs) as they may impact the product functions, safety, and efficacy.

Similar to endogenous protein therapeutics, RTPs also contain CTMs and/or PTMs. RTPs are mostly produced using in vitro cell culture systems or transgenic animals or plants. More recently, in vitro cell free system has been developed to produce recombinant proteins. Regardless of the technology used to produce RTPs, almost all the RTPs produced so far contain PTMs. The structure and the nature of PTMs of RTPs may vary with cell line, cell culture conditions, animal species used in transgenic animal technology, plant species used in transgenic plant technology, etc. Hence, CTMs or PTMs in RTPs may vary from lot-to-lot which might impact the product quality, safety, and efficacy (Jefferis 2016). Accordingly, it is necessary to characterize the CTMs or PTMs of RTPs and set proper control strategies and/or specifications for lot release of such proteins to establish their safety and efficacy.

Although proteins are susceptible for many CTM and/or PTM modifications but not every protein may contain every one of the PTMs listed in Table 1.2. Also, the list shown in Table 1.2 is a snap shot of a large number of CTMs or PTMs that modify proteins under different conditions, and it is nearly impossible to address each and every one of those CTMs or PTMs in one book. Hence, this book will address some of the very important CTMs or PTMs that are highly relevant to product quality, safety, and efficacy of protein therapeutics including the RTPs and monoclonal antibodies (mAbs). These important CTMs and/or PTMs are briefly described below in alphabetical order. These important modifications are discussed in detail in the individual chapters in this book in alphabetical order from Chapters 2 to 20. In addition, some modifications that are not covered in Chapters 2–20 are briefly discussed in Chapter 21.

## Brief Introductions to Individual Chapters

### Chapter 2: Acetylation of Proteins

Acetylation is a chemical or enzymatic acylation reaction in which hydroxyl (–OH) and/or amine (–$NH_2$) functional groups are derivatized by acetyl ($CH_3CO$–) groups and is an organic condensation reaction. Protein acetylation is mostly an enzymatic reaction and is catalyzed by acetyltransferases. Acetylation of proteins is both a CTM and a PTM, and the protein deacetylation is also an enzymatic process. Acetylation and deacetylation is a dynamic process which may impact different biological functions of proteins. This chapter discusses the mechanism of acetylation and deacetylation, biological impact of acetylation and deacetylation, and methods to detect and quantitate acetylation of proteins.

### Chapter 3: C-Terminal Lys or Arg Clipping of Proteins

Many cell-surface proteins and secreted proteins often contain C-terminal heterogeneity due to the presence or the absence of C-terminal Lys or Arg residues. This type of C-Terminal heterogeneity is very commonly observed in mAbs. Carboxypeptidases easily cleaves C-terminal Lys or Arg residues. The specificity of carboxypeptidases is often pH-dependent and the enzymatic reaction is mostly incomplete. Hence, many proteins have C-terminal heterogeneity. This chapter provides an overview of C-terminal Lys and Arg clipping and its impact on the biological properties of proteins.

### Chapter 4: Cysteinylation of Proteins

Cys residues present in proteins normally forms intra- and/or interchain disulfide bonds. Such Cys residues are called as paired Cys residues. However, some proteins contain free Cys residues that are often referred as unpaired Cys residues. The unpaired Cys residues often form disulfide bonds with free Cys molecules. Such a process is called cysteinylation. In some proteins, trisulfide bonds are observed in which a free Cys residue is disulfide bonded with two internal Cys residues of proteins. This chapter discusses the cysteinylation of proteins with disulfide and trisulfide bonds along with their biological significance.

### Chapter 5: Deamidation of Proteins

Deamidation is a chemical reaction in which organic amides undergo elimination reaction in the presence of base and/or metal ions or at elevated temperature to form carboxylic acid and ammonia. Amino acids such as Asn and Glu contain amide groups that may undergo deamidation in the presence of base and/or metal ions or at elevated temperature. This chapter describes the mechanism of deamidation, factors that influence deamidation, and effect of deamidation on biological functions of proteins. In addition, description of available methods to detect deamidation is discussed in this chapter. Methods using high performance liquid chromatography (HPLC), capillary electrophoresis (CE), and mass spectrometry are also briefly described.

## Chapter 6: Glycation of Proteins

Glycation is another type of PTM that occurs chemically when the proteins comes in contact with reducing sugars such as glucose in solution. Glycation is a chemical modification of proteins in which free amine groups present in basic amino acids such as Lys undergo chemical condensation reaction with reducing sugars such as glucose to form Schiff's base (aldimine linkage). The unstable aldimine linkage undergoes Amadori rearrangement to form a relatively stable covalent Amadori ketoamine linkage. As a result, an open chain glucose residue attaches to the side chain primary amine group of Lys residues. This open chain glucose residue undergoes multiple cleavages to form advanced glycation end (AGE) products through Maillard reactions. The AGEs have been shown to be toxic to humans and may cause many neurological diseases such as Parkinson, Alzheimer, etc., diseases.

This chapter describes the mechanism of glycation and impact of glycation on protein functions. AGE products and their toxic effects is also discussed in this chapter. Impact of glycation on human diseases is briefly discussed in this chapter. In addition, this chapter also contains a discussion on the available methods to detect protein glycation and their related AGE products.

## Chapter 7: Glycosylation of Proteins

Glycosylation is one of the major modifications of proteins. There are multiple types of glycosylation that includes N- and O-glycosylation. Glycosylation can be either a CTM or a PTM. However, majority of protein glycosylation is considered as a PTM. *O*-GlcNAcylation has been described as a CTM, whereas almost all other types of glycosylation have been described as PTMs. This chapter introduces types of protein glycosylation and briefly discusses their biological significance.

## Chapter 8: N-glycosylation of Proteins

This chapter describes the biosynthesis of protein N-glycosylation and structure of *N*-glycans. Microheterogeneity of *N*-glycans is also discussed in this chapter. Terminal residues of *N*-glycans and their effect on serum half-life of proteins is briefly discussed in this chapter. This chapter also contains a discussion on microheterogeneity of antibody glycosylation and the impact of Fc glycans on antibody functions. Available methods to detect and quantify protein N-glycosylation is also discussed. Immunogenicity of nonhuman type glyco-epitopes is briefly discussed in this chapter.

### Chapter 9: O-glycosylation of Proteins

Some proteins like erythropoietin (EPO) contain *O*-glycans in addition to *N*-glycans. This chapter is focused on the biosynthesis of *O*-glycans, mucin type O-glycosylation, their impact on protein structure and functions. The impact of O-glycosylation on serum half-life of proteins is also discussed. This chapter also contain a brief discussion on the methods to detect and quantitate *O*-glycans. Differences in O-glycosylation between human and animal cells including CHO and mouse hybridoma cells are also discussed in this chapter.

### Chapter 10: Hydroxylation of Proteins

Hydroxylation is an oxidation reaction in which —C—H bond converts into —C—OH bond. In chemistry, hydroxylation usually requires metal catalysts and heat. In biology, hydroxylation is normally catalyzed by hydroxylases. In many proteins, hydroxylation of Pro residue is very abundant, and both 3-hydroxyproline and 4-hydroxyproline residues are present in collagen. In addition, 5-hydroxylysine residues are also found in collagen. The hydroxyproline and hydroxylysine residues play important role in the structure and function of both collagen and gelatin. In addition, hydroxyphenylalanine and hydroxytyrosine residues are also found in some proteins. This chapter discusses the hydroxylation of proteins and the biological significance of hydroxylation on proteins. Hydroxylation of other amino acids (except Lys, Pro, Phe and Tyr) is not covered in this chapter.

### Chapter 11: Methylation of Proteins

Methylation of proteins is an alkylation reaction in which a methyl group is covalently linked to functional groups such as hydroxyl groups, carboxyl groups, and/or amine groups. In proteins, methylation is mediated by methyltransferases and *S*-adenosyl methionine is the primary donor of the methyl group. Different methyltransferases are involved in transferring methyl groups from *S*-adenosyl methionine to hydroxyl groups, carboxyl groups, and amine groups. Hence, different methyltransferases are also discussed in this chapter. In addition to amino acids, sugar residues can also get methylated and hence methylation of sugars is also briefly discussed in this chapter. In addition, discussion on mechanism of protein methylation, mechanism of O-methylation of sugar residues, physicochemical and biological significance of methylation, methylation in RTPs and methods to analyze protein methylation is also presented in this chapter.

### Chapter 12: Oxidation of Proteins

This chapter describes the types of protein oxidation, the amino acid residues that are susceptible for oxidation, mechanism of oxidation, and available methods to detect oxidation. The impact of oxidation on protein functions along with the human diseases associated with the protein oxidation are also discussed in this chapter. Impact of protein oxidation on human life and oxidation–reduction cycle is also briefly discussed in this chapter. Impact of protein oxidation on neurological diseases and age-related diseases are briefly discussed in this chapter.

### Chapter 13: Phosphorylation of Proteins

This chapter describes phosphorylation of proteins, amino acid residues that undergo phosphorylation, impact of phosphorylation on proteins functions and stability. A brief discussion on the methods to detect phosphorylation is included in this chapter. Biopharmaceuticals that contain phosphate residues are also discussed in this chapter. In addition to phosphorylation of amino acid residues of proteins, phosphorylated glycan residues of glycoproteins are also discussed, briefly. This chapter also contains a discussion on types of phosphorylation present in glycans and methods to detect such phosphorylated glycan epitopes.

### Chapter 14: Prenylation of Proteins

Prenylation is the covalent attachment of farnesyl or geranylgeranyl group as thioether linked to the thiol (–SH) group at the C-terminal Cys residues of proteins that contain a common CAAX consensus sequence in the C-termini. More than 120 human proteins are identified to be prenylated with either farnesyl or geranylgeranyl groups. Prenylation is an enzymatic reaction mediated by either farnesyltransferases or geranylgeranyltransferases. This chapter briefly discusses the mechanism of prenylation and biological significance of prenylation on proteins.

### Chapter 15: Proteolysis of Proteins

This chapter describes the in vitro and in vivo protein degradation by proteolysis, chemical mediated peptide bond cleavage, and autocleavage of proteins. Proteolysis mediated by serine proteases, cysteine proteases, etc., is also discussed, briefly. This chapter also contains a brief discussion on the proteolysis mediated by metalloproteinases and impact of protein degradation on biological activity of proteins, human disease, age, etc.

### Chapter 16: Selenylation of Proteins

Selenium (Se) is a nonmetal element and selenylation refers to covalently bound Se to amino acid residues of proteins. Both selenocysteine and selenomethionine are found in many endogenous proteins. Selenylated proteins are found in both prokaryotes and eukaryotes. This chapter briefly describes the selenylated proteins and their biological significance.

### Chapter 17: Signal Peptides of Proteins

Signal peptides are hydrophobic peptide moieties that are present in the N-terminus of membrane bound proteins and secreted proteins. Signal peptides are normally cleaved off from proteins by signal peptidases upon their exit through the endoplasmic reticulum. Signal peptidases are highly sequence specific enzymes and amino acid sequence of signal peptides needs to be highly complementary to the proteins sequences for the enzymes to act efficiently. Otherwise, the signal peptides are not completely cleaved by the peptidases and the uncleaved signal peptides might be potential safety risk in endogenous proteins and in RTPs. This chapter describes biochemical properties of signal peptides in endogenous proteins and in RTPs.

### Chapter 18: Sulfation of Proteins and Glycoproteins

Sulfation of biomolecules is a derivatization reaction of hydroxyl groups and/or unprotonated amine groups with sulfuryl group ($-SO_3$). Sulfation is mediated by sulfotransferases (SulTs) which transfers the $-SO_3$ group from 3′-phosphoadenosine-5′-phosphosulfate (PAPS). Protein sulfation mainly occurs on the hydroxyl group of Tyr residues. In addition, sulfation of glycans attached to proteins and also sulfation of glycosaminoglycans (GAGs) has been observed. This chapter discusses the biosynthesis of PAPS, sulfation reactions in the cytosol, sulfation of proteins in the Golgi, mechanism of Tyr sulfation, sulfation of GAGs, biological functions of GAGs, sulfation in RTPs, and methods to analyze sulfation.

### Chapter 19: SUMOylation

SUMOylation is a PTM in which small ubiquitin like modifier (SUMO) proteins are covalently linked through their C-terminal Gly residue with the Lys residues of target proteins similar to ubiquitination. SUMO proteins are small with ~100 amino acid residues with a molecular weight of ~12 kDa. Unlike some ubiquitinated proteins, SUMOylation is not a marker for protein degradation. SUMOylation and desumoylation are enzymatic modifications which affects many protein functions as well as some cellular functions. This chapter describes the mechanism of SUMOylation and its impact on protein and cellular functions.

### Chapter 20: Ubiquitination

Ubiquitin is a small protein with 76 amino acid residues with a molecular weight of ~8.5 kDa. Ubiquitin is present in eukaryotes but not in prokaryotes. Ubiquitination of proteins is an enzymatic process involving three different steps with three different groups of enzymes. During ubiquitination, activation, conjugation, and ligation reactions are mediated by ubiquitin-activating enzymes, ubiquitin-conjugating enzymes, and ubiquitin ligases, respectively. Through these three steps, an isopeptide bond is formed between ubiquitin with either Lys residues or with the N-terminal amino acid residues of the target proteins. Ubiquitination impacts many protein functions and some proteins may be marked for protein degradation through proteasome upon ubiquitination. This chapter discusses the mechanism of protein ubiquitination and the impact of ubiquitin moiety on the protein functions.

### Chapter 21: Other PTMs

Chapter 21 provides a brief discussion on other PTMs of proteins that are not covered in Chapters 2–20. Although it is very difficult to cover each and every PTM of proteins, attempts were made in Chapters 2–20 to describe some of the very important modifications of proteins that are highly relevant to the biopharmaceutical industry as well as to the academia. Many of the modifications described in this chapter are also highly useful to understand the structure and functions of proteins. The CTM and PTM modifications described in Chapter 21 are described in the alphabetical order. The discussion is mainly centered around the introduction and the biological significance of other CTMs and PTMs that are not covered in Chapters 2–20.

## References

Avila-Chávez, E., Torres-y-Torres, N., and Tovar-Palacio, A.R. (1997). New concepts in anionic and cationic amino acid transport. *Rev. Investig. Clin.* 49 (5): 411–424.

Bordes, R. and Holmberg, K. (2015). Amino acid-based surfactants – do they deserve more attention? *Adv. Colloid Interface Sci.* 222: 79–91. https://doi.org/10.1016/j.cis.2014.10.013.

Choudhary, C., Kumar, C., Gnad, F. et al. (2009). Lysine acetylation targets protein complexes and co-regulates major cellular functions. *Science* 325 (5942): 834–840. https://doi.org/10.1126/science.1175371.

Dolinnaya, N.G., Ogloblina, A.M., and Yakubovskaya, M.G. (2016). Structure, properties, and biological relevance of the DNA and RNA G-quadruplexes: overview 50 years after their discovery. *Biochemistry (Mosc)* 81 (13): 1602–1649. https://doi.org/10.1134/S0006297916130034.

Fischer, S., Marquart, K.F., Pieper, L.A. et al. (2017). miRNA engineering of CHO cells facilitates production of difficult-to-express proteins and increases success in cell line development. *Biotechnol. Bioeng.* 114 (7): 1495–1510. https://doi.org/10.1002/bit.26280.

Galeotti, C., Kaveri, S.V., and Bayry, J. (2017). IVIG-mediated effector functions in autoimmune and inflammatory diseases. *Int. Immunol.* https://doi.org/10.1093/intimm/dxx039.

Good, R.A. (1972). On the threshold of biologic engineering. *Am. J. Med. Technol.* 38 (5): 153–162.

Hertweck, C. (2011). Biosynthesis and charging of pyrrolysine, the 22nd genetically encoded amino acid. *Angew. Chem. Int. Ed* 50 (41): 9540–9541. https://doi.org/10.1002/anie.201103769.

Ishchenko, A., Abola, E.E., and Cherezov, V. (1607). Crystallization of membrane proteins: an overview. *Methods Mol. Biol.* 2017: 117–141. https://doi.org/10.1007/978-1-4939-7000-1_5.

Jefferis, R. (2016). Posttranslational modifications and the immunogenicity of biotherapeutics. *J. Immunol. Res.* 2016: https://doi.org/10.1155/2016/5358272.

Jia, H., Heymann, M., Bernhard, F. et al. (2017). Cell-free protein synthesis in micro compartments: building a minimal cell from biobricks. *New Biotechnol.* 39 (Pt B): 199–205. https://doi.org/10.1016/j.nbt.2017.06.014.

Kessler, B.M. and Edelmann, M.J. (2011). PTMs in conversation: activity and function of deubiquitinating enzymes regulated via post-translational modifications. *Cell Biochem. Biophys.* 60 (1-2): 21–38. https://doi.org/10.1007/s12013-011-9176-6.

Kinoshita, Y. and Saeki, H. (2017). A review of the active treatments for toxic epidermal necrolysis. *J. Nippon Med. Sch.* 84 (3): 110–117. https://doi.org/10.1272/jnms.84.110.

Kumar, R., Sanawar, R., Li, X., and Li, F. (2017). Structure, biochemistry, and biology of PAK kinases. *Gene* 605: 20–31. https://doi.org/10.1016/j.gene.2016.12.014.

Kurepa, J. and Smalle, J.A. (2008). Structure, function and regulation of plant proteasomes. *Biochimie* 90 (2): 324–335.

Marrack, J.R. and Hoch, H. (1949). Serum proteins: a review. *J. Clin. Pathol.* 2 (3): 161–192.

Nishimura, H. (2017). Renin-angiotensin system in vertebrates: phylogenetic view of structure and function. *Anat. Sci. Int.* 92 (2): 215–247. https://doi.org/10.1007/s12565-016-0372-8.

Palacín, M. (1994). A new family of proteins (rBAT and 4F2hc) involved in cationic and zwitterionic amino acid transport: a tale of two proteins in search of a transport function. *J. Exp. Biol.* 196: 123–137.

Parayath, N.N. and Amiji, M.M. (2017). Therapeutic targeting strategies using endogenous cells and proteins. *J. Control. Release* 258: 81–94. https://doi.org/10.1016/j.jconrel.2017.05.004.

Patil, R. and Walther, J. (2017). Continuous manufacturing of recombinant therapeutic proteins: upstream and downstream technologies. *Adv. Biochem. Eng. Biotechnol.* https://doi.org/10.1007/10_2016_58.

Perez, J.G., Stark, J.C., and Jewett, M.C. (2016). Cell-free synthetic biology: engineering beyond the cell. *Cold Spring Harb. Perspect. Biol.* 8 (12): https://doi.org/10.1101/cshperspect.a023853.

Sakami, W. and Harrington, H. (1963). Amino acid metabolism. *Annu. Rev. Biochem.* 32: 355–398.

Semenchenko, A., Oliveira, G., and Atman, A.P. (2016). Hybrid agent-based model for quantitative in-silico cell-free protein synthesis. *Biosystems* 150: 22–34.

Tanaka, K. (2009). The proteasome: overview of structure and functions. *Proc. Jpn. Acad. Ser. B Phys. Biol. Sci.* 85 (1): 12–36.

Voet, J.G. and Pratt Donald Voet, C.W. (2006). *Fundamentals of Biochemistry: Life at the Molecular Level*, 2e. Hoboken, NJ: Wiley.

Walsh, G. (2010). Post-translational modifications of protein biopharmaceuticals. *Drug Discov. Today* 15 (17–18): 773–780. https://doi.org/10.1016/j.drudis.2010.06.009.

Work, E. (1961). The mucopeptides of bacterial cell walls: a review. *J. Gen. Microbiol.* 25: 169–189.

Wu, H. and Fuxreiter, M. (2016). The structure and dynamics of higher-order assemblies: amyloids, signalosomes, and granules. *Cell* 165 (5): 1055–1066. https://doi.org/10.1016/j.cell.2016.05.004.

Wu, L., Candille, S.I., Choi, Y. et al. (2013). Variation and genetic control of protein abundance in humans. *Nature* 499 (7456): 79–82. https://doi.org/10.1038/nature12223.

Zamecnik, P.C. (1950). The use of labeled amino acids in the study of the protein metabolism of normal and malignant tissues: a review. *Cancer Res.* 10 (11): 659–667.

# 2

# Acetylation of Proteins

## Introduction

In organic chemistry, acetylation is a very common occurrence and is also called as ethanoylation. Acetylation is a part of the acylation reaction that involves a chemical condensation reaction between hydroxyl groups or amine groups of an organic molecule with acetic anhydride that act as a donor of acetyl groups (Fritz and Schenk 1959). Chemically, acylation is a chemical process in which hydrogen atom of hydroxyl group or amine group is replaced by an acyl carbonyl group such as acetyl ($CH_3CO$–) group. During organic synthesis, acylation can be achieved by heating a mixture of an acyl anhydride or acyl chloride and amine or hydroxyl group containing organic compounds in an organic solvent with or without a catalyst. The most commonly known acylation reaction is the acetylation of Gly in the presence of either acetic anhydride or acetyl chloride. Acetylation of Gly is an N-acetylation reaction to form *N*-acetylglycine as shown in Figure 2.1.

*N*-acetylglycine can be easily prepared by heating a mixture of glycine and acetic anhydride in benzene or glacial acetic acid in the presence of a catalyst (Curtius and Radenhausen 1895; Dakin 1929). During the acetylation reaction, benzene or glacial acetic acid is used as a solvent medium. *N*-acetylglycine is widely used in peptidomimetics and also in drug discovery research.

Similar to N-acetylation, O-acetylation reaction is also a very common reaction in organic chemistry. O-acetylation of hydroxyl groups is routinely used during organic synthesis (Martin 1971). A classic example of O-acetylation reaction in organic chemistry is the conversion of salicylic acid (monohydroxybenzoic acid) to acetylsalicylic acid (aspirin) as shown in Figure 2.2.

Conversion of salicylic acid into aspirin involves acetylation of hydroxyl group of salicylic acid with acetic anhydride in the acetic acid medium. Aspirin is a very commonly used painkiller, and recently, it is also being used as a blood thinner (Davison 1971; Costa et al. 2017). Acetylation reaction is also very commonly used during the synthesis of heroin (Odell et al. 2006).

*Co- and Post-Translational Modifications of Therapeutic Antibodies and Proteins*, First Edition. T. Shantha Raju.

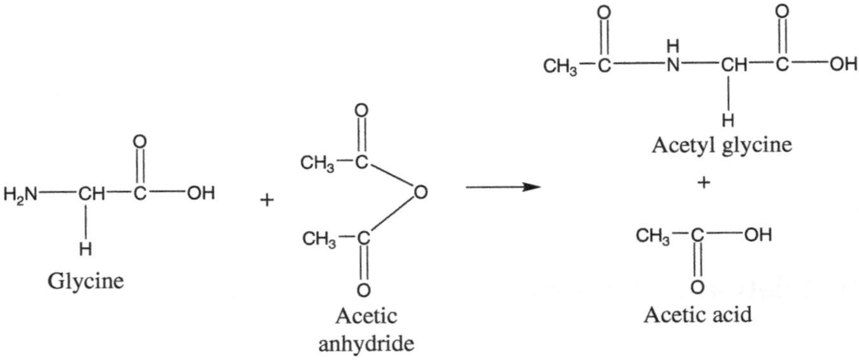

Figure 2.1 N-acetylation of glycine to form *N*-acetylglycine using acetic anhydride as an acetylating reagent. The final products of the reaction are *N*-acetyl glycine and acetic acid. (*See insert for color representation of this figure.*)

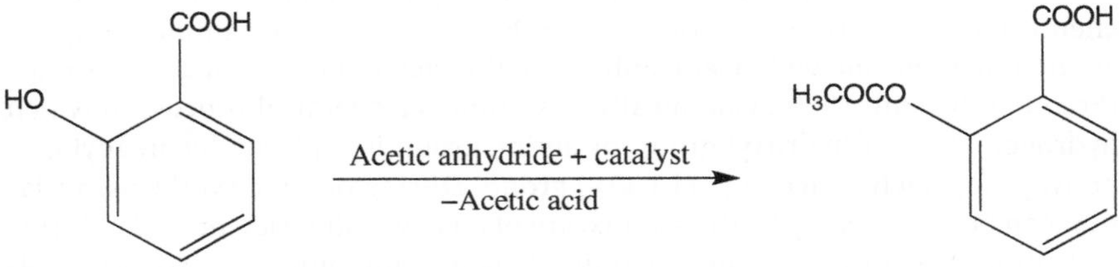

Figure 2.2 Conversion of salicylic acid into aspirin by O-acetylation.

O-acetylation reaction converts morphine into diacetylmorphine (DAM), which is also commonly known as heroin (Odell et al. 2006). Esterification of fatty acids is a common acylation process that involves the conversion of free fatty acids into fatty acid esters using acylation process. Acetylation and deacetylation reactions of hydroxyl groups and amine functional groups of organic compounds are routinely used in organic synthesis to protect the respective functional groups.

Acetylation in proteins is a very common occurrence, and thousands of mammalian proteins are found to be acetylated (Javaid and Choi 2017). Similar to organic compounds, acetylation of proteins is also commonly associated with hydroxyl or amine functional groups of constituent amino acid residues. Acetylation of proteins can occur both by enzymatic and nonenzymatic processes. Enzymatic acetylation in proteins is both a CTM and a PTM (Fritz et al. 2012). Enzymatic acetylation reactions of proteins are mediated by acetyltransferases. Acetyl-coenzyme A (Ac-CoA) is the most common donor of acetyl groups in biological systems (Choudhary et al. 2009). In proteins, enzymatic O-acetylation of hydroxyl groups is mediated by *O*-acetyltransferases, and N-acetylation of amine groups is mediated by

*N*-acetyltransferases (Choudhary et al. 2009). In addition, O-deacetylation of acetyl groups of proteins is mediated by *O*-deacetylases, and N-deacetylation is mediated by *N*-deacetylases (Choudhary et al. 2009). Acetylation and deacetylation is a dynamic biological process that may involve in turning "on and off" of protein functions/activities (Glozak et al. 2005; Glozak and Seto 2009). In addition, acetylation and deacetylation of biomolecules are part of gene regulation in living organisms (Zhao et al. 2010; Verdin and Ott 2015).

Recent proteomics studies have identified that thousands of mammalian proteins are acetylated (Arnesen et al. 2009; Van Damme et al. 2015). In Salmonella, about 90% of proteins that are involved in its central metabolism have been shown to be acetylated (Bernal et al. 2014). In biological systems, histones are the most commonly acetylated proteins (Suzuki and Luo 2017). In addition to histones, p53, tubulins, chromatin proteins, metabolic enzymes, etc., are also found to be acetylated (Cameron et al. 2016; Eckschlager et al. 2017; Javaid and Choi 2017). As described earlier, protein acetylation is both a CTM and a PTM and is an important protein modification in cell biology (Kurdistani 2014). O-acetylation may occur on the hydroxyl groups of Ser and Thr residues, whereas N-acetylation may occur on the amine side chains of Lys residues or amine functional groups of amino acids found in the N-termini of proteins. In proteins, O-acetylation is very rare and might be nonenzymatic. However, N-acetylation is a common occurrence that can be both enzymatic and nonenzymatic also. In addition to proteins, some sugar residues of oligosaccharides, polysaccharides, and several other glycoconjugates are found to be N- and O-acetylated. Both 4-*O*-acetylated sialic acids and 8-*O*-acetylated sialic acids are found to be present in mammalian tissue samples (Kamerling and Vliegenthart 1975). O-acetylated sialic acids are also being measured as cancer biomarkers (Krishna and Varki 1997; Khedri et al. 2017). N-acetylation is very common in glycoconjugates and proteoglycans where amine sugars such as glucosamine (GlcN), galactosamine (GalN), sialic acids are N-acetylated. Chitin is a biopolymer consisting of *N*-acetylglucosamine (GlcNAc) and is commonly found in the shells of shell fish and crabs. Sialic acids are N-acetylated and are very common sugar residues of glycoproteins and polysaccharides. In addition, heparin, heparin sulfate, and other glycosaminoglycans (GAGs) may also contain both N- and O-acetylated sugar residues.

## Mechanism of N-acetylation at the N-termini of Proteins

In humans, about 85% of proteins are N-acetylated at their N-termini (Silva and Martinho 2015; Drazic et al. 2016; Lee et al. 2016). Similarly, about 85% of proteins present in yeast are N-acetylated at their N-termini (Hirano et al. 2016). In addition, N-terminal acetylated proteins are also present in prokaryotes and archaea (Aksnes et al. 2015). N-acetylation of amine groups at the N-terminus of proteins is catalyzed by a set of enzyme complexes that are collectively called as N-terminal acetyltransferases (NATs) (Varland et al. 2015). These NAT enzymes mediate the transfer of acetyl groups from Ac-CoA to the $\alpha$-amino group of the N-terminal amino acids of proteins. Six different NAT enzymes have been found in humans (Starheim et al. 2012). These enzymes are labeled as NatA, NatB, NatC, NatD, NatE, and NatF. Each of these enzymes possesses different specificities for different amino acids or amino acid sequence in the N-termini of proteins. These six different NATs are responsible for the N-terminal acetylation of the nascent polypeptide chain synthesized within the cells. Unlike other acetylations, the existing data so far suggests that the N-terminal acetylation is irreversible (Starheim et al. 2012).

NatA consists of two subunits; Naa10 is a catalytic subunit and Naa15 is the auxiliary subunit (Starheim et al. 2012). These subunits of NatA are highly complex, and their complexity increases from lower eukaryotes to higher eukaryotes. NatA preferentially N-acetylates amine functional groups at the N-terminally situated Ser, Ala, Gly, Thr, Val, and Cys residues (Soppa 2010). The N-acetylation occurs only after the removal of initiator methionine residues by methionine amino-peptidases from the nascent proteins (Starheim et al. 2012).

Similar to NatA, two subunits are present in NatB enzyme complexes. In NatB, the catalytic subunit is Naa20p and the auxiliary subunit is Naa25p. Both of these subunits are found in yeast and humans (Soppa 2010). The Naa20p and Naa25p are ribosome-associated subunits of NatB in yeast. However, in humans both subunits are ribosome-associated and are nonribosomal forms. NatB substrates are N-terminal methionine residues with the sequence specificity of Met–Glu, Met–Asp, Met–Asn, and Met–Gln.

NatC consists of three subunits, one catalytic subunit (Naa30p), and two auxiliary subunits (Naa35p and Naa38p). Yeast ribosome contains all the three NatC subunits. All these three subunits are also found in nonribosomal forms. The substrates for NatC are N-terminal methionine residues with the starting sequence of Met–Leu, Met–Ile, Met–Trp, and Met–Phe.

NatD is different than NatA, NatB, and NatC as it contains only the catalytic subunit (Naa40p). Histones H2A and H4 are the substrates for NatD, but it is not a lysine *N*-acetyltransferase (KATs) as it only acetylates the N-terminal amino acids.

NatE enzyme complex is composed of Naa50p subunit and two NatA subunits (Naa10p and Naa15p). However, the substrate specificity of NatE subunits is different than the NatA subunits.

NatF is a newly discovered enzyme (discovered in 2011) and is composed of Naa60p enzyme subunit (Van Damme et al. 2011). Only higher eukaryotes contain NatF, but not the lower eukaryotes. Substrate specificity of NatF overlaps with NatC and NatE as it *N*-acetylates the N-terminal Met residues with the starting sequence of Met–Lys, Met–Leu, Met–Ile, Met–Trp, and Met–Phe. N-termini of cytosolic transmembrane proteins are preferentially N-acetylated by NatF. Humans contain a higher abundance of NatF acetylated proteins than yeast.

## Mechanism of N-acetylation and N-deacetylation of Lysine Residues

Histone acetylation and deacetylation are part of the gene regulation (Konsoula and Barile 2012). Lys residues present in the N-terminal tail of histones undergo acetylation and deacetylation processes (Wakeel et al. 2018). N-acetylation of Lys residues of histones are catalyzed by histone acetyltransferases (HATs). These HATs are now called as lysine acetyltransferases (KATs). This is because these enzymes also acetylate Lys residues present in nonhistone proteins as well (Suryanarayanan and Singh 2018). KATs mediate the transfer of acetyl group from Ac-CoA to $-NH_2$ group of Lys residues as shown in Figure 2.3. The N-deacetylation of Lys residues is catalyzed by histone deacetylases (HDACs) that are now termed as Lys deacetylases (KDACs) because these enzymes can deacetylate the acetyl groups of Lys residues present in nonhistone proteins as well (Su et al. 2018).

## Mechanism of O-acetylation of Sugar Residues

O-acetylated sugar residues are present as constituent monosaccharides of oligosaccharides and polysaccharides that are present in bacteria, parasites, human tissues, etc. (Klein and Roussel 1998; Mandal et al. 2015). In addition, N-acetylated sugars such as *N*-acetylglucosamine (GlcNAc), *N*-acetylgalactosamine (GalNAc), sialic acids (SAs such as *N*-acetylneuraminic acid [NANA], and *N*-glycolylneuraminic acid [NGNA]) are present in oligosaccharides, polysaccharides, glycoconjugates, etc. Some of these N-acetylated sugar residues may also contain O-acetyl groups covalently linked to the free hydroxyl groups. For example, O-acetylated sialic acids are found in bacteria, fungi, virus, humans, etc. (Klein and Roussel 1998). Both 4-O-acetylated sialic acids and 9-O-acetylated sialic acids are found in human tissues (Herrler et al. 1991; Mandal et al. 2000). The O-acetylated sialic acid residues are believed to be identified as biomarkers for cancer (Chowdhury and

$$\text{CoA–S–CO–CH}_3 + \text{H}_2\text{N–Lys–protein} \longrightarrow \text{CoA–SH} + \text{CH}_3\text{–CO–NH–Lys–protein}$$

**Figure 2.3** N-acetylation of lysine residues of proteins (example of histone proteins). CoA, coenzyme A.

Mandal 2009). O-acetylation of sialic acids is mediated by sialic acid specific *O*-acetyltransferases and *O*-acetylesterases (Mandal et al. 2015).

## Biological Significance of Protein Acetylation

Acetylation of organic molecules such as aspirin increases their ability to cross blood-brain barrier (Vasović et al. 2008; Bhatt and Addepalli 2012). As a result of acetylation, small molecule drugs reach brain more quickly thus increasing their effectiveness to treat the disease (Contestabile and Sintoni 2013). Acetylation is one of the main reasons for heroin to be far more potent narcotic than morphine.

Acetylation is a common CTM and PTM that might affect several functions of proteins (Huynh et al. 2017). Introduction of acetyl group ($CH_3CO-$) to amine group at the N-termini or at the side chain amine group of Lys residues may affect the physicochemical properties such as hydrophobicity and hydrophilicity of proteins including the overall charge of proteins. Hence, acetylation may affect the cellular functions such as protein synthesis, stability, localization, metabolism (Choudhary et al. 2014), etc.

N-acetylation of proteins in the N-termini may impact their stability. Acetylation in the N-termini of proteins blocks the N-terminal ubiquitination and hence prevents the subsequent degradation of proteins (see Chapter 20) (Starheim et al. 2012). Acetylation at the N-terminus may help the protein localization (Dhar et al. 2017). For example, N-terminal acetylation of Ar13p protein helps its localization to Golgi membrane (Dhar et al. 2017). Acetylation in the N-terminus of proteins is also involved in cell cycle regulation and apoptosis (Aksnes et al. 2015). For example, NatA acetylates the N-terminus of caspase-2 and the N-acetylated caspase-2 interacts with the adaptor protein associated death domain that could activate caspace-2 to induce apoptosis of cells (Feng et al. 2015). Acetylation of N-terminus of ribosome proteins may impact the biosynthesis of proteins (Kamita et al. 2011). With the deletion of NatA and NatB strains, a noticeable decrease in the protein biosynthesis was noticed (Katima et al. 2011). Hence, acetylation may impact the protein biosynthesis (Shi and Tu 2014). In human cancer, NATs may act as onco-proteins and tumor suppressors (Zeng et al. 2016) in certain type of human carcinoma such as cervical carcinoma, breast cancer, bladder cancer. Naa10p is overexpressed (Zeng et al. 2016).

Acetylation and deacetylation of lysine residues of proteins is a dynamic process similar to methylation and phosphorylation (see Chapter 11 for methylation and Chapter 13 for phosphorylation). Acetylation and deacetylation is a very significant post-translational regulatory mechanism of proteins (Baeza et al. 2016). Regulation of transcription factors, effector proteins, molecular chaperones, cytoskeletal proteins, etc. is controlled by acetylation and deacetylation of lysine residues. Protein acetylation and deacetylation may be involved in the crosstalk with phosphorylation, methylation, SUMOylation, ubiquitination, etc. Such crosstalk between acetylation and other PTMs may dynamically control cell signaling (Kontaxi et al. 2017).

In the field of epigenetics, the identification of histone acetylation and deacetylation is an important transcriptional regulation (Heuser et al. 2017). In addition to histones, acetylation and deacetylation of p53 is involved in its activation and deactivation. The p53 protein is termed as a "guardian of the genome" and is a tumor suppression gene that may play an important role in cells signal transactions (Janus et al. 1999). By doing so p53 prevents the mutation and hence stabilize the genome (Viadiu et al. 2014). The p53 protein also initiates the programmed cell death when there is a severe DNA damage in the cells (Mollereau and Ma 2014). The p53 proteins contain three acetylation sites at K164, K120 and at the amino acid in the C-termini. When cells undergo significant stress, increased acetylation of p53 protein is observed. In addition, O-acetylation of Ser and Thr residues is recently observed in eukaryotes (Seepersaud et al. 2017).

## Acetylation in Recombinant Therapeutic Proteins

Both N- and O-acetylation may occur in recombinant therapeutic proteins (RTPs) depending on the cell type and culture conditions used during the production (Soo et al. 2017). Often clonal selection may be a differentiating factor in N- and/or O-acetylation of RTPs. N-acetylation may occur at the N-termini that will pose problems in protein sequencing of RTPs using Edmund degradation method of sequencing. N-acetylation may also occur at the side chain amine groups of Lys residues, depending on the accessibility of such residues on the surface of RTPs.

In addition to N-acetylation, O-acetylation may occur at hydroxyl side chains of Ser or Thr residues of RTPs. Since the O-acetyl groups are very labile, they may be sensitive to change in pH of buffers used during process development and/or during formulation. Hence, O-acetyl groups may be lost during purification and/or storage of RTPs. O-acetylated sialic acid residues have been observed in several RTPs derived glycans (Moore et al. 1997; Gawlitzek et al. 1999; Gawlitzek et al. 2000). N-acetylated sugars such as GlcNAc, GalNAc are common constituents of both *N*- and *O*-glycans of RTPs (Varki 1993, 2017).

## Methods to Analyze Acetylation in Proteins and Carbohydrates

O-acetyl groups are acid and base labile, whereas the N-acetyl groups are relatively stable in both acidic and basic medium. Both O-acetyl and N-acetyl groups can be detected and monitored using high-resolution mass spectrometry (Gao et al. 2017; Kori et al. 2017). Additionally, the acetyl groups of proteins including RTPs can be detected by ELISA assays using specific antibodies that recognize either O-acetyl or N-acetyl or both O- and N-acetyl groups (Chen et al. 2017; Lin et al. 2017). However, ELISA assays have very high variability, and often, antibody cross reactivity is a problem. Since acetylation impacts the surface hydrophobicity and hydrophilicity along with the charge state, acetylated proteins can be separated by using HPLC techniques such as reversed phase (RP), ion-exchange chromatography (IEC), hydrophobic interaction chromatography (HIC). Even the electrophoretic methods such as capillary zone electrophoresis (CZE), capillary iso-electric focusing (cIEF) methods can also be used to detect protein acetylation. Both $^{1}H$ and $^{13}C$ NMR methods have been used to detect acetylation in biomolecules including proteins, carbohydrates, lipids, etc. The three hydrogen atoms linked to carbon atom of the methyl group give a distinct resonance for O-acetyl and N-acetyl groups. This is because the carbonyl group to which methyl group is linked affect the resonance of the hydrogen atoms. Thus $^{1}H$ NMR method is a very useful tool to detect both O-acetyl and N-acetyl groups. In addition, $^{13}C$ NMR method is also very useful to detect and quantitate O-acetyl and N-acetyl groups. The carbon atoms of $CH_3$- group and carbonyl group give distinct resonances that can be identified and quantitated. However, care must be taken to avoid deacetylation during sample preparation and also during analysis using acidic or basic buffer medium.

## References

Aksnes, H., Hole, K., and Arnesen, T. (2015). Molecular, cellular, and physiological significance of N-terminal acetylation. *Int. Rev. Cell Mol. Biol.* 316: 267–305. https://doi.org/10.1016/bs.ircmb.2015.01.001.

Arnesen, T., Van Damme, P., Polevoda, B. et al. (2009). Proteomics analyses reveal the evolutionary conservation and divergence of N-terminal acetyltransferases from yeast and humans. *Proc. Natl. Acad. Sci. U. S. A.* 106 (20): 8157–8162. https://doi.org/10.1073/pnas.0901931106.

Baeza, J., Smallegan, M.J., and Denu, J.M. (2016). Mechanisms and dynamics of protein acetylation in mitochondria. *Trends Biochem. Sci.* 41 (3): 231–244. https://doi.org/10.1016/j.tibs.2015.12.006.

Bernal, V., Castaño-Cerezo, S., Gallego-Jara, J. et al. (2014). Regulation of bacterial physiology by lysine acetylation of proteins. *New Biotechnol.* 31 (6): 586–595. https://doi.org/10.1016/j.nbt.2014.03.002.

Bhatt, L.K. and Addepalli, V. (2012). Potentiation of aspirin-induced cerebroprotection by minocycline: a therapeutic approach to attenuate exacerbation of transient focal cerebral ischaemia. *Diab. Vasc. Dis. Res.* 9 (1): 25–34. https://doi.org/10.1177/1479164111427753.

Cameron, A.M., Lawless, S.J., and Pearce, E.J. (2016). Metabolism and acetylation in innate immune cell function and fate. *Semin. Immunol.* 28 (5): 408–416. https://doi.org/10.1016/j.smim.2016.10.003.

Chen, S., Chen, S., Duan, Q., and Xu, G. (2017). Site-specific acetyl lysine antibodies reveal differential regulation of histone acetylation upon kinase inhibition. *Cell Biochem. Biophys.* 75 (1): 119–129. https://doi.org/10.1007/s12013-016-0777-y.

Choudhary, C., Kumar, C., Gnad, F. et al. (2009). Lysine acetylation targets protein complexes and co-regulates major cellular functions. *Science* 325 (5942): 834–840. https://doi.org/10.1126/science.1175371.

Chowdhury, S. and Mandal, C. (2009). O-acetylated sialic acids: multifaceted role in childhood acute lymphoblastic leukaemia. *Biotechnol. J.* 4 (3): 361–374. https://doi.org/10.1002/biot.200800253.

Choudhary, C., Weinert, B.T., Nishida, Y. et al. (2014 Aug). The growing landscape of lysine acetylation links metabolism and cell signaling. *Nat. Rev. Mol. Cell Biol.* 15 (8): 536–550. https://doi.org/10.1038/nrm3841.

Contestabile, A. and Sintoni, S. (2013). Histone acetylation in neurodevelopment. *Curr. Pharm. Des.* 19 (28): 5043–5050.

Costa, A.C., Reina-Couto, M., Albino-Teixeira, A., and Sousa, T. (2017). Aspirin and blood pressure: Effects when used alone or in combination with antihypertensive drugs. *Rev. Port. Cardiol.* 36 (7–8): 551–567. https://doi.org/10.1016/j.repc.2017.05.008.

Curtius, T. and Radenhausen, R. (1895). Hydrazide und Azide organischer Säuren. X Abhandlung. 35. Ueber Hydrazide substituirter Amidosäuren und das Hydrazid der Fumarsäure. *J. Prakt. Chem.* 52 (1): 433–454. https://doi.org/10.1002/prac.18950520134.

Dakin, H.D. (1929). The condensation of aromatic aldehydes with glycine and acetylglycine. *J. Biol. Chem.* 82 (2): 439–446.

Davison, C. (1971). Salicylate metabolism in man. *Ann. N. Y. Acad. Sci.* 179: 249–268.

Dhar, S., Gursoy-Yuzugullu, O., Parasuram, R., and Price, B.D. (2017). The tale of a tail: histone H4 acetylation and the repair of DNA breaks. *Philos. Trans. R. Soc. Lond. Ser. B Biol. Sci.* 372 (1731): https://doi.org/10.1098/rstb.2016.0284.

Drazic, A., Myklebust, L.M., Ree, R., and Arnesen, T. (2016 Oct). The world of protein acetylation. *Biochim. Biophys. Acta* 1864 (10): 1372–1401. https://doi.org/10.1016/j.bbapap.2016.06.007.

Eckschlager, T., Plch, J., Stiborova, M., and Hrabeta, J. (2017). Histone deacetylase inhibitors as anticancer drugs. *Int. J. Mol. Sci.* 18 (7): https://doi.org/10.3390/ijms18071414.

Feng, Y., Liu, T., Dong, S.Y. et al. (2015). Rotenone affects p53 transcriptional activity and apoptosis via targeting SIRT1 and H3K9 acetylation in SH-SY5Y cells. *J. Neurochem.* 134 (4): 668–676. https://doi.org/10.1111/jnc.13172.

Fritz, J.S. and Schenk, G.H. (1959). Acid-catalyzed acetylation of organic hydroxyl groups. *Anal. Chem.* 31 (11): 1808–1812.

Fritz, K.S., Galligan, J.J., Hirschey, M.D. et al. (2012). Mitochondrial acetylome analysis in a mouse model of alcohol-induced liver injury utilizing SIRT3 knockout mice. *J. Proteome Res.* 11 (3): 1633–1643. https://doi.org/10.1021/pr2008384.

Gao, J., Qin, X.J., Jiang, H. et al. (2017). Detecting serum and urine metabolic profile changes of CCl(4)-liver fibrosis in rats at 12 weeks based on gas chromatography-mass spectrometry. *Exp. Ther. Med.* 14 (2): 1496–1504. https://doi.org/10.3892/etm.2017.4668.

Gawlitzek, M., Papac, D.I., Sliwkowski, M.B., and Ryll, T. (1999). Incorporation of 15N from ammonium into the N-linked oligosaccharides of an immunoadhesin glycoprotein expressed in Chinese hamster ovary cells. *Glycobiology* 9 (2): 125–131.

Gawlitzek, M., Ryll, T., Lofgren, J., and Sliwkowski, M.B. (2000). Ammonium alters N-glycan structures of recombinant TNFR-IgG: degradative versus biosynthetic mechanisms. *Biotechnol. Bioeng.* 68 (6): 637–646.

Glozak, M.A., Sengupta, N., Zhang, X., and Seto, E. (2005). Acetylation and deacetylation of non-histone proteins. *Gene* 363: 15–23.

Glozak, M.A. and Seto, E. (2009). Acetylation/deacetylation modulates the stability of DNA replication licensing factor Cdt1. *J. Biol. Chem.* 284 (17): 11446–11453. https://doi.org/10.1074/jbc.M809394200.

Herrler, G., Szepanski, S., and Schultze, B. (1991). 9-*O*-acetylated sialic acid, a receptor determinant for influenza C virus and coronaviruses. *Behring Inst. Mitt.* (89): 177–184.

Heuser, M., Yun, H., and Thol, F. (2017). Epigenetics in MDS. *Semin. Cancer Biol.* (17): 30093–30097. https://doi.org/10.1016/j.semcancer.2017.07.009.

Hirano, H., Kimura, Y., and Kimura, A. (2016). Biological significance of co- and post-translational modifications of the yeast 26S proteasome. *J. Proteome* 134: 37–46. https://doi.org/10.1016/j.jprot.2015.11.016.

Huynh, N.C., Everts, V., and Ampornaramveth, R.S. (2017). Histone deacetylases and their roles in mineralized tissue regeneration. *Bone Rep.* 7: 33–40. https://doi.org/10.1016/j.bonr.2017.08.001.

Janus, F., Albrechtsen, N., Dornreiter, I. et al. (1999). The dual role model for p53 in maintaining genomic integrity. *Cell. Mol. Life Sci.* 55 (1): 12–27.

Javaid, N. and Choi, S. (2017). Acetylation- and methylation-related epigenetic proteins in the context of their targets. *Genes (Basel)* 8 (8): https://doi.org/10.3390/genes8080196.

Kamerling, J.P. and Vliegenthart, J.F. (1975). Identification of *O*-cetylated *N*-acylneuraminic acids by mass spectrometry. *Carbohydr. Res.* 41: 7–17.

Kamita, M., Kimura, Y., Ino, Y. et al. (2011). $N(\alpha)$-Acetylation of yeast ribosomal proteins and its effect on protein synthesis. *J. Proteome* 74 (4): 431–441. https://doi.org/10.1016/j.jprot.2010.12.007.

Khedri, Z., Xiao, A., Yu, H. et al. (2017). A chemical biology solution to problems with studying biologically important but unstable 9-*O*-acetyl sialic acids. *ACS Chem. Biol.* 12 (1): 214–224. https://doi.org/10.1021/acschembio.6b00928.

Klein, A. and Roussel, P. (1998 Jan). O-acetylation of sialic acids. *Biochimie* 80 (1): 49–57.

Konsoula, Z. and Barile, F.A. (2012). Epigenetic histone acetylation and deacetylation mechanisms in experimental models of neurodegenerative disorders. *J. Pharmacol. Toxicol. Methods* 66 (3): 215–220. https://doi.org/10.1016/j.vascn.2012.08.001.

Kontaxi, C., Piccardo, P., and Gill, A.C. (2017). Lysine-directed post-translational modifications of Tau protein in Alzheimer's disease and related tauopathies. *Front. Mol. Biosci.* 4: 56. https://doi.org/10.3389/fmolb.2017.00056.

Kori, Y., Sidoli, S., Yuan, Z.F. et al. (2017). Proteome-wide acetylation dynamics in human cells. *Sci. Rep.* 7 (1): 10296. https://doi.org/10.1038/s41598-017-09918-3.

Krishna, M. and Varki, A. (1997). 9-O-Acetylation of sialomucins: a novel marker of murine CD4 T cells that is regulated during maturation and activation. *J. Exp. Med.* 185 (11): 1997–2013.

Kurdistani, S.K. (2014). Chromatin: a capacitor of acetate for integrated regulation of gene expression and cell physiology. *Curr. Opin. Genet. Dev.* 26: 53–58. https://doi.org/10.1016/j.gde.2014.06.002.

Lee, K.E., Heo, J.E., Kim, J.M., and Hwang, C.S. (2016). N-terminal acetylation-targeted N-end rule proteolytic system: the Ac/N-end rule pathway. *Mol. Cells* 39 (3): 169–178. https://doi.org/10.14348/molcells.2016.2329.

Lin, Y.C., Lin, Y.C., Wu, C.C. et al. (2017). The immunomodulatory effects of TNF-α inhibitors on human Th17 cells via RORγt histone acetylation. *Oncotarget* 8 (5): 7559–7571. https://doi.org/10.18632/oncotarget.13791.

Mandal, C., Chatterjee, M., and Sinha, D. (2000). Investigation of 9-O-acetylated sialoglycoconjugates in childhood acute lymphoblastic leukaemia. *Br. J. Haematol.* 110 (4): 801–812.

Mandal, C., Schwartz-Albiez, R., and Vlasak, R. (2015). Functions and biosynthesis of O-acetylated sialic acids. *Top. Curr. Chem.* 366: 1–30. https://doi.org/10.1007/128_2011_310.

Martin, B.K. (1971). The formulation of aspirin. *Adv. Pharm. Sci.* 3: 107–171.

Mollereau, B. and Ma, D. (2014). The p53 control of apoptosis and proliferation: lessons from Drosophila. *Apoptosis* 19 (10): 1421–1429. https://doi.org/10.1007/s10495-014-1035-7.

Moore, A., Mercer, J., Dutina, G. et al. (1997). Effects of temperature shift on cell cycle, apoptosis and nucleotide pools in CHO cell batch cultues. *Cytotechnology* 23 (1–3): 47–54. https://doi.org/10.1023/A:1007919921991.

Odell, L.R., Skopec, J., and McCluskey, A. (2006). A 'cold synthesis' of heroin and implications in heroin signature analysis utility of trifluoroacetic/acetic anhydride in the acetylation of morphine. *Forensic Sci. Int.* 164 (2–3): 221–229.

Seepersaud, R., Sychantha, D., Bensing, B.A. et al. (2017). O-acetylation of the serine-rich repeat glycoprotein GspB is coordinated with accessory Sec transport. *PLoS Pathog.* 13 (8): https://doi.org/10.1371/journal.ppat.1006558.

Shi, L. and Tu, B.P. (2014). Protein acetylation as a means to regulate protein function in tune with metabolic state. *Biochem. Soc. Trans.* 42 (4): 1037–1042. https://doi.org/10.1042/BST20140135.

Silva, R.D. and Martinho, R.G. (2015). Developmental roles of protein N-terminal acetylation. *Proteomics* 15 (14): 2402–2409. https://doi.org/10.1002/pmic.201400631.

Soo, B.P.C., Tay, J., Ng, S. et al. (2017). Correlation between expression of recombinant proteins and abundance of $H_3K_4Me_3$ on the enhancer of human cytomegalovirus major immediate-early promoter. *Mol. Biotechnol.* 59 (8): 315–322. https://doi.org/10.1007/s12033-017-0019-6.

Soppa, J. (2010). Protein acetylation in archaea, bacteria, and eukaryotes. *Archaea* 16: https://doi.org/10.1155/2010/820681.

Starheim, K.K., Gevaert, K., and Arnesen, T. (2012). Protein N-terminal acetyltransferases: when the start matters. *Trends Biochem. Sci.* 37 (4): 152–161. https://doi.org/10.1016/j.tibs.2012.02.003.

Su, X.M., Ren, Y., Li, M.L. et al. (2018). Performance evaluation of histone deacetylases in lungs of mice exposed to ovalbumin aerosols. *J. Physiol. Pharmacol.* 69 (2): https://doi.org/10.26402/jpp.2018.2.12.

Suryanarayanan, V. and Singh, S.K. (2018). Deciphering the binding mode and mechanistic insights of pentadecylidenemalonate (1b) as activator of histone acetyltransferase PCAF. *J. Biomol. Struct. Dyn.* 25: 1–40. https://doi.org/10.1080/07391102.2018.1479658.

Suzuki, K. and Luo, Y. (2017). Histone acetylation and the regulation of major histocompatibility class II gene expression. *Adv. Protein Chem. Struct. Biol.* 106: 71–111. https://doi.org/10.1016/bs.apcsb.2016.08.002.

Van Damme, P., Hole, K., Pimenta-Marques, A. et al. (2011). NatF contributes to an evolutionary shift in protein N-terminal acetylation and is important for normal chromosome segregation. *PLoS Genet.* 7 (7): e1002169. https://doi.org/10.1371/journal.pgen.1002169.

Van Damme, P., Hole, K., Gevaert, K., and Arnesen, T. (2015). N-terminal acetylome analysis reveals the specificity of Naa50 (Nat5) and suggests a kinetic competition between N-terminal acetyltransferases and methionine aminopeptidases. *Proteomics* 15 (14): 2436–2446. https://doi.org/10.1002/pmic.201400575.

Varki, A. (2017). Biological roles of glycans. *Glycobiology* 27 (1): 3–49. https://doi.org/10.1093/glycob/cww086.

Varki, A. (1993). Biological roles of oligosaccharides: all of the theories are correct. *Glycobiology* 3 (2): 97–130.

Varland, S., Osberg, C., and Arnesen, T. (2015 Jul). N-terminal modifications of cellular proteins: the enzymes involved, their substrate specificities and biological effects. *Proteomics* 15 (14): 2385–2401. https://doi.org/10.1002/pmic.201400619.

Vasović, V., Banić, B., Jakovljević, V. et al. (2008). Effect of aminophylline on aspirin penetration into the central nervous system in rats. *Eur. J. Drug Metab. Pharmacokinet.* 33 (1): 23–30.

Verdin, E. and Ott, M. (2015). 50 years of protein acetylation: from gene regulation to epigenetics, metabolism and beyond. *Nat. Rev. Mol. Cell Biol.* 16 (4): 258–264. https://doi.org/10.1038/nrm3931. Epub.

Viadiu, H., Fronza, G., and Inga, A. (2014). Structural studies on mechanisms to activate mutant p53. *Subcell. Biochem.* 85: 119–132. https://doi.org/10.1007/978-94-017-9211-0_7.

Wakeel, A., Ali, I., Khan, A.R. et al. (2018). Involvement of histone acetylation and deacetylation in regulating auxin responses and associated phenotypic changes in plants. *Plant Cell Rep.* 37 (1): 51–59. https://doi.org/10.1007/s00299-017-2205-1.

Zhao, S., Xu, W., Jiang, W. et al. (2010). Regulation of cellular metabolism by protein lysine acetylation. *Science* 327 (5968): 1000–1004. https://doi.org/10.1126/science.1179689.

Zeng, Y., Zheng, J., Zhao, J. et al. (2016). High expression of Naa10p associates with lymph node metastasis and predicts favorable prognosis of oral squamous cell carcinoma. *Tumour Biol.* 37 (5): 6719–6728. https://doi.org/10.1007/s13277-015-4563-z.

# 3

# C-terminal Lys or Arg Clipping in Proteins

## Introduction

Very often C-terminal Lys or Arg residues are absent from cell surface or secreted proteins that are expected to contain these residues based on their gene sequence (Harris 1995). This is because C-terminal Lys or Arg residues are being removed by carboxypeptidases from these proteins (Matthews et al. 2004). Since clipping of C-terminal Lys or Arg residues by carboxypeptidases is often incomplete, the enzymatic digestion process generates additional heterogeneity in the cell surface-bound or secreted protein molecules. The C-terminal heterogeneity due to the presence or the absence of C-terminal Lys or Arg residues is very common in recombinant proteins produced using mammalian cells such as Chinese Hamster Ovary (CHO) cells (Harris 1995). C-terminal Lys heterogeneity is very commonly observed in mAbs produced using CHO cells (Hu et al. 2017). The observed C-terminal heterogeneity in mAbs produced using CHO cells is due to the activity of carboxypeptidase D in CHO cell culture media (Hu et al. 2016). In addition to C-terminal Lys heterogeneity, C-terminal Arg heterogeneity is also observed in mAbs produced using CHO cells (Zhang et al. 2015).

## Biological Significance of C-terminal Lys or Arg Clipping in Proteins

Although it is generally observed that the C-terminal Lys or Arg heterogeneity does not impact the antibody binding to antigen or its biological activity, recent studies suggest that these heterogeneities might impact the antibody binding to Fc receptors thus affecting antibody effector functions (Hintersteiner et al. 2016). The C-terminal heterogeneity may also impact the antibody serum half-life (Hintersteiner et al. 2016). In addition, C-terminal Lys residues has been shown to impact the complement-dependent cytotoxicity (CDC) activity

*Co- and Post-Translational Modifications of Therapeutic Antibodies and Proteins*, First Edition. T. Shantha Raju.

of mAbs (van den Bremer et al. 2015). IgG hexamerization is required for the efficient binding of antibody to C1q protein and hence CDC activity (Wang et al. 2016). C-terminal Lys residues have been shown to impact the IgG hexamerization process and hence may impact the antibody binding to C1q proteins (van den Bremer et al. 2015). Such an impact on mAbs might be due to the effect of C-terminal Lys or Arg residues on the surface charge of antibody molecules that might affect the avidity of antibodies to bind Fc receptors. In addition to mAbs, in some proteins, the internal Lys or Arg residues might become C-terminal residues upon proteolytic cleavages. This is more evident in zymogens in which preprotein or proproteins undergo cleavage to activate the protease (Foltmann 1988). For example, in the case of zymogen of rhtPA, proteolytic cleavage may result in C-terminal Lys residues that can be cleaved by carboxypeptidase D present in CHO cells thus generating additional heterogeneity of the protein molecules (Harris 1995). In addition, C-terminal heterogeneity may impact the overall charge heterogeneity of the proteins thus impact the p*I* of the molecule.

## Analysis of C-terminal Lys or Arg Clipping in Proteins

The C-terminal heterogeneity can be analyzed by charge separation-based methods such as IEC, IEF, cIEF methods. If the C-terminal heterogeneity impacts the surface hydrophobicity, then HIC method can be used to separate and quantiate such heterogeneity. In fact HIC method has been used to separate and quantiate C-terminal Lys heterogeneity in mAbs (Beck et al. 2005). Since Lys residues are charged residues, the mAbs samples with or without C-terminal Lys residues can be separated and quantitated by IEC, HIC, IEF, cIEF, etc., methods. The C-terminal heterogeneity can also be detected using high-resolution mass spectrometry and other complementary methods. The mass difference of the presence or the absence of Lys residue is good enough to detect the C-terminal Lys residue clipping in mAbs. If the C-terminal heterogeneity impacts the biological function of proteins, it may become critical quality attribute (CQA).

## References

Beck, A., Bussat, M.C., Zorn, N. et al. (2005). Characterization by liquid chromatography combined with mass spectrometry of monoclonal anti-IGF-1 receptor antibodies produced in CHO and NS0 cells. *J. Chromatogr. B Analyt. Technol. Biomed. Life Sci.* 819 (2): 203–218.

van den Bremer, E.T., Beurskens, F.J., Voorhorst, M. et al. (2015). Human IgG is produced in a pro-form that requires clipping of C-terminal lysines for maximal complement activation. *MAbs* 7 (4): 672–680. https://doi.org/10.1080/19420862.2015.1046665.

Foltmann, B. (1988). Structure and function of proparts in zymogens for aspartic proteinases. *Biol. Chem. Hoppe Seyler* 369 (Suppl): 311–314.

Harris, R.J. (1995). Processing of C-terminal lysine and arginine residues of proteins isolated from mammalian cell culture. *J. Chromatogr. A* 705 (1): 129–134.

Hintersteiner, B., Lingg, N., Zhang, P. et al. (2016). Charge heterogeneity: Basic antibody charge variants with increased binding to Fc receptors. *MAbs* 8 (8): 1548–1560.

Hu, Z., Zhang, H., Haley, B. et al. (2016). Carboxypeptidase D is the only enzyme responsible for antibody C-terminal lysine cleavage in Chinese hamster ovary (CHO) cells. *Biotechnol. Bioeng.* 113 (10): 2100–2106. https://doi.org/10.1002/bit.25977.

Hu, Z., Tang, D., Misaghi, S. et al. (2017). Evaluation of heavy chain C-terminal deletions on productivity and product quality of monoclonal antibodies in Chinese hamster ovary (CHO) cells. *Biotechnol. Prog.* 33 (3): 786–794. https://doi.org/10.1002/btpr.2444.

Matthews, K.W., Mueller-Ortiz, S.L., and Wetsel, R.A. (2004). Carboxypeptidase N: a pleiotropic regulator of inflammation. *Mol. Immunol.* 40 (11): 785–793.

Wang, G., de Jong, R.N., van den Bremer, E.T. et al. (2016). Molecular basis of assembly and activation of complement component C1 in complex with immunoglobulin G1 and antigen. *Mol. Cell* 63 (1): 135–145. https://doi.org/10.1016/j.molcel.2016.05.016.

Zhang, X., Tang, H., Sun, Y.T. et al. (2015). Elucidating the effects of arginine and lysine on a monoclonal antibody C-terminal lysine variation in CHO cell cultures. *Appl. Microbiol. Biotechnol.* 99 (16): 6643–6652. https://doi.org/10.1007/s00253-015-6617-y.

# 4

# Cysteinylation of Proteins

## Introduction

Cysteinylation refers to disulfide bond formation between free thiol groups of internal cysteine (Cys) residues of proteins with free Cys molecules (Kita et al. 2016; Liu et al. 2018). Cysteinylation is a post-translational modification (PTM) and is usually a reversible modification. Internal Cys residues of proteins normally form disulfide bonds with other free thiol groups of internal Cys residues of the same protein chain or another protein chain of the same protein (Patil et al. 2015). The disulfide bond formation between two Cys residue of the same and single protein chain is called as intra-chain disulfide bond (Hatahet et al. 2014). The disulfide bond formation between two Cys residues of two different chains of same protein is called as inter-chain disulfide (Moritz and Stracke 2017). The intra-chain or inter-chain disulfide bonded Cys residues are often referred to as paired Cys residues (Windsor et al. 1993). However, many proteins often contain internal Cys residues with free thiol groups that are often called as unpaired Cys residues (Wu et al. 2010). The unpaired Cys residues of proteins often form disulfide bonds with free cysteine molecules (Kita et al. 2016; Liu et al. 2018). Such a disulfide bond formation between protein-bound Cys and free Cys molecules is called as cysteinylation. In addition, recently, trisulfide bonds have been detected in which a free cysteine molecule forms sulfhydryl bonds with two protein-bound cysteine residues (Goldrick et al. 2017). The trisulfide bonds are commonly observed in monoclonal antibodies (mAbs), antibody fusion proteins, and antibody drug conjugates (Kshirsagar et al. 2012; Cumnock et al. 2013; Goldrick et al. 2017; Liu et al. 2017).

## Biological Significance of Cysteinylation of Proteins

Some bacteria such as *Bacillus subtilis* use cysteinylation as a mechanism to protect free thiol groups following oxidative stress (Hochgräfe et al. 2007).

*Co- and Post-Translational Modifications of Therapeutic Antibodies and Proteins*, First Edition. T. Shantha Raju.

During oxidative stress, free thiol groups of proteins are usually protected and regulated by S-thiolation (Loi et al. 2015). Glutathione enzymes are involved in S-thiolation to protect the free thiol groups of proteins. Many Gram-positive bacteria such as *B. subtilis* do not contain low molecular weight thiols like glutathione. Hence, *B. subtilis* and other Gram-positive bacteria use cysteinylation to protect the free thiol groups (Hochgräfe et al. 2007). Cu/Zn-superoxide-dismutase (SOD1) is a metalloenzyme that catalyzes the reduction of superoxide anions into molecular oxygen and hydrogen peroxide. The resulting hydrogen peroxide in turn can oxidize SOD1 that affect the protein conformation and activity. However, the Cys 111 of SOD is cysteinylated that prevents its oxidation (Auclair et al. 2013).

## Cysteinylation and Trisulfide Bonds in Recombinant Therapeutic Proteins

Cysteinylation and trisulfide bonds have been detected in a number of recombinant therapeutic proteins (RTPs) and mAbs (Kita et al. 2016). Cysteinylation and trisulfide bonds have been shown to impact the biological properties of RTPs (Seibel et al. 2017). For example, McSherry et al. detected zero, one, and two cysteinylation in the complementarity-determining region 3 (CDR3) of a mAb light chain that affects the binding of antibody to its antigen (McSherry et al. 2016). In addition, Kita et al. reported the presence of both cysteinylation and trisulfide bonds in a mAb (Kita et al. 2016). Further, removal of cysteinylation has been shown to improve the biological activity and stability along with reducing the heterogeneity of a mAb (Banks et al. 2008). In addition to impacting the biological activity and binding affinity, cysteinylation may also impact the higher-order structure of proteins (Banks et al. 2008). Sulfocysteine has been used to reduce the trisulfide bond formation in mAbs produced in Chinese Hamster Ovary (CHO) cells (Seibel et al. 2017).

## Analysis of Cysteinylation of Proteins

Cysteinylation of proteins results in the overall molecular weight shift of approximately +121 Da. This mass shift can be easily detected using high-resolution mass spectrometry (Lim et al. 2001). In addition, reversed-phase high performance liquid chromatography (RP-HPLC), IEC-HPLC (IEC, ion-exchange chromatography), mass spectrometry with or without limited proteolysis methods have also been used to detect cysteinylation and trisulfide bonds in mAbs (Lim et al. 2001; Gadgil et al. 2006; Banks et al. 2008; Kita et al. 2016; McSherry et al. 2016). Nonreducing peptide mapping is also very useful to detect and quantitate cysteinylation (Gadgil et al. 2006; Kita et al. 2016).

## References

Auclair, J.R., Brodkin, H.R., D'Aquino, J.A. et al. (2013). Structural consequences of cysteinylation of Cu/Zn-superoxide dismutase. *Biochemistry* 52 (36): 6145–6150. https://doi.org/10.1021/bi400613h.

Banks, D.D., Gadgil, H.S., Pipes, G.D. et al. (2008). Removal of cysteinylation from an unpaired sulfhydryl in the variable region of a recombinant monoclonal $IgG_1$ antibody improves homogeneity, stability, and biological activity. *J. Pharm. Sci.* 97 (2): 775–790.

Cumnock, K., Tully, T., Cornell, C. et al. (2013). Trisulfide modification impacts the reduction step in antibody-drug conjugation process. *Bioconjugate Chem.* 24 (7): 1154–1160. https://doi.org/10.1021/bc4000299.

Gadgil, H.S., Bondarenko, P.V., Pipes, G.D. et al. (2006). Identification of cysteinylation of a free cysteine in the Fab region of a recombinant monoclonal $IgG_1$ antibody using Lys-C limited proteolysis coupled with LC/MS analysis. *Anal. Biochem.* 355 (2): 165–174.

Goldrick, S., Holmes, W., Bond, N.J. et al. (2017). Advanced multivariate data analysis to determine the root cause of trisulfide bond formation in a novel antibody-peptide fusion. *Biotechnol. Bioeng.* 114 (10): 2222–2234. https://doi.org/10.1002/bit.26339.

Hatahet, F., Boyd, D., and Beckwith, J. (2014). Disulfide bond formation in prokaryotes: history, diversity and design. *Biochim. Biophys. Acta* 1844 (8): 1402–1414. https://doi.org/10.1016/j.bbapap.2014.02.014.

Hochgräfe, F., Mostertz, J., Pöther, D.C. et al. (2007). S-cysteinylation is a general mechanism for thiol protection of *Bacillus subtilis* proteins after oxidative stress. *J. Biol. Chem.* 282 (36): 25981–25985.

Kita, A., Ponniah, G., Nowak, C., and Liu, H. (2016). Characterization of cysteinylation and trisulfide bonds in a recombinant monoclonal antibody. *Anal. Chem.* 88 (10): 5430–5437. https://doi.org/10.1021/acs.analchem.6b00822.

Kshirsagar, R., McElearney, K., Gilbert, A. et al. (2012). Controlling trisulfide modification in recombinant monoclonal antibody produced in fed-batch cell culture. *Biotechnol. Bioeng.* 109 (10): 2523–2532. https://doi.org/10.1002/bit.24511.

Lim, A., Wally, J., Walsh, M.T. et al. (2001). Identification and location of a cysteinyl posttranslational modification in an amyloidogenic kappa1 light chain protein by electrospray ionization and matrix-assisted laser desorption/ionization mass spectrometry. *Anal. Biochem.* 295 (1, 1): 45–56.

Liu, R., Chen, X., Dushime, J. et al. (2017). The impact of trisulfide modification of antibodies on the properties of antibody-drug conjugates manufactured using thiol chemistry. *MAbs* 9 (3): 490–497. https://doi.org/10.1080/19420862.2017.1285478.

Liu, J., Zhang, J., Xu, F. et al. (2018). Structural characterizations of human periostin dimerization and cysteinylation. *FEBS Lett.* 592 (11): 1789–1803. https://doi.org/10.1002/1873-3468.13091.

Loi, V.V., Rossius, M., and Antelmann, H. (2015). Redox regulation by reversible protein S-thiolation in bacteria. *Front Microbiol.* 6: 187. https://doi.org/10.3389/fmicb.2015.00187.

McSherry, T., McSherry, J., Ozaeta, P. et al. (2016). Cysteinylation of a monoclonal antibody leads to its inactivation. *MAbs* 8 (4): 718–725. https://doi.org/10.1080/19420862.2016.1160179.

Moritz, B. and Stracke, J.O. (2017). Assessment of disulfide and hinge modifications in monoclonal antibodies. *Electrophoresis* 38 (6): 769–785. https://doi.org/10.1002/elps.201600425.

Patil, N.A., Tailhades, J., Hughes, R.A. et al. (2015). Cellular disulfide bond formation in bioactive peptides and proteins. *Int. J. Mol. Sci.* 16 (1): 1791–1805. https://doi.org/10.3390/ijms16011791.

Seibel, R., Maier, S., Schnellbaecher, A. et al. (2017). Impact of S-sulfocysteine on fragments and trisulfide bond linkages in monoclonal antibodies. *MAbs* 9 (6): 889–897. https://doi.org/10.1080/19420862.2017.1333212.

Windsor, W.T., Syto, R., Tsarbopoulos, A. et al. (1993). Disulfide bond assignments and secondary structure analysis of human and murine interleukin 10. *Biochemistry* 32 (34): 8807–8815.

Wu, S.L., Jiang, H., Hancock, W.S., and Karger, B.L. (2010). Identification of the unpaired cysteine status and complete mapping of the 17 disulfides of recombinant tissue plasminogen activator using LC-MS with electron transfer dissociation/collision induced dissociation. *Anal. Chem.* 82 (12): 5296–5303. https://doi.org/10.1021/ac100766r.

# 5

# Deamidation of Proteins

## Introduction

Organic molecules containing amide group undergoes an elimination reaction in which they lose an ammonia molecule or an ammonium ion under basic conditions and/or at elevated temperature (Keceli and Toscano 2012). Such an elimination reaction is called deamidation in which amide functional group of the molecule converts into carboxylic acid functional group as shown in Figure 5.1.

Amino acids such as Asn and Gln contain amide groups. The amide group in Asn and Gln may undergo an elimination reaction in which they lose an amide group (in the form of an ammonia molecule). Asn may undergo deamidation to convert into aspartic acid (Asp) and/or isoaspartic acid (IsoAsp), whereas deamidation of Gln converts it mostly into glutamic acid.

Deamidation is a chemical reaction and hence, it is a nonenzymatic post-translational modification (PTM) of proteins (Lindner and Helliger 2001). Almost all proteins containing Asn and Gln eventually undergo deamidation (Clarke 1987). Some proteins undergo spontaneous deamidation and some proteins undergo deamidation under basic conditions and/or at elevated temperature (Clarke 1987). The rate at which a protein undergoes deamidation may depend on the amino acid sequence and local environment that also depends on the 3D structure of proteins (Stephenson and Clarke 1989).

Deamidation is one of the factors that define the very useful lifetime of proteins (Lindner and Helliger 2001; Reissner and Aswad 2003). Deamidation may affect the physicochemical properties of proteins as it results in converting an amide moiety into a carboxylic acid moiety (Xie and Schowen 1999). Deamidation may affect the isoelectric point of a given protein in which a basic or neutral protein may convert into an acidic protein. As a result, deamidation may also affect the surface hydrophobicity and hydrophilicity of proteins. Hence, deamidation may affect the chromatographic properties of proteins (Stroop 2007).

Biological properties of proteins may also be affected by deamidation (Volkin et al. 1997; Qin et al. 2000). If the deamidation occurs in the binding domains

*Co- and Post-Translational Modifications of Therapeutic Antibodies and Proteins*, First Edition. T. Shantha Raju.

$$R—CO—NH_2 + H_2O \longrightarrow R—COOH + NH_3$$

**Figure 5.1** Deamidation of an organic amide molecule. R = an alkyl or aryl group (alkyl is an aliphatic organic group and aryl is an aromatic organic group). Deamidation mostly occurs in the presence of base and/or heat.

of antibodies, binding proteins, etc., it will affect the antigen–antibody binding of proteins to their receptors, etc. If the deamidation site is in the catalytic domain of enzymes, it may affect the enzymatic properties of such proteins. Deamidation in the Fc portion of IgGs may affect the antibody effector functions including their ability to kill the cells (Wooding et al. 2016; Zhao et al. 2016). Since deamidation affects the isoelectric properties of proteins, it may affect their resistance to proteases (Liu et al. 2016). Further, deamidation may impact the proteolytic activities of proteases (Qin et al. 2000).

Deamidation may also affect the pharmacokinetic properties of proteins (Huang et al. 2005). For example, Huang et al., reported that the deamidation may decrease serum half-life of proteins (Huang et al. 2005). Deamidation may also impact the immunogenicity of proteins (Verma et al. 2016). This is because, upon deamidation protein sequence may become unnatural and hence such proteins may become immunogenic (Verma et al. 2013, 2016; Zhao et al. 2016).

## Mechanism of Deamidation of Proteins

Proteins deamidation is a stepwise chemical reaction. Mechanism of deamidation of Asn residues is shown in Figure 5.2. During the initial reaction of deamidation, Asn first converts into succinimide intermediate by losing an ammonia molecule. This succinimide intermediate undergoes hydrolysis into either aspartic acid (Asp) or isoaspartic (IsoAsp) acid as shown in Figure 5.2. Although aspartic acid is a natural amino acid, isoaspartic acid is an unnatural amino acid and hence, proteins containing isoaspartic acid may be immunogenic to humans.

Similar to deamidation of Asn residues, Gln residues also undergoes deamidation in a stepwise fashion. In the first step, Gln converts into glutarimide intermediate during which an ammonia molecule is eliminated as shown in Figure 5.3. In the second step, the glutarimide intermediate undergoes hydrolysis and converts into either glutamic acid or isoglutamic acid. While glutamic acid is a natural amino acid, the isoglutamic acid is an unnatural amino acid. Hence, proteins with isoglutamic acid may create immunogenicity issues for humans. However, most often Gln converts into glutamic acid than isoglutamic acid (Li et al. 2010) unlike during Asn deamidation.

The chemical deamidation is a very common PTM of proteins and eventually occurs at all Asn and Gln residues of proteins with the exception of –NP- and

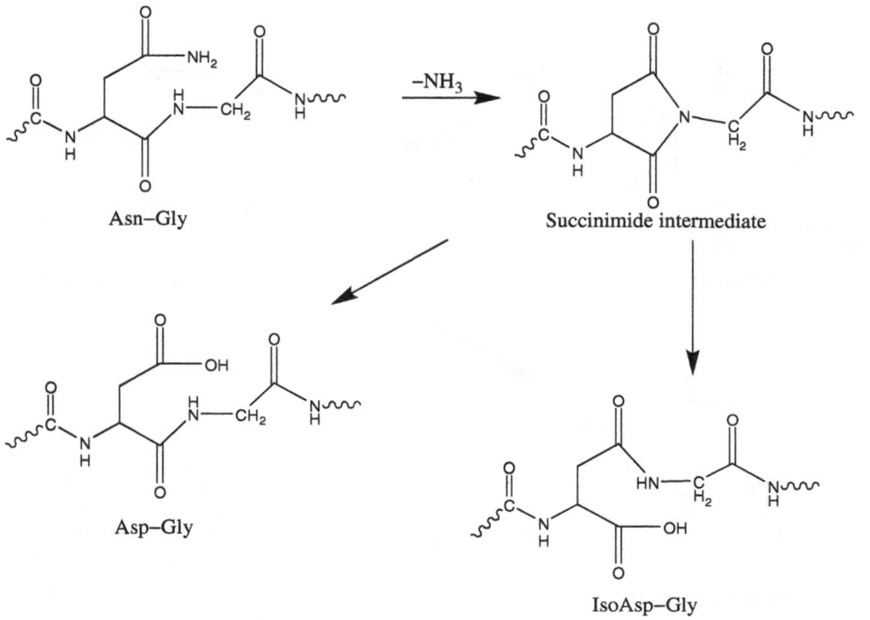

Asn–Gly

$-NH_3$

Succinimide intermediate

Asp–Gly

IsoAsp–Gly

**Figure 5.2** Deamidation of Asn into Asp or IsoAsp in a protein with the participation of neighboring Gly residue.

–QP- sequences. However, the rates of deamidation of Asn and Gln residues may depend on the local environment and neighboring group participation. For example, surface exposed Asn and Gln residues undergoes deamidation much more readily than the Asn and Gln residues that are not surface exposed (Gupta and Srivastava 2004). Similarly, deamidation may occur much more rapidly if the Asn and/or Gln residues are next to small and flexible amino acid residues (Bischoff and Kolbe 1994; Reissner and Aswad 2003). For instance, proteins containing Asn residue followed by a glycine residue undergo deamidation much more quickly than the other Asn residues in the protein (Washington et al. 2013).

Nature of solutions and salts in the buffer may also influence the rate of deamidation of proteins (Cleland et al. 1993). Deamidation may occur much more quickly at basic pH compared to neutral pH (Cleland et al. 1993; Wooding et al. 2016). Temperature of solutions may also impact the rates of deamidation. For instance, proteins stored at elevated temperature are observed to have more significant deamidation compared to proteins stored at lower temperature (Cleland et al. 1993). Buffer components such as metal ions may also influence the rate of deamidation (Brewer 1981). Storage containers

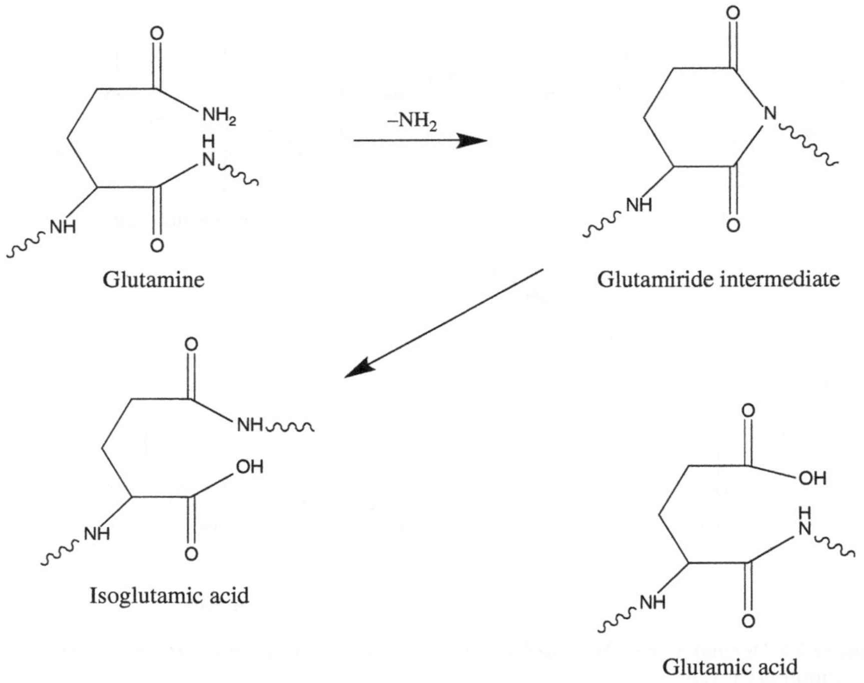

**Figure 5.3** Deamidation of glutamine residues of proteins.

and closers may also influence the rate of deamidation of proteins (Zhou et al. 2010).

## Physicochemical Characteristics of Deamidated Proteins

Deamidation results in a net mass change of approximately +0.984 amu that can be easily detectable by high-resolution mass spectrometric analysis. Protein deamidation results in changing amide functionality to carboxylic acid functionality. Hence, deamidation may also result in change in overall ionic charge that might affect the total charge of proteins. Because of this reason isoelectric point of proteins may also change upon deamidation. Hence, deamidation may convert a basic protein into an acidic protein. Deamidation may also result in changing the overall hydrophobicity and hydrophilicity of proteins. If the Asn and/or Gln residues are surface exposed, deamidation may change the surface hydrophobicity and hydrophilicity of proteins. Because of the change in charge, the electrophoretic mobility of deamidated proteins may be different from their native proteins. Because of the change in overall charge, the deamidated proteins may be more sensitive proteolysis.

## Biological Significance of Deamidation of Proteins

Deamidation may affect the proteins functions including their biological activity, enzymatic activity, binding affinity, etc. (Enyedi et al. 2015; Leblanc et al. 2017; Miao et al. 2017; Noguchi 2010). For example, Yu et al. reported that the deamidation of Gln40 blocks structural reconfiguration and activation of SCF (SKP, Cullin, F-box) ubiquitin ligase complex by Nedd8 (Yu et al. 2015). Additionally, single deamidation has been reported to impair the oncogenic properties of Bcl-xL proteins both in vivo and in vitro (Beaumatin et al. 2016). In addition, modulation of eukaryotic host cells by bacterial virulence factors is mediated by deamidation (Washington et al. 2013).

IgGs contain deamidation sites both in the Fab and Fc domains (Chelius et al. 2005). Deamidation can occur on the intact antibodies, antibody fragments, such as Fab and Fc, and also on the antibody drug conjugates as well as on other antibody-related products (Geuijen et al. 2017). Deamidation of antibodies and related products can occur both in vitro and in vivo (Liu et al. 2009). The deamidation in the Fc region of IgGs may impact the antibody effector functions (Liu et al. 2009; Sinha et al. 2009; Zhang et al. 2014). Deamidation in the complementarity-determining region (CDR) regions of IgGs may impact the antibody binding to antigens (Chelius et al. 2005; Huang et al. 2005; Wooding et al. 2016).

### Deamidation and Immunogenicity

Deamidation may affect the immunogenicity of proteins. Using a genetically deamidated protein, Verma et al. (2016) have shown that the deamidation decreases immunogenicity of recombinant protective antigen (rPA) thus impacting the effectiveness of the rPA vaccine. The loss of immunogenicity of rPA is not due to the presence of an adjuvant but is due to the inability of the deamidated rPA to elicit T-cell help for antibody production (Hassett et al. 2015; Verma et al. 2016). Deamidation may also generate proteins with isoaspartic acid residues which is not natural and hence may present immunogenicity issues in humans. Extreme deamidation may create super acidic proteins and such proteins may become highly immunogenic. Almost all proteins containing Asn and/or Glu residues eventually undergo deamidation. The eventual deamidation may lead to protein degradation by proteolysis or elimination due to immunogenicity issues. Deamidation is also shown to play a role in cataract formation and in aging (Lampi et al. 2014).

### Deamidation and Pharmacokinetics Properties of Proteins

Deamidation may also affect the pharmacokinetic properties of proteins (Esposito et al. 2015; Liu et al. 2009; Tesar et al. 2017). Deamidation of glutaminyl and asparaginyl residues found in peptides and proteins may occur both in vitro and in vivo (Liu et al. 2009). The nonenzymatic deamidation of Asn and Gln residues of proteins and peptides was observed in vitro as well as in vivo (Liu et al. 2009). Deamidated proteins and peptides have been shown to have reduced serum half-life compared to the native proteins (Liu et al. 2009; Møss and Bundgaard 1990; Zhang et al. 2014). Hence, protein engineering methods have been used to mutate the Asn residues to improve the serum-life of some proteins (Liu et al. 2009; Tesar et al. 2017).

### Deamidation in Recombinant Therapeutic Proteins

Deamidation has been observed in almost all the RTPs including mAbs (Zhao et al. 2016). Somatostatin was the very first recombinant mammalian protein produced in bacteria back in 1977 (Itakura et al. 1977). However, to date no information is available on the deamidation of somatostatin. Recombinant human insulin, Humulin, was first produced in *Escherichia coli* in 1978, which was first approved as a human therapeutic in 1982 (Brange et al. 1992; Cohn 1977, 1978; Crea et al. 1978; FDA Drug Bulletin 1982). In 1981, Fisher and Porter 1981 discovered the deamidation in bovine insulin. In 1991, deamidation of recombinant human insulin was observed (Wright 1991). Similarly, deamidation of recombinant human growth hormone was also observed by Skottner et al. (1988). To date deamidation of almost all RTPs has been observed (Harris 2005). In many of these RTPs, deamidation impacts the products quality, safety, and efficacy (Alt et al. 2016). If a deamidation site is in the CDR region of mAbs, it may impact the antibody binding to antigen (Ohkuri et al. 2013). Deamidation of Asn residues in the Fc domains may impact the antibody effector functions and FcRn binding of mAbs (Miao et al. 2017). Hence, deamidation is an important quality attribute for RTPs (Alt et al. 2016). Specifically, if deamidation impacts the antigen binding of an antibody, it will be a critical quality attribute (CQA) for RTPs. In such cases, it is necessary to monitor the extent of deamidation in each and every manufacturing lots of RTPs and mAbs.

### Methods for the Analysis of Deamidation in Proteins

Deamidation is an important PTM that affects physicochemical, biological, and pharmacokinetic properties of proteins. For many RTPs including mAbs, deamidation might be a CQA as it may affect their safety and efficacy profiles. Hence, major advances have been made to develop highly sensitive and reproducible methods to detect as well as to quantitate deamidation of proteins. In proteins, deamidation not only results in the net mass shift of approximately +0.984 amu but also results in change in the overall charge of the corresponding amino acid residue and the associated proteins. The mass difference from deamidation in proteins may be easily detected using high-resolution mass spectrometry. In addition, the change in charge is very useful to separate the deamidated proteins from the native proteins using ion-exchange chromatography (IEC), hydrophobic interaction chromatography (HIC), iso-electric focusing (IEF), and capillary electrophoretic methods (Lew et al. 2015). The site of deamidation can be easily detected using peptide mapping methods such as reversed-phase high performance liquid chromatography (RP-HPLC) and RP-HPLC combined with online (or inline) MS methods (Formolo et al. 2014; Lew et al. 2015). Deamidation results in proteins containing either Asp or isoAsp. These two isomers are relatively difficult to differentiate because the mass and charge are the same for these two variants. Hence, many of the methods described above are not very useful to distinguish these two isomers. However, the electron capture dissociation mass spectral (ECD-MS) fragmentation method is very useful to distinguish Asp from isoAsp residues (Zhang et al. 2015; Mulagapati et al. 2017). The ECD-MS method is also very useful to quantify the Asp and isoAsp residues in proteins (Mukherjee et al. 2010; Zhang et al. 2015). Additionally, enzyme linked immunosorbent assay (ELISA) methods have also been developed to detect and quantitate isoAsp residues in certain proteins (Wang et al. 2013; Masiri et al. 2016). ELISA assays have also been developed to detect and quantitate deamidation in proteins and in RTPs (Tranquet et al. 2015). However, the ELISA methods have very high variability compared to the chromatographic or electrophoretic methods.

## References


Alt, N., Zhang, T.Y., Motchnik, P. et al. (2016). Determination of critical quality attributes for monoclonal antibodies using quality by design principles. *Biologicals* 44 (5): 291–305. https://doi.org/10.1016/j.biologicals.2016.06.005.

Beaumatin, F., El Dhaybi, M., Lasserre, J.P. et al. (2016). N52 monodeamidated Bcl-xL shows impaired oncogenic properties in vivo and in vitro. *Oncotarget* 7 (13): 17129–17143. https://doi.org/10.18632/oncotarget.7938.

Bischoff, R. and Kolbe, H.V. (1994). Deamidation of asparagine and glutamine residues in proteins and peptides: structural determinants and analytical methodology. *J. Chromatogr. B Biomed. Appl.* 662 (2): 261–278.

Brange, J., Hallund, O., and Sørensen, E. (1992). Chemical stability of insulin. 5. Isolation, characterization and identification of insulin transformation products. *Acta Pharm Nord.* 4 (4): 223–232.

Brewer, J.M. (1981). Yeast enolase: mechanism of activation by metal ions. *CRC Crit. Rev. Biochem.* 11 (3): 209–254.

Chelius, D., Rehder, D.S., and Bondarenko, P.V. (2005). Identification and characterization of deamidation sites in the conserved regions of human immunoglobulin gamma antibodies. *Anal. Chem.* 77 (18): 6004–6011.

Clarke, S. (1987). Propensity for spontaneous succinimide formation from aspartyl and asparaginyl residues in cellular proteins. *Int. J. Pept. Protein Res.* 30 (6): 808–821.

Cleland, J.L., Powell, M.F., and Shire, S.J. (1993). The development of stable protein formulations: a close look at protein aggregation, deamidation, and oxidation. *Crit. Rev. Ther. Drug Carrier Syst.* 10 (4): 307–377. Review. Erratum in: (1994). *Crit. Rev. Ther. Drug Carrier Syst.* 11 (1): 60.

Cohn, V. (1977). Scientists duplicate rat insulin gene. *Washington Post* (24 May): A1+.

Cohn, V. (1978). Scientists in California create gene to make human insulin. *Washington Post* (7 September): A1+.

Crea, R., Kraszewski, A., Hirose, T., and Itakura, K. (1978). Chemical synthesis of genes for human insulin. *Proc. Natl. Acad. Sci. U.S.A.* 75 (12): 5765–5769.

Enyedi, K.N., Czajlik, A., Knapp, K. et al. (2015). Development of cyclic NGR peptides with thioether linkage: structure and dynamics determining deamidation and bioactivity. *J. Med. Chem.* 58 (4): 1806–1817. https://doi.org/10.1021/jm501630j.

Esposito, S., Deventer, K., Geldof, L., and Van Eenoo, P. (2015). In vitro models for metabolic studies of small peptide hormones in sport drug testing. *J. Pept. Sci.* 21 (1): 1–9.

FDA Drug Bulletin (1982). Human insulin receives FDA approval. *FDA Drug Bull.* 12 (3): 18–19.

Fisher, B.V. and Porter, P.B. (1981). Stability of bovine insulin. *J. Pharm. Pharmacol.* 33 (4): 203–206.

Formolo, T., Heckert, A., and Phinney, K.W. (2014). Analysis of deamidation artifacts induced by microwave-assisted tryptic digestion of a monoclonal antibody. *Anal. Bioanal. Chem.* 406 (26): 6587–6598.

Geuijen, K.P.M., Oppers-Tiemissen, C., Egging, D.F. et al. (2017). Rapid screening of IgG quality attributes - effects on Fc receptor binding. *FEBS Open Bio.* 7 (10): 1557–1574. https://doi.org/10.1002/2211-5463.12283.

Gupta, R. and Srivastava, O.P. (2004). Deamidation affects structural and functional properties of human alphaA-crystallin and its oligomerization with alphaB-crystallin. *J. Biol. Chem.* 279 (43): 44258–44269.

Harris, R.J. (2005). Heterogeneity of recombinant antibodies: linking structure to function. *Dev. Biol. (Basel)* 122: 117–127.

Hassett, K.J., Vance, D.J., Jain, N.K. et al. (2015). Glassy-state stabilization of a dominant negative inhibitor anthrax vaccine containing aluminum hydroxide and glycopyranoside lipid A adjuvants. *J. Pharm. Sci.* 104 (2): 627–639.

Huang, L., Lu, J., Wroblewski, V.J. et al. (2005). In vivo deamidation characterization of monoclonal antibody by LC/MS/MS. *Anal. Chem.* 77 (5): 1432–1439.

Itakura, K., Hirose, T., Crea, R. et al. (1977). Expression in *Escherichia coli* of a chemically synthesized gene for the hormone somatostatin. *Science* 198 (4321): 1056–1063.

Keceli, G. and Toscano, J.P. (2012). Reactivity of nitroxyl-derived sulfinamides. *Biochemistry* 51 (20): 4206–4216. https://doi.org/10.1021/bi300015u.

Lampi, K.J., Wilmarth, P.A., Murray, M.R., and David, L.L. (2014). Lens β-crystallins: the role of deamidation and related modifications in aging and cataract. *Prog. Biophys. Mol. Biol.* 115 (1): 21–31. https://doi.org/10.1016/j.pbiomolbio.2014.02.004.

Leblanc, Y., Ramon, C., Bihoreau, N., and Chevreux, G. (2017). Charge variants characterization of a monoclonal antibody by ion exchange chromatography coupled on-line to native mass spectrometry: case study after a long-term storage at +5°C. *J. Chromatogr. B Anal. Technol. Biomed. Life Sci.* 1048: 130–139. https://doi.org/10.1016/j.jchromb.2017.02.017.

Lew, C., Gallegos-Perez, J.L., Fonslow, B. et al. (2015). Rapid level-3 characterization of therapeutic antibodies by capillary electrophoresis electrospray ionization mass spectrometry. *J. Chromatogr. Sci.* 53 (3): 443–449.

Li, X., Lin, C., and O'Connor, P.B. (2010). Glutamine deamidation: differentiation of glutamic acid and gamma-glutamic acid in peptides by electron capture dissociation. *Anal. Chem.* 82 (9): 3606–3615. https://doi.org/10.1021/ac9028467.

Lindner, H. and Helliger, W. (2001). Age-dependent deamidation of asparagine residues in proteins. *Exp. Gerontol.* 36 (9): 1551–1563.

Liu, Y.D., van Enk, J.Z., and Flynn, G.C. (2009). Human antibody Fc deamidation in vivo. *Biologicals* 37 (5): 313–322. https://doi.org/10.1016/j.biologicals.2009.06.001.

Liu, S., Moulton, K.R., Auclair, J.R., and Zhou, Z.S. (2016). Mildly acidic conditions eliminate deamidation artifact during proteolysis: digestion with endoprotease Glu-C at pH 4.5. *Amino Acids* 48 (4): 1059–1067. https://doi.org/10.1007/s00726-015-2166-z.

Masiri, J., Benoit, L., Katepalli, M. et al. (2016). Novel monoclonal antibody-based immunodiagnostic assay for rapid detection of deamidated gluten residues. *J. Agric. Food. Chem.* 64 (18): 3678–3687. https://doi.org/10.1021/acs.jafc.5b06085.

Miao, S., Xie, P., Zou, M. et al. (2017). Identification of multiple sources of the acidic charge variants in an $IgG_1$ monoclonal antibody. *Appl. Microbiol. Biotechnol.* 101 (14): 5627–5638. https://doi.org/10.1007/s00253-017-8301-x.

Møss, J. and Bundgaard, H. (1990). Kinetics and pattern of degradation of thyrotropin-releasing hormone (TRH) in human plasma. *Pharm. Res.* 7 (7): 751–755.

Mukherjee, R., Adhikary, L., Khedkar, A., and Iyer, H. (2010). Probing deamidation in therapeutic immunoglobulin gamma ($IgG_1$) by 'bottom-up' mass spectrometry with electron transfer dissociation. *Rapid Commun. Mass Spectrom.* 24 (7): 879–884. https://doi.org/10.1002/rcm.4464.

Mulagapati, S., Koppolu, V., and Raju, T.S. (2017). Decoding of O-linked glycosylation by mass spectrometry. *Biochemistry* 56 (9): 1218–1226. https://doi.org/10.1021/acs.biochem.6b01244.

Noguchi, S. (2010). Structural changes induced by the deamidation and isomerization of asparagine revealed by the crystal structure of Ustilago sphaerogena ribonuclease U2B. *Biopolymers* 93 (11): 1003–1010. https://doi.org/10.1002/bip.21514.

Ohkuri, T., Murase, E., Sun, S.L. et al. (2013). Characterization of deamidation at Asn138 in L-chain of recombinant humanized Fab expressed from *Pichia pastoris*. *J. Biochem.* 154 (4): 333–340. https://doi.org/10.1093/jb/mvt061.

Qin, K., Yang, D.S., Yang, Y. et al. (2000). Copper(II)-induced conformational changes and protease resistance in recombinant and cellular PrP. Effect of protein age and deamidation. *J. Biol. Chem.* 275 (25): 19121–19131.

Reissner, K.J. and Aswad, D.W. (2003). Deamidation and isoaspartate formation in proteins: unwanted alterations or surreptitious signals? *Cell. Mol. Life Sci.* 60 (7): 1281–1295.

Sinha, S., Zhang, L., Duan, S. et al. (2009). Effect of protein structure on deamidation rate in the Fc fragment of an $IgG_1$ monoclonal antibody. *Protein Sci.* 18 (8): 1573–1584. https://doi.org/10.1002/pro.173.

Skottner, A., Forsman, A., Skoog, B. et al. (1988). Biological characterization of charge isomers of human growth hormone. *Acta Endocrinol. (Copenh)* 118 (1): 14–21.

Stephenson, R.C. and Clarke, S. (1989). Succinimide formation from aspartyl and asparaginyl peptides as a model for the spontaneous degradation of proteins. *J. Biol. Chem.* 264 (11): 6164–6170.

Stroop, S.D. (2007). A modified peptide mapping strategy for quantifying site-specific deamidation by electrospray time-of-flight mass spectrometry. *Rapid Commun. Mass Spectrom.* 21 (6): 830–836.

Tesar, D., Luoma, J., Wyatt, E.A. et al. (2017). Protein engineering to increase the potential of a therapeutic antibody Fab for long-acting delivery to the eye. *MAbs* 9 (8): 1297–1305. https://doi.org/10.1080/19420862.2017.1372078.

Tranquet, O., Lupi, R., Echasserieau-Laporte, V. et al. (2015). Characterization of antibodies and development of an indirect competitive immunoassay for detection of deamidated gluten. *J. Agric. Food. Chem.* 63 (22): 5403–5409. https://doi.org/10.1021/acs.jafc.5b00923.

Verma, A., McNichol, B., Domínguez-Castillo, R.I. et al. (2013). Use of site-directed mutagenesis to model the effects of spontaneous deamidation on the immunogenicity of *Bacillus anthracis* protective antigen. *Infect Immun.* 81 (1): 278–284. https://doi.org/10.1128/IAI.00863-12.

Verma, A., Ngundi, M.M., and Burns, D.L. (2016). Mechanistic analysis of the effect of deamidation on the immunogenicity of anthrax protective antigen. *Clin. Vaccine Immunol.* 23 (5): 396–402. https://doi.org/10.1128/CVI.00701-15.

Volkin, D.B., Mach, H., and Middaugh, C.R. (1997). Degradative covalent reactions important to protein stability. *Mol. Biotechnol.* 8 (2): 105–122.

Wang, X., Li, Q., and Davies, M. (2013). Development of antibody arrays for monoclonal antibody higher order structure analysis. *Front. Pharmacol.* 4: 103. https://doi.org/10.3389/fphar.2013.00103.

Washington, E.J., Banfield, M.J., and Dangl, J.L. (2013). What a difference a Dalton makes: bacterial virulence factors modulate eukaryotic host cell signaling systems via deamidation. *Microbiol. Mol. Biol. Rev.* 77 (3): 527–539. https://doi.org/10.1128/MMBR.00013-13.

Wooding, K.M., Peng, W., and Mechref, Y. (2016). Characterization of pharmaceutical IgG and biosimilars using miniaturized platforms and LC-MS/MS. *Curr. Pharm. Biotechnol.* 17 (9): 788–801.

Wright, H.T. (1991). Sequence and structure determinants of the nonenzymatic deamidation of asparagine and glutamine residues in proteins. *Protein Eng.* 4 (3): 283–294.

Xie, M. and Schowen, R.L. (1999). Secondary structure and protein deamidation. *J. Pharm. Sci.* 88 (1): 8–13.

Yu, C., Mao, H., Novitsky, E.J. et al. (2015). Gln40 deamidation blocks structural reconfiguration and activation of SCF ubiquitin ligase complex by Nedd8. *Nat. Commun.* 6: 10053.

Zhang, Y.T., Hu, J., Pace, A.L. et al. (2014). Characterization of asparagine 330 deamidation in an Fc-fragment of $IgG_1$ using cation exchange chromatography and peptide mapping. *J. Chromatogr. B Anal. Technol. Biomed. Life Sci.* 965: 65–71. https://doi.org/10.1016/j.jchromb.2014.06.018.

Zhang, J., Reza Malmirchegini, G., Clubb, R.T., and Loo, J.A. (2015). Native top-down mass spectrometry for the structural characterization of human hemoglobin. *Eur. J. Mass Spectrom. (Chichester)* 21 (3): 221–231. https://doi.org/10.1255/ejms.1340.

Zhao, J., Li, J., Xu, S., and Feng, P. (2016). Emerging roles of protein deamidation in innate immune signaling. *J. Virol.* 90 (9): 4262–4268. https://doi.org/10.1128/JVI.01980-15.

Zhou, S., Zhang, B., Sturm, E. et al. (2010). Comparative evaluation of disodium edetate and diethylenetriaminepentaacetic acid as iron chelators to prevent metal-catalyzed destabilization of a therapeutic monoclonal antibody. *J. Pharm. Sci.* 99 (10): 4239–4250. https://doi.org/10.1002/jps.22141.

# 6

# Glycation of Proteins

## Introduction

Protein glycation is a nonenzymatic post-translational modification (PTM) of primary amine groups of proteins (Anguizola et al. 2013). Glycation primarily occurs at the α-amino termini of proteins and also at epsilon amino group of the Lys side chain (Raghav and Ahmad 2014). Glycation is a chemical modification in which susceptible primary amine groups reacts with carbonyl groups to form a highly reactive and unstable Schiff base that undergoes multistep rearrangement called Amadori rearrangement to form ketoamine (Cao et al. 2015; Li et al. 2016; Taghavi et al. 2017; Wei et al. 2017). The ketoamine formed as a result of glycation is also called as Amadori ketoamine. This Amadori ketoamine contains a more stable covalent bond than the reactive Schiff's base.

Reducing sugars such as glucose, galactose, and mannose contain a hemiacetal group that is in equilibrium with aldose form in solutions. The aldose form of reducing sugars contains a reactive aldehyde group. This reactive aldehyde group of reducing sugars such as hexoses reacts with the available primary amine groups of organic molecules to form a Schiff's base. In the presence of water, Schiff base undergoes Amadori rearrangement to form a more stable Amadori ketoamine product (Smith and Thornalley 1992).

Positively charged primary amine groups are present on the surface of proteins and many of them are solvent exposed (Cherian and Abraham 1993). However, not all the primary amine groups that are present on the surface of proteins are prone to glycation (Silva et al. 2014; Paudel et al. 2016; Tu et al. 2017). Unlike N-glycosylation no specific protein sequence has been observed that signals the glycation hotspots on the surface of proteins (Rabbani et al. 2016). However, three-dimensional structural studies of proteins suggest that the local environments with neighboring histidine residues or other basic amino acid residues such as arginine residues have been observed to influence the glycation of primary amine groups of Lys residues (Clark et al. 2013). Additionally, neighboring Lys residues may also influence the glycation of amine functional groups of surface exposed Lys residues. In addition, primary

*Co- and Post-Translational Modifications of Therapeutic Antibodies and Proteins*, First Edition. T. Shantha Raju.

amine groups that are close to negatively charged amino acid residues along with the neighboring aromatic amino acid residues are also observed to be most reactive to reducing sugars for glycation of proteins. Hence, reactivity of primary amine groups of proteins is dependent on the local environment that facilitates the initial deprotonation of amine group to form Schiff's base (with reducing sugars) that undergoes multiple rearrangement reactions initiated by acid/base catalysis to form a relatively stable Amadori ketoamine product (Lima et al. 2009).

The Amadori ketoamine product of glycation contains a keto group that is a reactive group (Venkatraman et al. 2001). The keto group ultimately undergoes further rearrangement and/or degradation to form advanced glycation end products (AGEs). The presence of reactive oxygen species and nitric oxide radicals may also influence the degradation of Amadori ketoamine product and formation of AGEs. These AGE products are recognized by a receptor called receptor of advanced glycation end products (RAGEs). The complex formed by AGE products with RAGE are immunogenic and are found to be toxic to humans (Chikazawa et al. 2013; Sessa et al. 2014).

Glycation of human serum proteins such as albumin, hemoglobin, immunoglobulins have been studied extensively (Rabbani et al. 2016). Glycation of human serum albumin (HSA) has been shown to impact the FcRn binding and hence affects the pharmacokinetic properties of albumin (Neelofar and Ahmad 2015; Wagner et al. 2016). Glycation of serum IgGs may impact their ability to bind their respective antigens and receptors including Fcγ receptors (Rabbani et al. 2016). Measuring the glycation levels of hemoglobin is one of the tests used to detect and diagnose diabetes in humans (Raghav and Ahmad 2014). Protein glycation may vary with the age and may be indicative of disease status (Svacina et al. 1990; Rabbani et al. 2016). Protein glycation may also vary with the diet and sex. In addition to diabetes, protein glycation has been linked to many autoimmune and neurologic diseases. The neurological diseases that are attributed to protein glycation include Alzheimer, Parkinson, etc. (Chen et al. 2017; Du et al. 2017; Somensi et al. 2017).

## Mechanism of Protein Glycation

Glycation is primarily a chemical condensation reaction between primary amine groups of amino acid residues of proteins and aldehyde groups of reducing sugars. Primary amine groups are basic functional groups that are highly reactive to carbonyl groups. Amino acid residues in the N-terminus of proteins and internal Lys residues contain primary amine groups. Aldose sugars such as glucose (Glc), galactose (Gal), etc. contain hemiacetal group

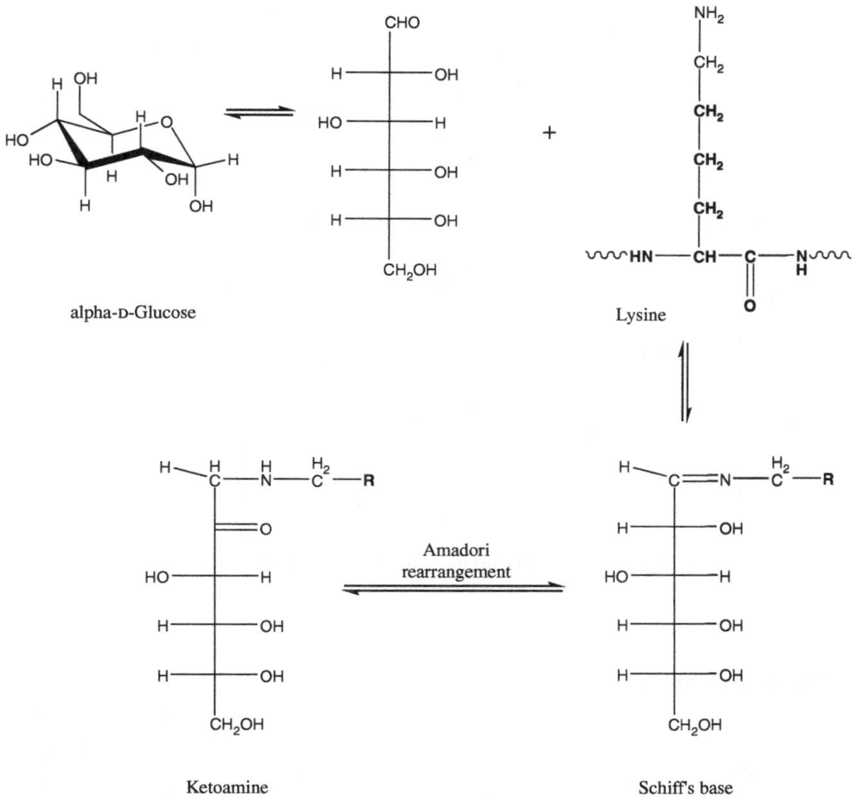

alpha-D-Glucose

Lysine

Amadori rearrangement

Ketoamine

Schiff's base

**Figure 6.1** Mechanism of protein glycation. (*See insert for color representation of this figure.*)

that acts as an aldehyde functionality. The open chain form of hexoses is in equilibrium with pyranose ring forms in solutions. Accordingly, the aldehyde groups present in the open chain form of hexoses is in equilibrium with the hemiacetal groups in solutions. Primary amine groups under basic conditions are positively charged that makes nucleophilic contact with the aldehyde groups of aldoses. Hence, primary amine groups undergo condensation reaction with carbonyl groups of reducing sugars to form reactive Schiff's base (see Figure 6.1). The Schiff base, highly reactive and very unstable, quickly undergoes Amadori rearrangement as shown in Figure 6.1 to form relatively stable Amadori ketoamine functionality (Mori and Manning 1986). The Amadori ketoamine that eventually undergoes degradation into small organic molecules that are collectively known as AGEs (Khalifah et al. 1999). Chemical structure of few examples of AGEs are shown in Figure 6.2.

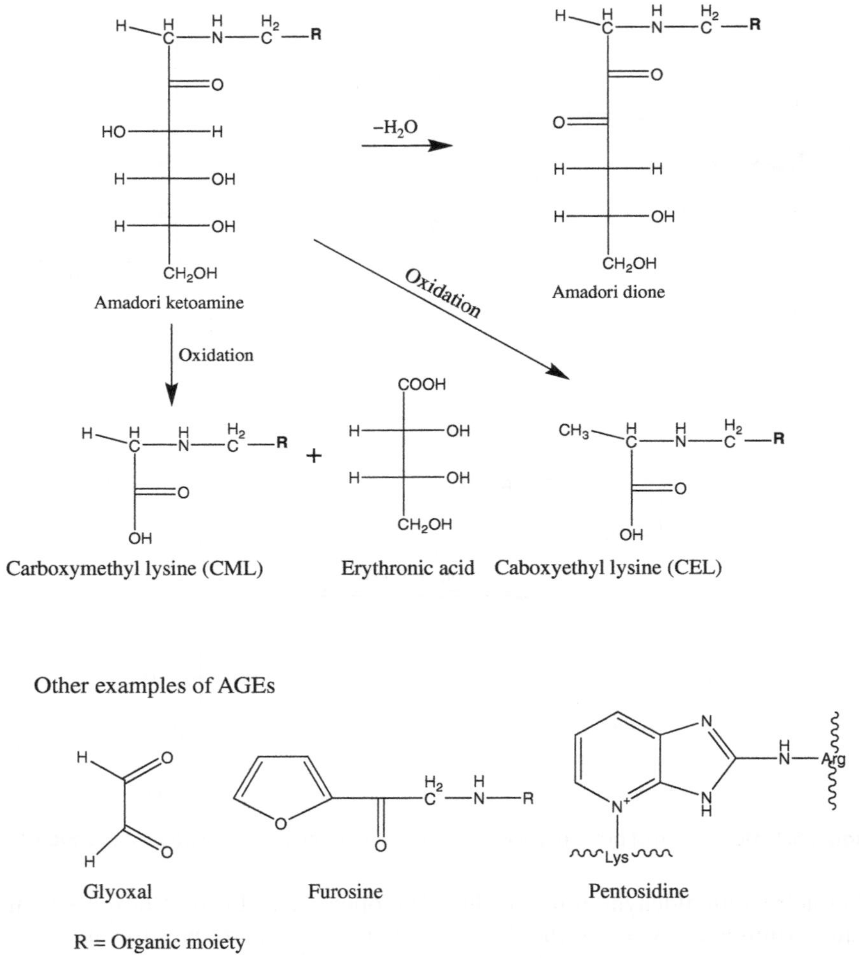

Amadori ketoamine

$-H_2O$

Amadori dione

Oxidation

Oxidation

Carboxymethyl lysine (CML) + Erythronic acid

Caboxyethyl lysine (CEL)

Other examples of AGEs

Glyoxal

Furosine

Pentosidine

R = Organic moiety

**Figure 6.2** Mechanism and chemical structure of few examples of advanced glycation end products (AGEs). (*See insert for color representation of this figure.*)

## Glycation of Proteins in Human

Human serum contains circulating proteins such as albumin, immunoglobulins, glycoproteins. Human serum also carries glucose that derives from food, drinks, drugs, vitamins, etc. The amount of glucose present in the human serum is dependent on the type of diet and the time of food intake. Proteins are large molecules and have longer serum half-life, whereas glucose is a small molecule, which does not stay longer in the serum. However, due to the continuous process of food intake and digestion, some amounts of reducing sugars such as glucose will be present in the human serum most of the time. In healthy human subjects, the presence of normal amount of glucose and other reducing sugars do not create a problem. However, the glucose content in the serum fluctuates a lot in patients with diabetes, and other autoimmune and neurological diseases (Pareek 2017; Zhang et al. 2017).

In humans, glycation of primary amine groups present in the serum proteins and glycoproteins with the circulating glucose (or other sugars) occurs in the serum. Accordingly, many serum proteins have been reported to be glycated (Neelofar and Ahmad 2015; Ishida et al. 2017). For example, glycated albumin, glycated immunoglobulins, glycated glycoproteins, glycated lipoproteins, etc. have been extensively characterized from human serum (Rabbani et al. 2016; Ishida et al. 2017). In addition, it has been shown that the level of glycation may vary with the age, gender, and disease status of human subjects (Du et al. 2017).

## Protein Glycation and Human Diseases

Protein glycation is very prevalent in mammals, specifically in humans (Bitensky et al. 1989). Glycation of human proteins may dependent on the disease status of human subjects (Du et al. 2017; Ishida et al. 2017; Isoda et al. 2017). In diabetic patients, increased glycation of serum albumin and to some extent other serum proteins including serum IgGs has been observed (Neelofar and Ahmad 2015). In addition to albumin and IgGs, other circulating proteins such as hemoglobin (Hb) are also glycated. In fact, measuring the levels of glycated hemoglobin (HbA1C) is one of the clinical diagnostic tests used to detect diabetes in humans (Salinas et al. 2017; Saxena et al. 2017). Recently, measuring the levels of glycation in serum albumin is suggested to be more indicative of progression of disease status in diabetic patients compared to hemoglobin tests (Neelofar and Ahmad 2015; Ghosh et al. 2017; Ishida et al. 2017). Many of the serum proteins such as albumin and IgGs have very long serum half-life. However, glycation of these proteins might affect their serum half-life. While in the serum, the more stable Amadori ketoamine product of glycated proteins may undergo degradation to form AGEs. The AGEs are immunogenic and toxic to humans that might be the cause of complications commonly observed in patients with diabetes such as diabetic neuropathy. AGEs are a heterogeneous group of organic molecules that are abundant in patients with age-related diseases including diabetes, atherosclerosis, liver cirrhosis. It has been observed by Chikazawa et al., C1q protein binds AGEs and the complex is involved in mediating the proteolytic cleavage of complement protein 5 to release C5a (Chikazawa et al. 2016). Additionally, the AGEs might be taken up by vascular tissues that may crosslink them to extracellular matrix. Such modifications may account for cytotoxicity, tissue necrosis, and vascular pathologies that is very commonly seen in diabetic complications. In addition, AGE-related cytotoxicity is considered as one of the main mechanism for neurological disorders such as Alzheimer, Parkinson, etc. diseases (Ahmed 1991).

## Glycation in Recombinant Therapeutic Proteins

Sugars such as lactose, trehalose, maltose are being used as components in formulation buffers to formulate recombinant therapeutic proteins (RTPs). These formulation components may contain residual amounts of Glc, Gal, Man, etc., as contaminants. In addition, sugars such as Glc, Gal, Man, etc. are being used as energy sources as feeds during cell culture fermentation process to manufacture RTPs. Since many of the RTPs contain surface-exposed Lys residues, they may undergo glycation during cell culture fermentation process, purification, and/or during storage in formulation buffers that contain reducing sugars. For example, rhDNase formulated in lactose containing buffer and stored in freeze-dried powdered state was found to be glycated at five of the available six Lys residues and to some extent at the amino terminus as well (Quan et al. 1999). Three-dimensional structural analysis of the rhDNase revealed that the requirements for glycation such as specificity of site reactivity, site accessibility and the nearby proton donor or acceptor groups are all maintained (Quan et al. 1999).

Glycation of recombinantly produced monoclonal antibodies (mAbs) has been extensively reported (Andya et al. 1999; Harris 2005; Quan et al. 2008; Zhang et al. 2008, 2009, 2012; Kaschak et al. 2011; Yuk et al. 2011; Butko et al. 2014; Wei et al. 2017). Glycation of mAbs has been shown to affect their stability, bioactivity, immunogenicity, etc. (Wei et al. 2017). Glycation in the complementarity-determining region (CDR) may impact antibody binding to its antigen. Similarly, glycation in the Fc region may impact antibody binding to Fc receptors including FcRn. Hence, glycation may impact antibody effector functions as well as its serum half-life. mAbs can also be glycated during manufacturing, storage and during in vivo circulation upon treatment (Wei et al. 2017). The AGEs derived from glycation may impact the color of mAbs solutions during storage (Butko et al. 2014). The yellow-brown color often observed in many RTPs and mAbs in solution during storage has been attributed to AGEs derived from glycation of the protein product (Butko et al. 2014).

Overall, glycation in RTPs and mAbs is very prevalent because of the presence of reducing sugars during the cell culture process, purification, formulation, storage, etc. The glycation may affect the solubility, stability, affinity, bioactivity, pharmacokinetics, etc., properties of RTPs and mAbs. Glycation may also affect the color of solutions of biopharmaceuticals due to AGEs. If the glycation affects the functions of RTPs and mAbs, it may become critical quality attribute (CQA). Hence, it may be necessary not only to control the exposure of RTPs and mAbs to reducing sugars, it may be a good idea to completely avoid reducing sugars as excipients of formulation.

Although glycation may impact the bioactivity of RTPs and mAbs, some RTPs and mAbs may be used to treat diseases that are derived from abnormal protein glycation and AGEs. For example, Yu et al. showed that rhEPO protects Schwann cells from AGE-related oxidative stress and apoptosis (Yu et al. 2015). Brederson et al. identified a mAb that binds RAGE and attenuates inflammatory and neuropathic pain in mouse (Brederson et al. 2016). Although these investigations are in early stages, it is possible to project that the use of RTPs and mAbs and other biotechnology derived products may become available to treat human diseases derived due to protein glycation and its toxic derivatives such as AGEs.

## Methods to Analyze Protein Glycation

There are several analytical methods available to measure protein glycation. High performance liquid chromatography (HPLC), enzyme linked immunosorbent assay (ELISA), capillary electrophoresis (CE), and mass spectrometry methods have been successfully used to measure protein glycation (Turpeinen et al. 1995; Harris 2005; Wei et al. 2017). HPLC method using boronate affinity column is a more specific method to measure glycation of purified proteins (Quan et al. 1999, 2008). This method has been more extensively used in biopharmaceutical industry to measure glycation of therapeutic recombinant proteins including mAbs (Quan et al. 1999, 2008). Figure 6.3 shows the separation of glycated mAb from non-glycated mAb using boronate affinity chromatography. The method is very useful to separate and quantitate the glycation in RTPs including mAbs. An iso-electric focusing (IEF) method has also been used to detect protein glycation in RTPs (Andya et al. 1999).

Bio-Rad laboratories has developed an automated analyzer that is also called Diamat to measure HbA1C levels and is based on cation-exchange chromatographic method (Turpeinen et al. 1995). Turpeinen et al. compared the Bio-Rad's method and Abbott's IMx analyzer method to their own HPLC method using PolyCat A column from Poly LC, Inc. (Turpeinen et al. 1995). According to these authors, no method is perfect in quantifying the glycation of haemoglobin but would be good enough to detect the protein glycation (Turpeinen et al. 1995). In addition to HPLC, ELISA, and mass spectral methods, chemical and computational approaches have also been used to detect and quantitate protein glycation (Saleem et al. 2015). For identifying potential protein glycation sites using conventional collision-induced dissociation-mass spectrometry (CID-MS), Saleem et al. employed sodium borohydride-based reduction chemistry to reduce the Amadori product and used CID-MS method along with the computational search engines to detect sites of glycation in

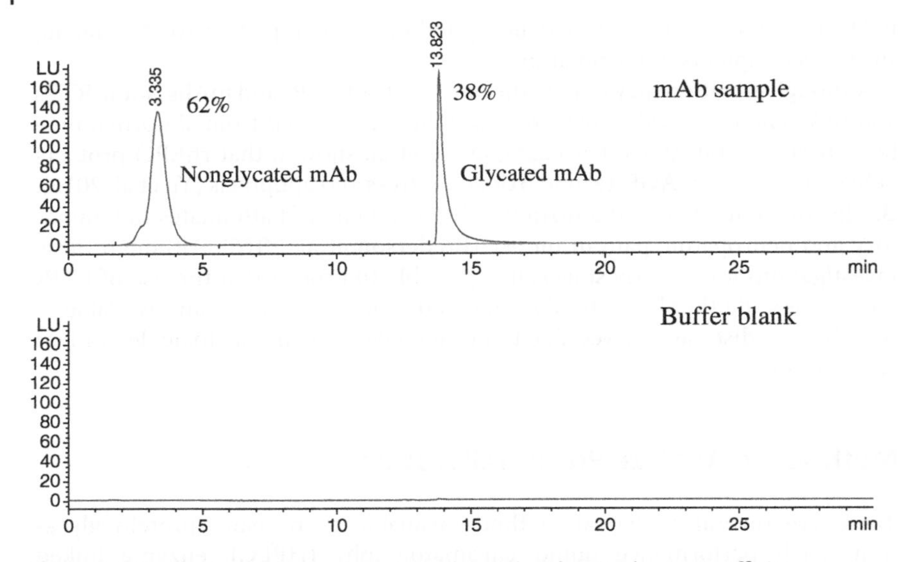


**Figure 6.3** Separation of glycated and non-glycated mAb using boronate affinity chromatography. (*See insert for color representation of this figure.*)

mAbs (Saleem et al. 2015). Accordingly, there are various methods available to predict, detect, and quantitate protein glycation including classical HPLC, CE, ELISA, etc. methods along with computational approaches. However, boronate affinity methods are still very useful to purify and/or enrich glycated samples for further structure-function studies of RTPs and mAbs.

## References

Ahmed, N. (1991). Glycation and diabetic complications. *J. Pak. Med. Assoc.* 41 (7): 171–174.

Andya, J.D., Maa, Y.F., Costantino, H.R. et al. (1999). The effect of formulation excipients on protein stability and aerosol performance of spray-dried powders of a recombinant humanized anti-IgE monoclonal antibody. *Pharm. Res.* 16 (3): 350–358.

Anguizola, J., Matsuda, R., Barnaby, O.S. et al. (2013). Review: glycation of human serum albumin. *Clin. Chim. Acta* 425: 64–76. https://doi.org/10.1016/j.cca.2013.07.013.

Bitensky, M.W., Kowluru, A., and Kowluru, R.A. (1989). Non-enzymatic glycation and protein recognition. *Prog. Clin. Biol. Res.* 304: 185–203.

Brederson, J.D., Strakhova, M., Mills, C. et al. (2016). A monoclonal antibody against the receptor for advanced glycation end products attenuates inflammatory and neuropathic pain in the mouse. *Eur. J. Pain* 20 (4): 607–614. https://doi.org/10.1002/ejp.775.

Butko, M., Pallat, H., Cordoba, A., and Yu, X.C. (2014). Recombinant antibody color resulting from advanced glycation end product modifications. *Anal. Chem.* 86 (19): 9816–9823. https://doi.org/10.1021/ac5024099.

Cao, H., Chen, T., and Shi, Y. (2015). Glycation of human serum albumin in diabetes: impacts on the structure and function. *Curr. Med. Chem.* 22 (1): 4–13.

Chen, Y.S., Wang, X.J., Feng, W., and Hua, K.Q. (2017). Advanced glycation end products decrease collagen I levels in fibroblasts from the vaginal wall of patients with POP via the RAGE, MAPK and NF-κB pathways. *Int. J. Mol. Med.* 40 (4): 987–998. https://doi.org/10.3892/ijmm.2017.3097.

Cherian, M. and Abraham, E.C. (1993). In vitro glycation and acetylation (by aspirin) of rat crystallins. *Life Sci.* 52 (21): 1699–1707.

Chikazawa, M., Otaki, N., Shibata, T. et al. (2013). Multispecificity of immunoglobulin M antibodies raised against advanced glycation end products: involvement of electronegative potential of antigens. *J. Biol. Chem.* 288 (19): 13204–13214. https://doi.org/10.1074/jbc.M113.452177.

Chikazawa, M., Shibata, T., Hatasa, Y. et al. (2016). Identification of C1q as a binding protein for advanced glycation end products. *Biochemistry* 55 (3): 435–446. https://doi.org/10.1021/acs.biochem.5b00777.

Clark, S.L., Santin, A.E., Bryant, P.A. et al. (2013). The initial noncovalent binding of glucose to human hemoglobin in nonenzymatic glycation. *Glycobiology* 23 (11): 1250–1259. https://doi.org/10.1093/glycob/cwt061.

Du, Q., Qian, M.M., Liu, P.L. et al. (2017). Glycation of high-density lipoprotein triggers oxidative stress and promotes the proliferation and migration of vascular smooth muscle cells. *J. Geriatr. Cardiol.* 14 (7): 473–480. https://doi.org/10.11909/j.issn.1671-5411.2017.07.003.

Ghosh, S., Datta, D., Cheema, M. et al. (2017). Aptasensor based optical detection of glycated albumin for diabetes mellitus diagnosis. *Nanotechnology* https://doi.org/10.1088/1361-6528/aa893a.

Harris, R.J. (2005). Heterogeneity of recombinant antibodies: linking structure to function. *Dev. Biol. (Basel)* 122: 117–127.

Ishida, Y.I., Kayama, T., Kibune, Y. et al. (2017). Identification of an argpyrimidine-modified protein in human red blood cells from schizophrenic patients: a possible biomarker for diseases involving carbonyl stress. *Biochem. Biophys. Res. Commun.* pii: S0006-291X(17)31681-9. doi: https://doi.org/10.1016/j.bbrc.2017.08.150.

Isoda, H., Takahashi, H., Eguchi, Y. et al. (2017). Re-evaluation of glycated hemoglobin and glycated albumin with continuous glucose monitoring system as markers of glycemia in patients with liver cirrhosis. *Biomed. Rep.* 7 (3): 286. https://doi.org/10.3892/br.2017.928.

Kaschak, T., Boyd, D., and Yan, B. (2011). Characterization of glycation in an $IgG_1$ by capillary electrophoresis sodium dodecyl sulfate and mass spectrometry. *Anal. Biochem.* 417 (2): 256–263. https://doi.org/10.1016/j.ab.2011.06.024.

Khalifah, R.G., Baynes, J.W., and Hudson, B.G. (1999). Amadorins: novel post-Amadori inhibitors of advanced glycation reactions. *Biochem. Biophys. Res. Commun.* 257 (2): 251–258.

Li, Z., Beeram, S.R., Bi, C. et al. (2016). High-performance affinity chromatography: applications in drug-protein binding studies and personalized medicine. *Adv. Protein Chem. Struct. Biol.* 102: 1–39. https://doi.org/10.1016/bs.apcsb.2015.09.007.

Lima, M., Moloney, C., and Ames, J.M. (2009). Ultra performance liquid chromatography-mass spectrometric determination of the site specificity of modification of beta-casein by glucose and methylglyoxal. *Amino Acids* 36 (3): 475–481. https://doi.org/10.1007/s00726-008-0105-y.

Mori, N. and Manning, J.M. (1986). Studies on the Amadori rearrangement in a model system: chromatographic isolation of intermediates and product. *Anal. Biochem.* 152 (2): 396–401.

Neelofar, K. and Ahmad, J. (2015). Amadori albumin in diabetic nephropathy. *Indian J. Endocrinol. Metab.* 19 (1): 39–46. https://doi.org/10.4103/2230-8210.146863.

Pareek, M. (2017). The interplay between fasting glucose, echocardiography, and biomarkers: pathophysiological considerations and prognostic implications. *Dan Med. J.* 64 (9): pii: B5400.

Paudel, G., Bilova, T., Schmidt, R. et al. (2016). Osmotic stress is accompanied by protein glycation in *Arabidopsis thaliana*. *J. Exp. Bot.* 67 (22): 6283–6295.

Quan, C.P., Wu, S., Dasovich, N. et al. (1999). Susceptibility of rhDNase I to glycation in the dry-powder state. *Anal. Chem.* 71 (20): 4445–4454.

Quan, C., Alcala, E., Petkovska, I. et al. (2008). A study in glycation of a therapeutic recombinant humanized monoclonal antibody: where it is, how it got there, and how it affects charge-based behavior. *Anal. Biochem.* 373 (2): 179–191.

Rabbani, N., Ashour, A., and Thornalley, P.J. (2016). Mass spectrometric determination of early and advanced glycation in biology. *Glycoconjugate J.* 33 (4): 553–568. https://doi.org/10.1007/s10719-016-9709-8.

Raghav, A. and Ahmad, J. (2014). Glycated serum albumin: a potential disease marker and an intermediate index of diabetes control. *Diabetes Metab. Syndr.* 8 (4): 245–251. https://doi.org/10.1016/j.dsx.2014.09.017.

Saleem, R.A., Affholter, B.R., Deng, S. et al. (2015). A chemical and computational approach to comprehensive glycation characterization on antibodies. *MAbs* 7 (4): 719–731. https://doi.org/10.1080/19420862.2015.1046663.

Salinas, M., López-Garrigós, M., Flores, E., Leiva-Salinas, C., and Pilot Group of the Appropriate Utilization of Laboratory Tests (2017). Glycated hemoglobin: a powerful tool not used enough in primary care. *J. Clin. Lab. Anal.* https://doi.org/10.1002/jcla.22310.

Saxena, P., Verma, P., and Goswami, B. (2017). Comparison of diagnostic accuracy of non-fasting DIPSI and HbA1C with fasting WHO criteria for diagnosis of gestational diabetes mellitus. *J. Obstet. Gynaecol. India* 67 (5): 337–342. https://doi.org/10.1007/s13224-017-0962-y.

Sessa, L., Gatti, E., Zeni, F. et al. (2014). The receptor for advanced glycation end-products (RAGE) is only present in mammals, and belongs to a family of cell adhesion molecules (CAMs). *PLoS ONE* 9 (1): e86903. https://doi.org/10.1371/journal.pone.0086903.

Silva, A.M., Sousa, P.R., Coimbra, J.T. et al. (2014). The glycation site specificity of human serum transferrin is a determinant for transferrin's functional impairment under elevated glycaemic conditions. *Biochem. J.* 461 (1, 1): 33–42. https://doi.org/10.1042/BJ20140133.

Smith, P.R. and Thornalley, P.J. (1992). Mechanism of the degradation of non-enzymatically glycated proteins under physiological conditions. Studies with the model fructosamine, N epsilon-(1-deoxy-D-fructos-1-yl)hippuryl-lysine. *Eur. J. Biochem.* 210 (3): 729–739.

Somensi, N., Brum, P.O., de Miranda Ramos, V. et al. (2017). Extracellular HSP70 activates ERK1/2, NF-kB and Pro-inflammatory gene transcription through binding with RAGE in A549 human lung cancer cells. *Cell. Physiol. Biochem.* 42 (6): 2507–2522. https://doi.org/10.1159/000480213.

Svacina, S., Hovorka, R., and Skrha, J. (1990). Computer models of albumin and haemoglobin glycation. *Comput. Meth. Programs Biomed.* 32 (3–4): 259–263.

Taghavi, F., Habibi-Rezaei, M., Amani, M. et al. (2017). The status of glycation in protein aggregation. *Int. J. Biol. Macromol.* 100: 67–74. https://doi.org/10.1016/j.ijbiomac.2015.12.085.

Tu, Z.C., Zhong, B.Z., and Wang, H. (2017). Identification of glycated sites in ovalbumin under freeze-drying processing by liquid chromatography high-resolution mass spectrometry. *Food Chem.* 226: 1–7. https://doi.org/10.1016/j.foodchem.2017.01.038.

Turpeinen, U., Karjalainen, U., and Stenman, U.H. (1995). Three assays for glycohemoglobin compared. *Clin. Chem.* 41 (2): 191–195.

Venkatraman, J., Aggarwal, K., and Balaram, P. (2001). Helical peptide models for protein glycation: proximity effects in catalysis of the Amadori rearrangement. *Chem. Biol.* 8 (7): 611–625.

Wagner, M.C., Myslinski, J., Pratap, S. et al. (2016). Mechanism of increased clearance of glycated albumin by proximal tubule cells. *Am. J. Physiol. Renal. Physiol.* 310 (10): F1089–F1102. https://doi.org/10.1152/ajprenal.00605.2015.

Wei, B., Berning, K., Quan, C., and Zhang, Y.T. (2017). Glycation of antibodies: modification, methods and potential effects on biological functions. *MAbs* 9 (4): 586–594. https://doi.org/10.1080/19420862.2017.1300214.

Yu, T., Li, L., Chen, T. et al. (2015). Erythropoietin attenuates advanced glycation endproducts-induced toxicity of Schwann cells in vitro. *Neurochem. Res.* 40 (4): 698–712. https://doi.org/10.1007/s11064-015-1516-2.

Yuk, I.H., Zhang, B., Yang, Y. et al. (2011). Controlling glycation of recombinant antibody in fed-batch cell cultures. *Biotechnol. Bioeng.* 108 (11): 2600–2610. https://doi.org/10.1002/bit.23218.

Zhang, B., Yang, Y., Yuk, I. et al. (2008). Unveiling a glycation hot spot in a recombinant humanized monoclonal antibody. *Anal. Chem.* 80 (7): 2379–2390. https://doi.org/10.1021/ac701810q.

Zhang, B., Mathewson, S., and Chen, H. (2009). Two-dimensional liquid chromatographic methods to examine phenylboronate interactions with recombinant antibodies. *J. Chromatogr. A* 1216 (30): 5676–5686. https://doi.org/10.1016/j.chroma.2009.05.084.

Zhang, J., Zhang, T., Jiang, L. et al. (2012). Rapid identification of low level glycation sites in recombinant antibodies by isotopic labeling with 13C6-reducing sugars. *Anal. Chem.* 84 (5): 2313–2320. https://doi.org/10.1021/ac202995x.

Zhang, X.D., Pechter, D., Yang, L. et al. (2017). Decreased complexity of glucose dynamics preceding the onset of diabetes in mice and rats. *PLoS ONE* 12 (9): e0182810. https://doi.org/10.1371/journal.pone.0182810.

# 7

# Glycosylation of Proteins

## Introduction

Glycans are monomeric, oligomeric, and/or polymeric organic biomolecules containing carbohydrate residues that are present in nature (Stone 1893; Flitsch and Ulijn 2003). The word "carbohydrates" originally defined as the naturally occurring organic molecules containing carbon, hydrogen, and oxygen as constituent atoms with 2 : 1 ratio of hydrogen to oxygen atoms (Stone 1893). Hence, carbohydrates have a very general empirical formula of $C_m(H_2O)_n$, where "$m$" could be same or different from "$n$" (Stone 1893). Accordingly, glucose has a molecular formula of $C_6(H_2O)_6$ that is same as the elemental formula of $C_6H_{12}O_6$. However, there are some exceptions to this empirical formula. For example, deoxyribose has an elemental formula of $C_5H_{10}O_4$ because one oxygen atom is missing in this molecule, and hence, it is called deoxyribose to represent missing oxygen, whereas ribose has an elemental formula of $C_5H_{10}O_5$ (Stone 1893). Deoxyribose and ribose are the constituent sugar moieties of DNA and RNA, respectively.

Accordingly, carbohydrates are multihydroxyl organic biomolecules containing aldehyde or keto functional group at either carbon 1 (C1) or carbon 2 (C2), respectively. These carbohydrate residues are also called as sugar molecules/residues because some of the carbohydrate molecules such as glucose, fructose, sucrose are sweet to taste to humans. Hence, carbohydrates or sugars are also referred as saccharides meaning sweet to taste. Further, some carbohydrates residues may also contain additional functional groups such as carboxyl, amine, acetyl, etc. groups. The monomeric carbohydrate residues such as glucose, galactose, mannose, etc., are termed as monosaccharides. Dimeric molecules such as lactose are called disaccharide, trimeric molecules such as chitotriose are called trisaccharides, and chitotetrose is a tetrasaccharide. Higher saccharides are termed as oligosaccharides, and the polymeric carbohydrates are called as polysaccharides (Stone 1893).

*Co- and Post-Translational Modifications of Therapeutic Antibodies and Proteins*, First Edition. T. Shantha Raju.

## Glycans and Aglycans

Glycans are saccharide/carbohydrate moieties and are usually covalently linked to each other as well as to other noncarbohydrate moieties. The noncarbohydrate moieties that are linked to glycans are usually called as aglycan moieties. The aglycan moieties usually are proteins, lipids, nucleic acids, etc. In plants glycans are also often found covalently linked to small organic molecules and such molecules are termed as glycosides (Eyjólfsson 1970; Heftmann 1974; Peters 1982). The plant-derived glycosides have been shown to exhibit medicinal properties (Peters 1982). Glycosides are also present in microorganisms, and many microbial-derived antibiotics are usually glycosides (Saribalis 1946; Scott 1946; Ganesan and Xu 2017; Hayes and Pietruszka 2017). In addition, glycans are also found as free saccharide units. For example, mammalian milk contains complex mixture of free saccharides that include monosaccharides, disaccharides, oligosaccharides, and polysaccharides (Wiegandt and Egge 1970; Jenness 1974; Peaker and Faulkner 1983; Fiat and Jollès 1989).

As described earlier, the glycans are present in nature not only as covalently linked to biomolecules such as proteins, lipids, and other organic molecules but are also present in nature as free saccharide molecules. The biomolecules in which the glycans are linked to proteins are called glycoproteins in which the glycans are the minor partners to proteins (Bourrillon and Got 1963; Winzler 1965). The biomolecules with glycans linked to lipids are called glycolipids in which glycans are minor components to lipids (Carter et al. 1965). Most often in glycoproteins and glycolipids, the glycan moieties are the most heterogeneous partners (Gabius 2017). Additionally, the molecules in which the glycans are the major components in proteins are termed as proteoglycans (Rosenberg 1973; Lamberg and Stoolmiller 1974). Similarly, the molecules in which the glycans are major components in lipids are called as lipoglycans as in lipooligosaccharides and lipopolysaccharides (Osborn et al. 1964).

## Glycosidic Bonds

Many of the plant-derived polysaccharides such as cellulose, starch, xylan, etc. are free saccharides and are not linked to aglycan moieties such as proteins, lipids, etc. Cellulose and starch are the two most commonly known polymers of glucose. In cellulose, the glucose molecules are covalently linked to each other by O-glycosidic bonds, and hence, it is a homopolymer of D-glucose. O-glycosidic bond is a covalent bond between two hydroxyl groups of two sugar residues in which a stable bond between two carbon atoms (at least one anomeric carbon atom) of two different sugar moieties is involved in bonding with an oxygen atom to form a C—O—C bond. During O-glycosidic bond formation a water molecule is eliminated as shown in Figures 7.1–7.3.

Starch is also a homopolymer of D-glucose but differs in its properties from cellulose. This is because in cellulose, the O-glycosidic bonds

$$R_3C\text{—}OH + R'_3C\text{—}OH \longrightarrow R_3\text{—}C\text{—}O\text{—}CR'_3 + H_2O$$

**Figure 7.1** Mechanism of O-glycosidic bond formation. R and R′, hydrogen atom or other molecules covalently linked to carbon atom.

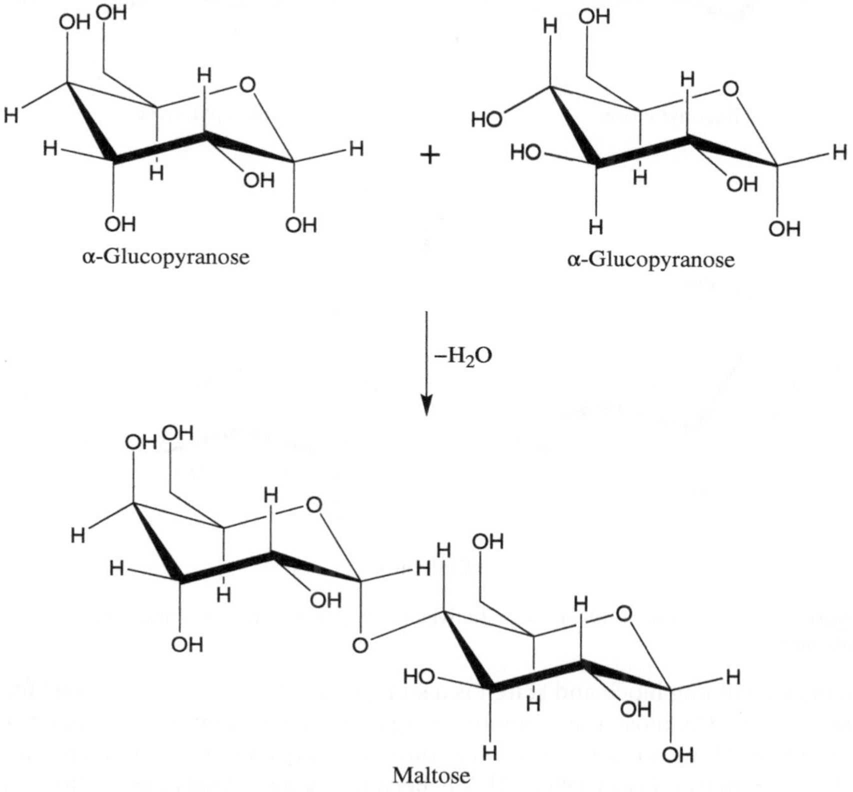

α-Glucopyranose + α-Glucopyranose

$-H_2O$

Maltose

**Figure 7.2** α-Glycosidic bond formation between two α-glucopyranose residues to form Maltose.

are in β-configuration, whereas in starch, the O-glycosidic bonds are in α-configuration. Cellobiose is a disaccharide derived from cellulose, whereas maltose is a disaccharide derived from starch (see Figures 7.2 and 7.3, respectively). Cellulose is mostly not soluble in water, whereas starch is somewhat water soluble and often forms gel-like solutions in water. This is because starch has α-glycosidic bonds and can form hydrogen bonds with water molecule, whereas cellulose has β-glycosidic bonds, which forms hydrogen bonds between themselves, and hence, the hydroxyl groups in cellulose may not be available for hydrogen bond formation with water molecules in solution (Zografi and Kontny 1986). Cellulose and starch are the two most abundant polymeric organic molecules available in nature (Akazawa and Tanaka 1967; Hassid 1969; Ward and Moo-Young 1989). Cellulose is present in almost all

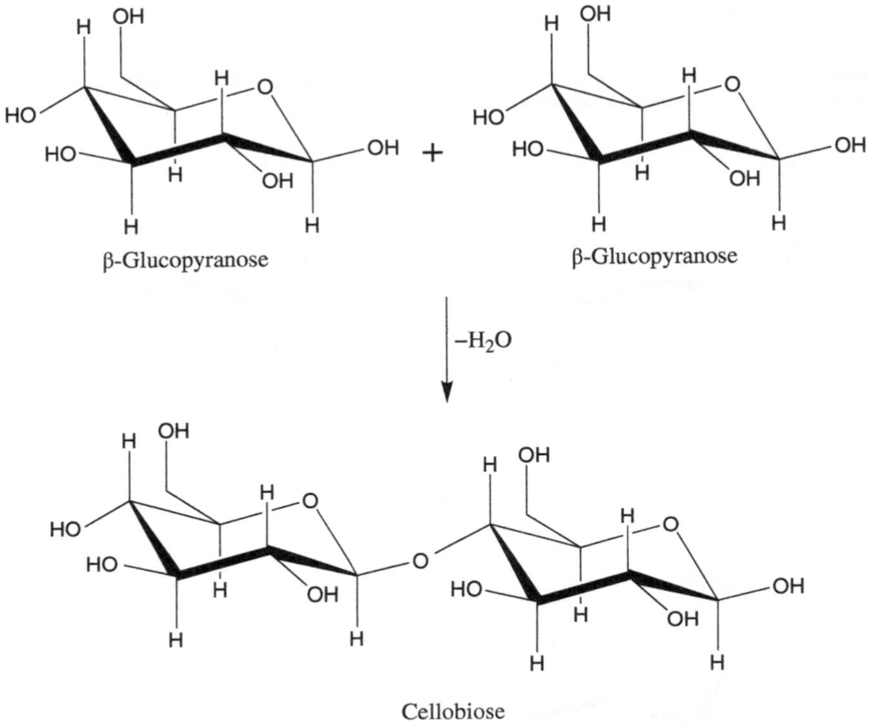

**Figure 7.3** β-Glycosidic bond formation between to β-glucopyranose residues to form Cellobiose.

plants as building block and starch is a storage polymer (energy reservoir) for many plants. Cellulose is a common energy source for many animals but not to humans. However, starch is a very common energy source for humans and other mammals (Hassid 1969). This is because many animals have cellulases to digest cellulose, whereas humans lack cellulases for digestion of cellulose.

## Aldoses and Ketoses

The monomeric units of carbohydrates, i.e. monosaccharides are similar to monomeric units of proteins, i.e. amino acid residues. Glucose is a very common monosaccharide and is found abundantly in nature (Ebert 1881). Besides glucose, there are several other monosaccharides, and some of these monosaccharides are shown in Table 7.1. Monosaccharides are categorized based on the number of carbon atoms present in their molecule along with the functional group present at carbon 1 (C1) or carbon 2 (C2), i.e. aldehyde

**Table 7.1** Partial list of monosaccharides and their derivatives found in living organisms.

| Name | Major forms | Source | Comments |
|---|---|---|---|
| Allose | Hexose, pyranose | Plants | A rare monosaccharide |
| Altrose | Hexose, pyranose | Plants | A rare monosaccharide |
| Arabinose | Pentose, furanose | Plants and bacteria | Rarely found in animals |
| Deoxy ribose | Deoxy pentose, furanose | Animals, plants and bacteria | Constituent of DNA |
| Fructose | Hexose, furanose | Plants | May also found in animals and bacteria |
| Fucose | Deoxy hexose, pyranose | Animals, plants, and bacteria | Abundant in animal glycoproteins |
| Galactosamine | Hexose, pyranose | Animals, plants, and bacteria | Often N-acetylated and/or sulfated |
| Galactose | Hexose, pyranose | Animals, plants, and bacteria | Highly abundant sugar present in both plants and animals |
| Galacturonic acid | Hexose, pyranose | Animals, plants, and bacteria | Abundant in plant polysaccharides |
| Glucosamine | Hexose, pyranose | Animals, plants, and bacteria | Often N-acetylated and/or sulfated |
| Glucose | Hexose, pyranose | Animals, plants, and bacteria | Constituent monosaccharide of cellulose, starch, and glycogen |
| Glucuronic acid | Hexose, pyranose | Plants, animals, and bacteria | Abundant in plant polysaccharide |
| Gulose | Hexose, pyranose | Plants | A rare monosaccharide |
| Heptose | Heptose, pyranose | Bacteria | Constituent sugar of lipopolysaccharides |
| Idose | Hexose, pyranose | Pants | Rarely found in animals |
| Iduronic acid | Hexose, pyranose | Animals, plants, and bacteria | Constituent sugar of glycosaminoglycans (GAGs) |
| Keto-deoxy-octulosonic acid (KDO) | Octose, pyranose | Bacteria | Constituent sugar of lipopolysaccharides |
| Mannose | Hexose, pyranose | Animals, plants, and bacteria | Abundant sugar in yeast |
| *N*-Acetyl neuraminic acid (NANA) | Nanose, pyranose | Animals, plants, and bacteria | Negatively charged monosaccharides commonly found in animal glycoproteins |

*(Continued)*

**Table 7.1** (Continued)

| Name | Major forms | Source | Comments |
|---|---|---|---|
| *N*-Acetylgalactosamine | Hexose, pyranose | Animals, plants, and bacteria | Derivative of galactose |
| *N*-Acetylglucosamine | Hexose, pyranose | Animals, plants, and bacteria | Derivative of glucose and a constituent monosaccharide of chitin |
| *N*-Glycolyl neuraminic acid (NGNA) | Nanose, pyranose | Animals, plants and bacteria | A derivative of NANA |
| Rhamnose | Deoxy hexose, pyranose | Plants | Often found in animals and bacteria |
| Ribose | Pentose, furanose | Animals, plants, and bacteria | Constituent of RNA |
| Xylose | Pentose, pyranose, and furanose | Animals, plants, and bacteria | Common linkage sugar of GAGs |

or keto group, respectively. Monosaccharides containing six carbon atoms are called hexoses or hexuloses, and monosaccharides containing five carbon atoms are called as pentoses or pentuloses. Accordingly, monosaccharides containing aldehyde functional group are called aldoses, and the corresponding monosaccharides containing keto functional group are called ketoses or alduloses. Glucose is in the category of hexose as it contains six carbon chains and contains an aldehyde functionality at C1 and hence it is an aldose, whereas fructose is a hexulose since it carries six carbon chain along with a keto functionality at C2 and hence it is a ketose. The C1 carbon in aldoses is termed as anomeric carbon and the aldehyde group in solution is in equilibrium with hemiacetal group as shown in Figure 7.4. The hydroxyl group in the hemiacetal group at C1 of aldoses and C2 of ketoses is very reactive similar to aldehyde or ketone functionalities and hence can form glycosidic bonds with hydroxyl groups present in other monosaccharides or in other molecules including aglycan moieties. The highly reactive hydroxyl group at C1 position of aldoses is also referred to as anomeric hydroxyl group, and the carbon atom at C1 position of aldoses is referred to as anomeric carbon atom. Similarly, in ketoses, the keto group can also form a hemiacetal group, and the resulting hydroxyl group at C2 is reactive with other hydroxyl groups to form glycosidic bonds. The hydroxyl group at C2 position of ketoses is also referred to as anomeric hydroxyl group and the carbon atom at C2 position of ketoses is termed as anomeric carbon atom.

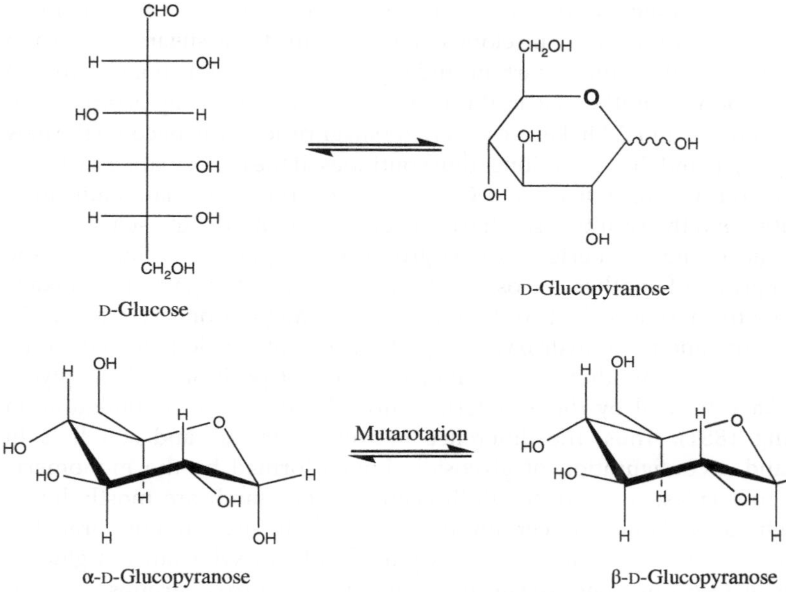

CHO, H—OH, HO—H, H—OH, H—OH, $CH_2OH$

D-Glucose

$CH_2OH$, O, OH, OH, OH, OH

D-Glucopyranose

α-D-Glucopyranose

Mutarotation

β-D-Glucopyranose

**Figure 7.4** Glucose forms open chain aldehyde and pyranose forms.

## Anomeric Groups: α- and β-Configurations

Lactose is a naturally occurring disaccharide consisting of glucose and galactose as its monosaccharide constituents, whereas sucrose is also a naturally occurring disaccharide of glucose and fructose. Milk from mammals contains lactose, and sugarcane is one of the many natural sources for sucrose (Paton and Cathcart 1911; Ritchie 1914; Virtanen and Nordlund 1934). Lactose is a reducing disaccharide, whereas sucrose is a nonreducing disaccharide (Doudoroff et al. 1947). In lactose, the anomeric hydroxyl group of Gal residue is O-linked to the hydroxyl group of Glc residue at O-4 position (hydroxyl group at C4). Hence, the anomeric hydroxyl group of Glc residue in lactose is free and very reactive. However, in sucrose, the anomeric hydroxyl group of Glc is O-linked to the anomeric hydroxyl group of Fru, and hence, sucrose has no reactive anomeric hydroxyl group. Trehalose is another example of a nonreducing sugar as it contains two glucose molecules linked to each other at their anomeric C1 positions. The saccharides containing free and reactive anomeric hydroxyl group are also termed as reducing sugars. Accordingly, lactose is a reducing disaccharide, whereas sucrose is a nonreducing disaccharide. Lactose can be further elongated as it contains a reducing group to form glycosidic bonds, whereas in sucrose, two sugars are linked to each other at their anomeric positions so that they cannot be elongated further. In aldose sugars, the carbon position 1 is called anomeric carbon, and it can form glycosidic bonds with the hydroxyl group of another sugar. If the O-linked hydroxyl group is other than the C1 in aldoses and C2 in ketoses, then the sugar residue will become the new reducing sugar and the chain elongation continues. If the participating hydroxyl group is from C1 for aldoses and C2 for ketoses, then the chain elongation terminates, and the resulting saccharide becomes a nonreducing sugar.

In cellulose, the anomeric hydroxyl group of one glucose is linked to the hydroxyl group of another glucose at O-4 position, and the glycosidic linkage formed by the anomeric hydroxyl group is in β-configuration (Armsby 1885). In starch, the anomeric hydroxyl group of glucose molecule is linked to the hydroxyl group of another glucose molecule at O-4 position and the glycosidic linkage formed by the anomeric hydroxyl group is in α-configuration (Pickering 1881). Thus, the difference between cellulose and starch is in the β- and α-configuration of glycosidic bonds formed by the monomeric glucose molecules, respectively. Both cellulose and starch are mostly linear polysaccharides. However, certain types of starch also contain branches wherein some glucose molecules are α-linked to hydroxyl groups of glucose moieties at O-6 positions. Accordingly, anomeric hydroxyl groups can also form glycosidic bonds with hydroxyl groups present at different positions (Montreuil 1984). Hence, glycans are very heterogeneous and highly complex because they can form glycosidic bonds with different hydroxyl groups in either α- or β-configurations (Montreuil 1984).

## Natural Diversity of Glycans

Carbohydrates are ubiquitous in nature and are present in all living organisms (McPherson 1911). In plants, they are present not only in the form of cellulose and starch, but they are also present as glycosides, glycoproteins, etc. In animals, they are present in the form of glycoproteins, glycolipids, proteoglycans, lipoglycans, free monosaccharides, oligosaccharides, polysaccharides, etc. In humans also, carbohydrates are present as constituents of glycoproteins, glycolipids, proteoglycans, lipoglycans, etc. Ribose and deoxy ribose are sugar residues that are the constituents of nucleic acids (Bagatell et al. 1959) which are present in almost all living beings. Additionally, in humans, free saccharides including monosaccharides, disaccharides, higher oligosaccharides, and polysaccharides are also present. For example, glycogen is a polysaccharide that is present in humans (Dowler and Mottram 1918). In microorganisms, carbohydrates are present as monosaccharides, oligosaccharides, and polysaccharides. Lipopolysaccharides, lipooligosaccharides, and capsular polysaccharides are present in many Gram-positive and Gram-negative bacteria (Bruner and Edwards 1948; Ewing and Kauffmann 1950).

## Glycans and Enzymes

It is now clear that the glycans form glycosidic bonds between themselves as well as with aglycan moieties. Most of the glycosidic bonds formation between glycans themselves and between glycans and aglycans are mediated by enzymes (Raju et al. 1996). The nature of linkages between glycans and aglycan moieties may vary with the type of aglycan moieties. The biosynthesis of glycans and glycans on aglycan moieties involves many different enzymes, and many of these enzymes are highly specific to anomericity, linkage types, and aglycan moieties (Raju et al. 1996). The enzymes involved in the glycosylation include glycosyltransferases (GTs), exoglycosidases, endoglycosidases, activated sugar synthetases, sugar transporters, etc. It has been postulated that more than 200 different enzymes are involved in the biosynthesis of glycans. GTs are involved in transferring the sugar moieties from activated sugars such as UDP-Glc to other sugar residues or to aglycan moieties (Shur and Roth 1975). These enzymes are highly specific to anomericity and to the linkage types (Strous 1986). For example, β-1,4-galactosyltransferase is involved in transferring Gal residue from UDP-Gal to GlcNAc residues at O-4 position in β-configuration (Lee et al. 2001). The enzyme β-1,4-galactosyltransferase is highly specific to the linkage type and anomericity and only mediates the formation of β-linkages. Sugar synthetases are involved in synthesizing the activated sugars. For example, cytidine monophosphate (CMP)-sialic acid synthetase is involved in the synthesis of CMP-sialic acid (Daunter and Newlands 1981). Exoglycosidases are involved in cleaving the terminal sugars from the nonreducing termini. These enzymes are also highly specific to anomericity and linkage types (Kobata 2013). For example, β-galactosidases cleave only β-linked Gal residues from the nonreducing termini, whereas α-galactosidases cleave only α-linked Gal residues from nonreducing termini (Lederberg 1950; Li et al. 1963).

In mammalian cells, most of the biosynthesis including processing of glycans occurs in the Golgi apparatus. However, synthesis of activated sugars mostly occurs in the cytoplasm, and the activated sugars are transported from cytoplasm to Golgi complex by sugar transporters. In addition, cleavage of sugars by exoglycosidases within the cells occurs in the acidic compartments. Accordingly, glycosylation of proteins is highly complex process and involves many different types of enzymes along with many different types of sugar residues. A few different types of protein glycosylation are briefly discussed below.

## N-glycosylation

Many proteins contain glycans covalently linked through the nitrogen atom of amide side chain of Asn residues. Such glycans types are collectively called as *N*-glycans or N-linked glycans or N-linked oligosaccharides and are described in detail in Chapter 8. N-glycosylation is a very major post-translational modification (PTM) of proteins that is very diverse and highly heterogeneous. *N*-glycans play major role in size, heterogeneity, solubility, and functions of proteins.

## O-glycosylation

In many proteins, glycans are also found to be covalently linked through hydroxyl groups of Ser and/or Thr residues. Such types of glycosylation is called O-glycosylation, and these types of glycans are often collectively referred as *O*-glycans or O-linked oligosaccharides. The *O*-glycans are also referred to as mucin glycans because mucin glycoproteins are heavily O-glycosylated. The biosynthesis of *O*-glycans is mediated by enzymes that are collectively called as O-glycosyltransferases (Lorenz et al. 1992). In lipids also, the glycans are mostly O-linked through hydroxyl groups to form glycolipids or lipoglycans (Dixon and Herbertson 1951). In collagen, glycans are found to be O-linked to the hydroxyl group of hydroxylysine residues. The glycans linked to hydroxylysine residues play an important role in the triple helical structure of collagen. In yeast, O-linked Man containing glycans are common. The most common types of O-glycosylation of proteins and their functions are discussed in detail in Chapter 9.

## Phospho-Serine Glycosylation

Phospho-serine is a derivative of Ser in which phosphate group is O-linked to the hydroxyl group of Ser residues. Phosphorylation of Ser is catalyzed by kinases that is discussed in detail in Chapter 13. Xyl, Fuc, Man, and GlcNAc residues are found to be O-linked to phosphate group of phosphoserine residues of proteins such as α-dystroglycan (Yoshida-Moriguchi et al. 2010). The glycosylation of phospho-serine residues is mediated by an enzyme called "like-acetylglucosaminyltransferase" (LARGE) protein (Yoshida-Moriguchi et al. 2010). The phospho-serine glycosylation may play a role in congenital muscular dystrophy (Yoshida-Moriguchi et al. 2010).

## GPI-Anchors (Glypiation)

Glycosylphosphatidylinositol (GPI)-anchor is a glycolipid consisting of phosphoethanolamine linker, glycan core, and phospholipid tail that is added to C-terminus of proteins during PTM. Many eukaryotic proteins contain

GPI-anchors that anchors the protein to the outer leaflet of the cell membrane (Paulick and Bertozzi 2008). Proteins containing GPI-anchors are diverse in nature in terms of their structure and functions. The functions of GPI-anchors include lipid raft partitioning, signal transduction, targeting to the apical membrane, and prion disease pathogenesis (Paulick and Bertozzi 2008; Krawitz et al. 2013).

## C-mannosylation

Thrombospondins (TSPs) are a family of five proteins with multifunctional properties (Morris and Kyriakides 2014). The TSPs are secreted glycoproteins and exhibits antiangiogenic properties (Baenziger et al. 1971). The C-linked mannose residues are attached to Trp residues in the consensus sequence of W-X-X-W, where X is any amino acid residue and the C-linked mannose residues are commonly found in TSPs (Baenziger et al. 1971). Although C-mannosylation is very unusual, it has been found in other proteins as well (Okamoto et al. 2017). For C-mannosylation, guanidine diphosphate (GDP)-Man is the sugar donor, and C-mannosyltransferase is the enzyme involved in transferring the mannose residue from GDP-mannose to indole moiety of Trp residues (Otani et al. 2018). C-mannosylation is also found in mammalian RNase 2 (Krieg et al. 1998) but it is not present in yeast- and bacteria-derived RNase 2.

## References


Akazawa, T. and Tanaka, Y. (1967). Biosynthesis of plant polysaccharides. *Tanpakushitsu Kakusan Koso* 12 (14): 1472–1478.

Armsby, H.P. (1885). The digestibility of cellulose. *Science* 5 (100): 11–12.

Baenziger, N.L., Brodie, G.N., and Majerus, P.W. (1971). A thrombin-sensitive protein of human platelet membranes. *Proc. Natl. Acad. Sci. U.S.A.* 68 (1): 240–243.

Bagatell, F.K., Wright, E.M., and Sable, H.Z. (1959). Biosynthesis of ribose and deoxyribose in *Escherichia coli*. *J. Biol. Chem.* 234 (6): 1369–1374.

Bourrillon, R. and Got, R. (1963). Glycopeptides and structure of glycoproteins. *Expos. Annu. Biochim. Med.* 24: 25–48.

Bruner, D.W. and Edwards, P.R. (1948). Changes induced in the O-antigens of Salmonella. *J. Bacteriol.* 55 (3): 449.

Carter, H.E., Johnson, P., and Weber, E.J. (1965). Glycolipids. *Annu. Rev. Biochem.* 34: 109–142.

Daunter, B. and Newlands, J. (1981). Seminal plasma biochemistry II: seminal plasma and spermatozoal cytidine monophosphate-sialic acid synthetase and sialyltransferase activities. *Andrologia* 13 (3): 215–224.

Dixon, K.C. and Herbertson, B.M. (1951). Cytoplasmic glycolipids of brain. *J. Pathol. Bacteriol.* 63 (1): 175.

Doudoroff, M., Hassid, W.Z., and Barker, H.A. (1947). Studies with bacterial sucrose phosphorylase; enzymatic synthesis of a new reducing and of a new non-reducing disaccharide. *J. Biol. Chem.* 168 (2): 733–746.

Dowler, V.B. and Mottram, V.H. (1918). The distribution of blood, glycogen and fat in the lobes of the liver. *J. Physiol.* 52 (2–3): 166–174.

Ebert, A.E. (1881). Glucose. *Science* 2 (75): 567–568.

Ewing, W.H. and Kauffmann, F. (1950). A new coli O-antigen group. *Public Health Rep.* 65 (41): 1341–1343.

Eyjólfsson, R. (1970). Recent advances in the chemistry of cyanogenic glycosides. *Fortschr. Chem. Org. Naturst.* 28: 74–108.

Fiat, A.M. and Jollès, P. (1989). Caseins of various origins and biologically active casein peptides and oligosaccharides: structural and physiological aspects. *Mol. Cell. Biochem.* 87 (1): 5–30.

Flitsch, S.L. and Ulijn, R.V. (2003). Sugars tied to the spot. *Nature* 421 (6920): 219–220.

Gabius, H.J. (2017). The sugar code: why glycans are so important. *Biosystems* 164: 102–111. https://doi.org/10.1016/j.biosystems.2017.07.003.

Ganesan, K. and Xu, B. (2017). Molecular targets of vitexin and isovitexin in cancer therapy: a critical review. *Ann. N. Y. Acad. Sci.* 1401 (1): 102–113. https://doi.org/10.1111/nyas.13446.

Hassid, W.Z. (1969). Biosynthesis of oligosaccharides and polysaccharides in plants. *Science* 165 (3889): 137–144.

Hayes, M.R. and Pietruszka, J. (2017). Synthesis of glycosides by glycosynthases. *Molecules* 22 (9): 1434. https://doi.org/10.3390/molecules22091434.

Heftmann, E. (1974). Recent progress in the biochemistry of plant steroids other than sterols (saponins, glycoalkaloids, pregnane derivatives, cardiac glycosides, and sex hormones). *Lipids* 9 (8): 626–639.

Jenness, R. (1974). Proceedings: biosynthesis and composition of milk. *J. Invest. Dermatol.* 63 (1): 109–118.

Kobata, A. (2013). Exo- and endoglycosidases revisited. *Proc. Jpn. Acad. Ser. B Phys. Biol. Sci.* 89 (3): 97–117.

Krawitz, P.M., Höchsmann, B., Murakami, Y. et al. (2013). A case of paroxysmal nocturnal hemoglobinuria caused by a germline mutation and a somatic mutation in PIGT. *Blood* 122 (7): 1312–1315. https://doi.org/10.1182/blood-2013-01-481499.

Krieg, J., Hartmann, S., Vicentini, A. et al. (1998). Recognition signal for C-mannosylation of Trp-7 in RNase 2 consists of sequence Trp-*x*-*x*-Trp. *Mol. Biol. Cell* 9 (2): 301–309.

Lamberg, S.I. and Stoolmiller, A.C. (1974). Glycosaminoglycans. A biochemical and clinical review. *J. Invest. Dermatol.* 63 (6): 433–449.

Lederberg, J. (1950). The beta-D-galactosidase of *Escherichia coli*, strain K-12. *J. Bacteriol.* 60 (4): 381–392.

Lee, J., Sundaram, S., Shaper, N.L. et al. (2001). Chinese hamster ovary (CHO) cells may express six beta 4-galactosyltransferases (β-4GalTs). Consequences of the loss of functional beta 4GalT-1, β-4GalT-6, or both in CHO glycosylation mutants. *J. Biol. Chem.* 276 (17): 13924–13934.

Li, Y.T., Li, S.C., and Shetlar, M.R. (1963). Alpha-galactosidase from *Diplococcus pneumoniae*. *Arch. Biochem. Biophys.* 103: 436–442.

Lorenz, C., Strahl-Bolsinger, S., and Ernst, J.F. (1992). Specific in vitro O-glycosylation of human granulocyte-macrophage colony-stimulating-factor-derived peptides by O-glycosyltransferases of yeast and rat liver cells. *Eur. J. Biochem.* 205 (3): 1163–1167.

McPherson, W. (1911). The formation of carbohydrates in the vegetable kingdom. *Science* 33 (839): 131–142.

Montreuil, J. (1984). Spatial conformation of glycans and glycoproteins. *Biol. Cell.* 51 (2): 115–131.

Morris, A.H. and Kyriakides, T.R. (2014). Matricellular proteins and biomaterials. *Matrix Biol.* 37: 183–191. https://doi.org/10.1016/j.matbio.2014.03.002.

Okamoto, S., Murano, T., Suzuki, T. et al. (2017). Regulation of secretion and enzymatic activity of lipoprotein lipase by C-mannosylation. *Biochem. Biophys. Res. Commun.* 486 (2): 558–563. https://doi.org/10.1016/j.bbrc.2017.03.085.

Osborn, M.J., Rosen, S.M., Rothfield, L. et al. (1964). Lipopolysaccharide of the Gram-negative cell wall. *Science* 145 (3634): 783–789.

Otani, K., Niwa, Y., Suzuki, T. et al. (2018). Regulation of granulocyte colony-stimulating factor receptor-mediated granulocytic differentiation by C-mannosylation. *Biochem. Biophys. Res. Commun.* 498 (3): 466–472. https://doi.org/10.1016/j.bbrc.2018.02.210.

Paton, D.N. and Cathcart, E.P. (1911). On the mode of production of lactose in the mammary gland. *J. Physiol.* 42 (2): 179–188.

Paulick, M.G. and Bertozzi, C.R. (2008). The glycosylphosphatidylinositol anchor: a complex membrane-anchoring structure for proteins. *Biochemistry* 47 (27): 6991–7000. https://doi.org/10.1021/bi8006324.

Peaker, M. and Faulkner, A. (1983). Soluble milk constituents. *Proc. Nutr. Soc.* 42 (3): 419–425.

Peters, U. (1982). Pharmacokinetic review of digitalis glycosides. *Eur. Heart J.* 3 (Suppl D): 65–78.

Pickering, S.U. (1881). The detection of starch and dextrin. *Science* 2 (32): 44–45.

Raju, T.S., Lerner, L., and O'Connor, J.V. (1996). Glycopinion: biological significance and methods for the analysis of complex carbohydrates of recombinant glycoproteins. *Biotechnol. Appl. Biochem.* 24 (Pt 3): 191–194.

Ritchie, J. (1914). Note on the non-lactose fermenters in fresh milk. *J. Hyg. (Lond).* 14 (3): 393–394.

Rosenberg, L. (1973). Cartilage proteoglycans. *Fed. Proc.* 32 (4): 1467–1473.

Saribalis, C. (1946). Streptomycin: a review of the literature. *Marquette Med. Rev.* 11: 179–190.

Scott, E.G. (1946). Streptomycin: a review of the literature. *Del. Med. J.* 18: 15.

Shur, B.D. and Roth, S. (1975). Cell surface glycosyltransferases. *Biochim. Biophys. Acta* 415 (4): 473–512.

Stone, W.E. (1893). The use of the term “carbohydrates”. *Science* 21 (528): 149–150. https://doi.org/10.1126/science.ns-21.528.149.

Strous, G.J. (1986). Golgi and secreted galactosyltransferase. *CRC Crit. Rev. Biochem.* 21 (2): 119–151.

Virtanen, A.I. and Nordlund, M. (1934). Synthesis of sucrose in plant tissue. *Biochem. J.* 28 (5): 1729–1732.

Ward, O.P. and Moo-Young, M. (1989). Enzymatic degradation of cell wall and related plant polysaccharides. *Crit. Rev. Biotechnol.* 8 (4): 237–274.

Wiegandt, H. and Egge, H. (1970). Oligosaccharides of human milk. *Fortschr. Chem. Org. Naturst.* 28: 404–428.

Winzler, R.J. (1965). Metabolism of glycoproteins. *Clin. Chem.* 11 (Suppl): 339–347.

Yoshida-Moriguchi, T., Yu, L., Stalnaker, S.H. et al. (2010). O-mannosyl phosphorylation of α-dystroglycan is required for laminin binding. *Science* 327 (5961): 88–92. https://doi.org/10.1126/science.1180512.

Zografi, G. and Kontny, M.J. (1986). The interactions of water with cellulose- and starch-derived pharmaceutical excipients. *Pharm. Res.* 3 (4): 187–194. https://doi.org/10.1023/A:1016330528260.

# 8

# N-glycosylation of Proteins

## Introduction

N-glycosylation is a very prevalent and one of the most complex post-translational modification (PTM) of proteins (Shrimal et al. 2015). Initial step of N-glycosylation of proteins involves the en bloc transfer of pre-assembled oligosaccharide chain by oligosaccharyltransferase in the endoplasmic reticulum (ER) (Aebi 2013). The enzyme, oligosaccharyltransferase, is very highly sequence-specific and mediates the transfer of preassembled oligosaccharide moiety to amide group of asparagine residue in a specific sequence of Asn-X-Ser/Thr/Cys, where X can be any amino acid residue except Pro (Mellquist et al. 1998). However, recently it has been shown that the N-glycosylation may also occur at Asn-X-Ser/Thr/Cys sequence, where X can be a Pro residue (Sun and Zhang 2015; Dutta et al. 2017). The multistep N-glycosylation of proteins occurs in the ER and Golgi apparatus that involves many different enzymes including glycosidases along with the glycosyltransferases (GTs) (Wang et al. 2015). Many endogenous proteins are N-glycosylated and the *N*-glycan moieties of these proteins are highly complex that are also highly heterogeneous (Opdenakker et al. 1993). The heterogeneity of *N*-glycans is often referred as microheterogeneity, which may affect the biological functions of proteins including the pharmacokinetic properties (Rudd and Dwek 1997). Endogenous proteins that are N-glycosylated include α-1-acid glycoprotein, glycophorin, fibrinogen, fibronectin, fetuin, etc. (Henschen-Edman 2001; Clerc et al. 2016). In many of these glycoproteins, the *N*-glycans affect their biological functions and also their serum half-life (Varki 1993, 2017). Many therapeutic proteins produced in mammalian cells using recombinant DNA technology are glycosylated (Butler and Spearman 2014). These include recombinant human erythropoietin (rhEPO), recombinant human tissue plasminogen (rhtPA), recombinant human DNase (rhDNase), etc. (Raju et al. 2000). *N*-Glycans of rhEPO are required for its biological activity and terminal sialylation affects the serum half-life of rhEPO (Elliott et al. 2004). In addition, the recombinant therapeutic proteins (RTPs) are

*Co- and Post-Translational Modifications of Therapeutic Antibodies and Proteins*, First Edition. T. Shantha Raju.

produced in insect cells, and yeast cells are also glycosylated (Sethuraman and Stadheim 2006).

Almost all the immunoglobulins isolated from mammalian species are N-glycosylated (Epp et al. 2016). The microheterogeneity of *N*-glycans present in the Fc region of IgGs is much more significant than the *N*-glycans present in the other serum glycoproteins (Jakab 2016). This is because *N*-glycans found in the Fc region of IgGs are truncated and may contain different terminal sugars (Raju 2008). The different terminal sugars impact the protein functions very differently. The terminal sugars present in the Fc glycan repertoire include sialic acids, Gal, GlcNAc, Man, etc. Fc glycans repertoire of IgGs include high mannose, hybrid, and complex biantennary structures. Structure and heterogeneity of high mannose and complex biantennary type *N*-glycans found in the Fc region of IgGs are shown in Figures 8.1 and 8.2, respectively.

*N*-Glycans of IgGs present in the Fc region affect the antibody effector functions including complement-dependent cytotoxicity (CDC) and antibody-dependent cellular cytotoxicity (ADCC) (Dorokhov et al. 2016). Also, the *N*-glycans of Fc region are required for antibodies to bind Fc receptors

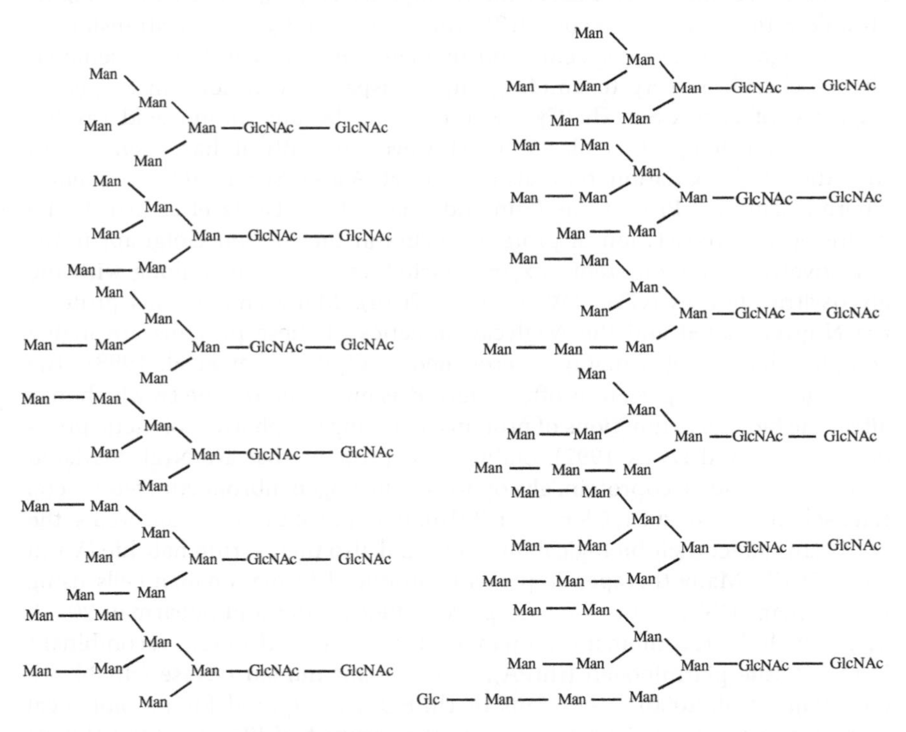


**Figure 8.1** Structure and heterogeneity of high-mannose type *N*-glycans found in many IgGs.

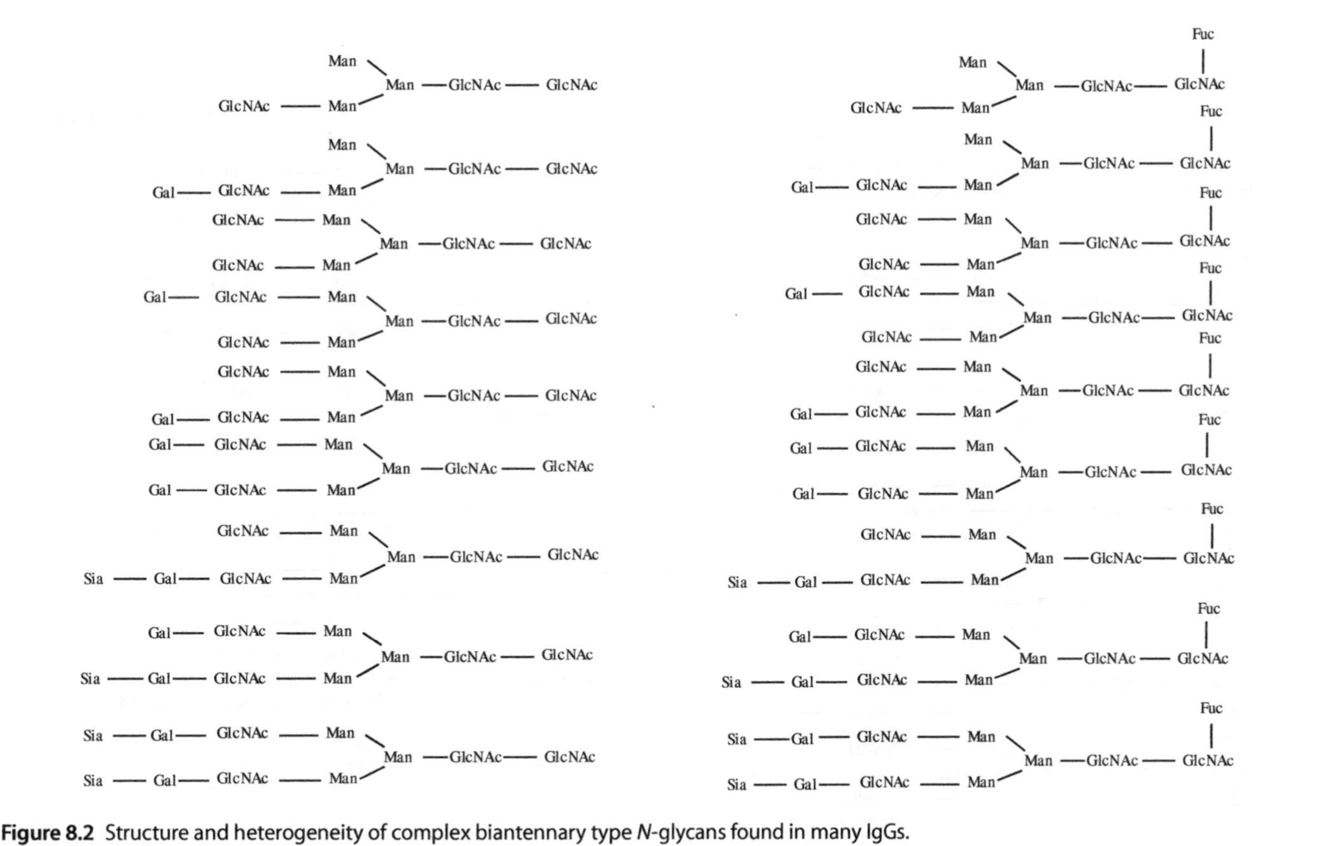


**Figure 8.2** Structure and heterogeneity of complex biantennary type *N*-glycans found in many IgGs.

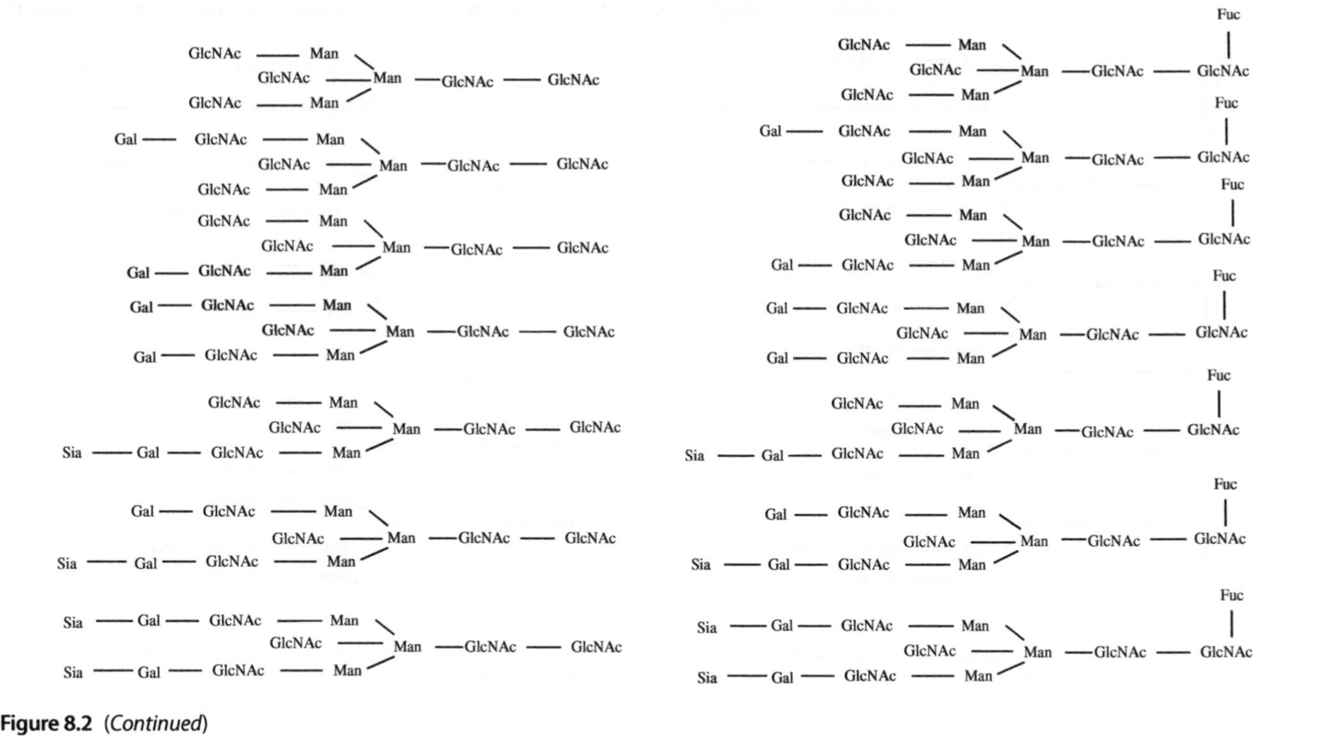

Figure 8.2 (*Continued*)

including FcγRI, FcγRII, FcγRIII, etc. (Quast et al. 2017). Fc glycans of IgGs may also impact the complement binding and hence the CDC activity of antibodies (Raju 2008). Recombinant human IgGs produced in mammalian cells are glycosylated and the *N*-glycans have been shown to play major role in ADCC, CDC, antibody-dependent cellular phagocytosis (ADCP), apoptosis, trogocytosis, etc. (Raju 2008; Plesner and Krejcik 2018). Further, Fc glycans may also impact the antibody stability and resistance to proteases (Raju and Scallon 2006, 2007).

## Mechanism of N-glycosylation of Proteins

### Biosynthesis of *N*-Glycans

Biosynthesis of *N*-glycans involves multiple steps including (i) biosynthesis of lipid-linked precursor oligosaccharide; (ii) en bloc transfer of precursor oligosaccharide to nascent polypeptide chain; (iii) trimming of the precursor oligosaccharide; and (iv) additional processing of oligosaccharide unit for chain elongation and/or modifications (Butters et al. 1981; Stanley et al. 2017). The entire biosynthetic pathway of *N*-glycans occurs in the ER and also in the Golgi apparatus (Kornfeld and Kornfeld 1985). The lipid moiety involved in the biosynthesis of the precursor oligosaccharide is dolichol phosphate which is present in the lipid bilayer of ER membrane (Butters et al. 1981).

#### Biosynthesis of Lipid-linked Precursor Oligosaccharide

Dolichol is an unsaturated lipid molecule containing repeating units of isoprene moiety (Waechter and Lennarz 1976). Chemically, isoprene is a 2-methyl-1,3-butadiene and is a colorless volatile organic liquid. The chemical composition of isoprene is $C_5H_8$ with a molecular formula of $CH_2{=}C(CH_3){-}CH{=}CH_2$ and its covalently linked monomeric units (~20 units) are present in dolichol. The elemental composition of dolichol is $C_{100}H_{164}O$ and its molecular weight is ~1382 amu. Dolichol is linked to phosphate, and the dolichol-phosphate is anchored to the phospholipid bilayer of ER membrane (Waechter and Lennarz 1976). The process of assembling the precursor oligosaccharide starts with the transfer of *N*-acetylglucosamine-1-phosphate (GlcNAc-1-P) from UDP-GlcNAc thus forming the pyrophosphate linkage on dolichol phosphate. Following this addition, a second GlcNAc to the first GlcNAc is added forming GlcNAc-O-GlcNAc disaccharide on dolichol pyrophosphate. Five Man residues are subsequently added, sequentially, by mannosyltransferase that mediates the transfer of Man residues from GDP-Man to GlcNAc residue and then to Man residues on dolichol phosphate. Upon the assembly of Dolichol-P-P-GlcNAc$_2$-Man$_5$, the entire molecule is translocated from cytoplasmic side of the ER membrane to luminal side of the ER membrane and the process is presumably mediated by the enzyme called flippase (Sanyal and Menon 2009).

The translocation of the dolichol-linked oligosaccharide allows it to be accessible for the enzymes in the lumen of ER. Following the translocation, four additional Man residues and three Glc residues are added by mannosyltransferases and glucosyltransferases from GDP-Man and UDP-Glc, respectively (Hubbard and Robbins 1980). The addition of four Man residues and three Glc residues occurs sequentially resulting in the formation of Dol-P-P-$GlcNAc_2Man_9Glc_3$ molecule (Hanover and Lennarz 1980).

#### En Bloc Transfer of the Precursor Oligosaccharide to Nascent Polypeptide Chain

The en bloc transfer of the preassembled $GlcNAc_2Man_9Glc_3$ glycan chain to the Asn residues of nascent polypeptide chains occurs in the lumen of the ER. The transfer of oligosaccharide is mediated by an enzyme called oligosaccharyltransferase and the enzymatic transfer reaction is supported by the energy released from the breaking of pyrophosphate bond between dolichol phosphate and glycan chain (Chalifour and Spiro 1988). The enzyme, oligosaccharyltransferase, is highly specific to sequence on the polypeptide chain and usually mediates the transfer of oligosaccharide moiety to Asn residues situated in the consensus sequence of Asn-X-Ser/Thr, where X is any amino acid except Pro (Welply et al. 1983). However, there are some exceptions to this sequence specificity such as in some cases, the enzyme can also transfer the oligosaccharide to Asn in the sequence of Asn-X-Cys (Mellquist et al. 1998), where X can be Pro also (Mazumder et al. 2012). The oligosaccharyltransferase efficiently transfer the oligosaccharide to Asn residues that are on the surface of the protein located on the lumen side of the ER (Kelleher and Gilmore 2006). However, in IgGs, the conserved N-glycosylation site is located in the $CH_2$ domain and the oligosaccharides in this region are buried within the space between the two $CH_2$ domains in the Fc portion. The *N*-glycan occupancy at this conserved N-glycosylation site is more than 90% and hence, the enzyme appears to efficiently transfer the oligosaccharide moiety to Asn residues that are not easily accessible (Yang et al. 2016).

#### Processing of the Glycan

Following the transfer of *N*-glycan to protein, trimming of the three Glc residues by glucosidase I and II takes place in the lumen of the ER (Hammond et al. 1994). Glucosidase I and II are exoglycosidases that removes the terminal sugars from the nonreducing termini of the oligosaccharide. If the proteins are folded correctly, glucosidase I and II removes the three Glc residues from the glycan moiety. If the protein is not folded properly, the glucosidases do not act, and the unfolded or partially folded proteins are recognized by chaperone proteins such as calnexin and calreticulin that helps them undergo recirculation for proper folding (Hebert et al. 1995). Proteins that remain unfolded even after the recirculation will be headed for degradation (Gardner et al. 2013). Following the removal of the three Glc residues, a single Man residue is removed by mannosidase 1 to form $GlcNAc_2Man_8$ glycan moiety in the ER (Elbein 1984).

The removal of the Man residue to form $GlcNAc_2Man_8$ glycan moiety signals the transfer of the glycoprotein from ER to Golgi. Upon the removal of the Man residue by mannosidase I, the glycoprotein with $GlcNAc_2Man_8$ glycan translocate into *cis*-Golgi from the lumen of the ER. In the *cis*-Golgi, the glycan moiety undergoes further modification or processing depending on the nature of the protein to which the glycan is attached. The glycans on some proteins such as DNase undergo phosphorylation which then transfers into acidic compartments. Majority of proteins with phosphorylated glycans are acidic enzymes (Friedman and Higgins 1995). In an alternative pathway, many glycoproteins in the *cis*-Golgi undergo further trimming by mannosidases to form $GlcNAc_2Man_5$ moiety (Tulp et al. 1986).

**Additional Processing of Oligosaccharide Unit for Chain Elongation and/or Modifications**

The formation of $GlcNAc_2Man_5$ glycan moiety on proteins signals the transfer of such glycoproteins from *cis*-Golgi to medial-Golgi (Bora de Oliveira et al. 2017). In the medial-Golgi, *N*-acetylglucosaminyltransferase-I (GnT-I) mediates the transfer of GlcNAc residue from UDP-GlcNAc to Man residue in the α-1,6-branch of glycan at O-2 position of Man residues (Stanley and Chaney 1985). At this stage, some protein molecules are transferred to *trans*-Golgi and such molecules will eventually end up forming hybrid glycan structures (Kobata 2000). Hybrid glycans contain more than three Man residues along with three GlcNAc residues and may contain Gal and sialic acid residues (Schachter and Jaeken 1999). Alternatively, many glycoproteins undergo further trimming by mannosidases in the *cis*-Golgi following GnT1 modification. Upon the addition of GlcNAc residue by GnT-I, mannosidases trim the two Man residues from the α-1,3-branch to form the $GlcNAc_2Man_3GlcNAc$ moiety on glycoproteins in *cis*-Golgi. The removal of the two mannose residues allows *N*-acetylglucosaminyltransferase-II (GnT-II) to transfer GlcNAc from UDP-GlcNAc to Man residue in the α-1,3-branch at O-2 position. Following the action by GnT-II, other GnTs such as GnT-III, GnT-IV, GnT-V will act (Schachter 2014). The action of GnTs will trigger the transfer of proteins from medial-Golgi to *trans*-Golgi. Addition of Fuc, Gal, Sialic acid residues by fucosyltransferases (FucTs), galactosyltransferases (GalTs), sialyltransferases (STs), etc., respectively, occurs in the *trans*-Golgi compartment (Gabius 2017).

## Microheterogeneity of *N*-Glycans

As described above, biosynthesis of *N*-glycans involves glycosidases, glycosyltransferases, nucleotide sugars, activated sugar transporters, etc. (Barel and Charbit 2017). It has been postulated that the biosynthesis machinery of *N*-glycans involves more than 200 different enzymes within a given living cell or organisms (Raju et al. 1996). Glycosidases and glycosyltransferases are highly linkage-specific, and for each glycosidic bonds, multiple enzymes may exists (Raju 2008; Kobata 2013). For example, at least six β-galactosyltransferases have been identified in Chinese hamster ovary (CHO) cells that are involved in the formation of β-1,4-Gal linkages to GlcNAc residues (Lee et al. 2001). Similarly, multiple STs, multiple FucTs, etc., have been identified (Ma et al. 2006; Huang et al. 2017). In addition, endoglycosidases and exoglycosidases are also highly linkage-specific (Kobata 2013). The glycosidases and GTs compete with each other for the available substrate/acceptor molecules, respectively, within the Golgi compartments. In addition to their competition to the substrate/acceptor molecules, the expression patterns of these enzymes may also vary with the species, organs, cellular location, etc. In addition to these factors involving glycosidases and GTs, the availability of activated sugars in the respective cellular compartments and the efficiency of the nucleotide sugar transporters to transfer the activated sugars may also vary with the species, organs, cellular location, etc. Hence, *N*-glycans of glycoproteins are very complex and highly heterogeneous (Raju 2008). The heterogeneity of *N*-glycans exists in terms of terminal sugar residues, branching pattern, chain length, etc. Hence, glycoproteins often contain a highly heterogeneous mixture of high mannose, hybrid and complex glycans that may contain multiple terminal end-groups including sialic acid, Gal, GlcNAc, Man, etc. Such heterogeneity of glycans in glycoproteins is also referred as microheterogeneity (Iwase 1988). The microheterogeneity of *N*-glycans may vary with protein sequence, site of attachment of glycans on proteins, and may also vary with the termini (Raju and Jordan 2012). The microheterogeneity of *N*-glycans may also vary with species, organisms, cell types, etc. (Raju et al. 2000). Within a given species, the microheterogeneity of *N*-glycans may vary with the age, gender, disease status, diet, etc. (Zeck et al. 2011).

Many glycoproteins contain multiple N-glycosylation sites (Howard et al. 1991). In a given protein, different glycosylation sites may contain different types of *N*-glycans. For instance, in a given glycoprotein that contain more than two or more glycosylatin sites, one glycosylation site may contain high mannose type *N*-glycans, another site may contain hybrid type *N*-glycans, and the other site may contain complex type *N*-glycans. Such a type of glycosylation pattern has been observed in rhtPA (Briggs et al. 2009). Hybrid type glycans are abundant in vertebrates compared to microorganisms, and hence, lower organisms may contain less diverse *N*-glycans compared to vertebrates. In addition, all three types of glycans, i.e. high mannose, hybrid and complex glycans may be present in a single glycosylation site. For example, immunoglobulins derived from chicken serum contain a mixture of high mannose, hybrid and complex glycan structure (Raju et al. 2000). IgGs derived from sheep serum contain a highly heterogeneous mixture of *N*-glycans compared to IgGs derived from other animal species (Raju et al. 2000). Also, bisecting GlcNAc residue containing glycans are more abundant in sheep IgGs compared to IgGs from other species (Raju et al. 2000).

### Species-Specific N-glycosylation

In humans, a vast majority of serum glycoproteins contain complex type *N*-glycans (Hughes 1975). However, minor amounts of high mannose and hybrid glycan structures are also found in human-derived glycoproteins (Shiyan and Bovin 1997). Most *N*-glycans found in human-derived glycoproteins are sialylated, and the type of sialic acid present in human glycoproteins is mostly *N*-acetylneuraminic acid (NANA). Accordingly, *N*-glycolylneuraminic acid (NGNA) residues are mostly absent in human-derived glycoproteins (Okerblom and Varki 2017). However, the glycoproteins derived from other animal species may contain a mixture of both NANA and NGNA (Raju et al. 2000). Some animal species exclusively express only NGNA containing sialylated *N*-glycans. This is because some species like humans are missing an active enzyme called cytidine monophosphate (CMP)-sialic acid hydroxylase that is responsible for the formation of N-glycolyl group on the sialic acid residues (Suzuki 2006). Although, humans contain the gene that express the CMP-sialic acid hydroxylase, this gene has a mutation that makes the enzyme inactive in humans (Chou et al. 1998). Accordingly, expression of sialylated *N*-glycans is species-specific because the active CMP-sialic acid hydroxylase is present in many animals but not in humans (Raju et al. 2000).

Mouse and other animal-derived glycoproteins often contain *N*-glycans with α-Gal residues (Flint et al. 1986). However, the human-derived glycoproteins do not contain *N*-glycans with α-Gal residues. α-Galactosyltransferases (α-GalTs) are responsible for the transfer of α-Gal residues to the terminal β-Gal residues on glycoproteins (Basu and Basu 1973). The active α-GalT enzyme is present in mouse and other animal species, but the active enzyme is absent in humans (Byrne et al. 2015). Mouse myeloma cells express active α-GalT, but the active enzyme is not present in CHO cells (Raju 2008). Hence, glycoproteins derived from CHO cells do not contain α-Gal epitopes. However, some mutant CHO cells have been shown to express active α-GalT (Cummings and Mattox 1988). Both human cells and normal CHO cells contain a mutant gene that expresses inactive α-GalT enzyme (Ashford et al. 1993). Active α-GalT enzyme is present in old world monkeys and chimpanzees, but not in humans (Galili 2015).

Some human glycoproteins contain *N*-glycans with bisecting GlcNAc residues (Raju et al. 2000). Further, mouse and other animal species also contain variable amounts of *N*-glycans with bisecting GlcNAc residues. Furthermore, *N*-glycans derived from sheep glycoproteins appears to contain higher amounts of bisecting GlcNAc residues but not the *N*-glycans derived from goat glycoproteins. Similarly, glycoproteins derived from cells of human, mouse, bovine, etc., origin contain variable amounts of *N*-glycans with bisecting GlcNAc residues. However, glycoproteins derived from normal CHO cells do not appear to contain *N*-glycans with bisecting GlcNAc residues. The addition of bisecting GlcNAc to *N*-glycans is mediated by an enzyme called *N*-acetylglucosaminyltransferase-III (GnT-III) (Stanley 2002). The gene responsible for the expression of GnT-III is mutated in wild-type CHO cells and this mutated gene expresses an inactive GnT-III enzyme in normal CHO cells (Stanley 2002). However, a mutant CHO cell line has been isolated that expresses active GnT-III enzyme (Campbell and Stanley 1984). The glycoproteins derived from this mutant CHO cell line contain *N*-glycans with bisecting GlcNAc residues (Stanley et al. 1991).

Human glycoproteins contain core fucosylated *N*-glycans in which fucose is α-1,6-linked to core GlcNAc residue that is attached to Asn residue of proteins. Similarly, glycoproteins derived from other animal species also contain core fucosylated *N*-glycans. However, the amount of core fucosylation in *N*-glycans vary from species to species (Raju et al. 2000). Also, the degree of core fucosylation may vary from organ to organ, tissue type, cell type, location of cells, etc. In humans, core fucosylation of *N*-glycans may also vary with age, gender, disease status, etc. Additionally, although human and other animal species contain α-1,6-fucose in the *N*-glycan core region, plants- and insects-derived glycoproteins contain *N*-glycans with core fucose-linked at α-1,3-position of GlcNAc attached to Asn residues (Raju 2008). Some glycoproteins may contain difucosylated *N*-glycans in which both α-1,3- and α-1,6-fucose residues are attached to the core GlcNAc residue which is linked to Asn residues of proteins (Geisler et al. 2015). In such glycoproteins, both α-1,3- and α-1,6-linked residues are present in the *N*-glycan core regions (Staudacher et al. 1999). However, such *N*-glycans are very rare and may present in some mutant cells-derived glycoproteins or in glycoproteins derived from microorganisms and plants (Wilson and Altmann 1998).

In addition to the core fucose, many glycoproteins contain fucose in the antennae of *N*-glycans. In such glycoproteins, the fucose residue/s are linked mostly to GlcNAc residues in the antennae (Lowe 2005). Such fucose residues in the antennae may be part of Lewis X and sialyl Lewis X epitopes. The formation of Lewis X and sialyl Lewis X epitopes on *N*-glycans is due to the transfer of Fuc from GDP-Fuc by FucT to GlcNAc residues present in the *N*-glycan antennae. Multiple FucTs have been shown to exist in human and other animal species (Becker and Lowe 2003). The expression pattern of FucTs may vary with species, and within species, the FucTs expression may vary with tissue, organ, cell type, etc. Hence, *N*-glycans with Lewis X and silayl Lewis X epitopes may vary with species, tissue, organ, cell type, etc.

Glycoproteins from plants and insects contain *N*-glycans with xylose (Xyl) linked at β-1,2-position to core Man residue attached to chitobiose saccharide (Cabanes-Macheteau et al. 1999). However, such types of glycoproteins with Xyl in the core region are not found in humans and other animal species. Plants and insects express xylosyltransferase (XylT) that is responsible for the biosynthesis of Xyl containing *N*-glycans of glycoproteins. This specific XylT is not present in humans and other animal species (Raju 2008).

Many mammalian glycoproteins contain *N*-glycans with repeating units of Galβ-1,4-GlcNAc residues. The saccharides with repeating units of Galβ-1,4-GlcNAc residues are referred as polylactosamine (poly LacNAc) residues. The poly LacNAc saccharides are linear polysaccharides present in some glycoproteins. In addition, in some glycoproteins, the poly LacNAc structures may be branched (Kinoshita et al. 2014). The poly LacNAc saccharides may also fucosylated and/or sialylated (Terada et al. 2005). Hence, in some glycoproteins, the poly LacNAc glycans may contain multiple Lewis X and/or sialyl Lewis X epitopes. The expression of poly LacNAc glycans containing glycoproteins is highly specific to species, tissue, organ, etc.

## Functions of *N*-glycans

It is now very well known that the *N*-glycans impact the structure and functions of proteins (Raju 2008). The impact of *N*-glycans on the structure and functions of proteins can be broadly categorized as follows: physicochemical, biological, and pharmacokinetic functions that are discussed below in different sections.

### Physicochemical Functions of *N*-glycans

During the biosynthesis of glycoproteins, the *N*-glycans undergo initial trimming process in the ER mediated by glucosidases and mannosidases thus generating intermediate glycan structures. These intermediate glycans are involved in performing specific functions. In the early stages of glycoprotein synthesis, different glycan structures containing terminal Glc and/or Man residues are required for proper folding. The glycoproteins that are not folded properly are recirculated for reglycosylation and refolding. Improperly glycosylated proteins are subject to deglycosylation and degradation within the ER. Hence, trimming of Glc and Man residues is required for intracellular transport of glycoproteins from ER to Golgi complex and targeting of these glycoproteins in *trans*-Golgi network. Accordingly, *N*-glycans play important roles in the transport and targeting of glycoproteins to different cellular compartments within the cells.

The protein backbone and the covalently linked *N*-glycans may exhibit different physicochemical properties. For example, protein backbone might be hydrophobic, but the covalently linked *N*-glycans are hydrophilic. Further, the *N*-glycans can be neutral or acidic, and the acidity of glycans may be due to sialylation or sulfation or phosphorylation of sugar residues. Accordingly, covalently linked *N*-glycans may impact the hydrophobicity and/or overall charge of glycoproteins. Hence, *N*-glycans may impact the solubility of glycoproteins in aqueous solutions. Very often the deglycosylated proteins are insoluble compared to their glycosylated counterparts (Woodward et al. 1987; Raju and Davidson 1994). The *N*-glycans may be short chain oligosaccharides or very long polylactosamine chains, but they can be bulky as they are able to form hydrogen bonds with water molecules in solution and hence the surface-exposed *N*-glycans may mask amino acid residues from surface exposure. Accordingly, the *N*-glycans may impact the protein–protein interactions and may also increase resistance of proteins to proteases (Raju and Scallon 2005).

**Biological Functions of *N*-glycans**

*N*-Glycans may impact the biological activities of proteins. Protein–protein interactions, enzymatic activities, interactions with receptors, etc., have been shown to be affected by *N*-glycans (Varki 2017). In IgGs, the *N*-glycans present in the Fc portion are required for antibody effector functions including ADCC activity and *N*-glycans affect the antibody binding to Fc receptors such as FcγRI, FcγRII, etc. Additionally, Fc glycans may also impact CDC activity of IgGs because *N*-glycans impact the antibody binding to C1q protein (Raju 2008). In addition, *N*-glycans present in the Fab domains may impact antibody binding to antigens.

Lectins are carbohydrate binding proteins, and many lectins are expressed in living organisms. *N*-Glycans binds to lectins and nonreducing terminal end groups such as sialic acids, Gal, GlcNAc, bisecting GlcNAc, Man, Fuc, etc., are recognized by different lectins. Carbohydrate–lectin interactions have been studied extensively and such interactions has been shown to impact many functions of living organisms (Dam et al. 2016). For example, carbohydrate–lectin interactions are involved in neutrophil rolling (Lowe 2005). In addition, carbohydrate–lectin interactions have been shown to be involved in alternative complement activity of IgGs (Malhotra et al. 1995).

Carbohydrate–lectin interactions have been shown to be involved in microbial interactions. For example, cell surface sialic acid residues are involved in host–pathogen interactions of several viruses (Byrd-Leotis et al. 2017). In addition, *N*-glycans has been shown to inhibit interactions between antigenic peptide region and antibody in Ebola (Mire and Geisbert 2017). In the case of Ebola, the *N*-glycans have been shown to protect the antigenic protein from proteolysis (Raju 2017). *N*-Glycans also protects the cell surface glycoproteins of viruses from antibody recognition by masking the antibody binding domains (Raju 2017). Further, *N*-glycans containing blood group epitopes may be involved in host–pathogen interactions, neutrophil activation, etc. (Varki 2017).

**Impact of *N*-Glycans on Pharmacokinetic Properties of Proteins**

*N*-Glycans have been shown to impact the serum half-life of glycoproteins (Varki 2017). For example, glycoproteins containing sialylated *N*-glycans have been shown to exhibit longer serum half-life compared to their asialylated counterparts (Raju et al. 2001). This is because asialylated glycoproteins are being recognized by asialoglycoprotein receptor present in the liver (Blasko et al. 2013). In addition to sialic acid, many glycoproteins contain Gal, GlcNAc, Man, etc., as their terminal sugar residues due to microheterogeneity (Briggs et al. 2009). Similar to glycoproteins containing terminal Gal residues, the glycoproteins containing terminal GlcNAc and Man residues are also cleared rapidly from the serum (Jones et al. 2007). This is because glycoproteins containing terminal GlcNAc and Man residues are being recognized by mannose receptor present in the liver (Steer and Ashwell 1986).

### *N*-Glycans and Human Diseases

In addition to impacting the physicochemical, biological, and pharmacokinetic properties of proteins, *N*-glycans are also involved in causing human diseases like metabolic diseases. The human metabolic diseases caused by *N*-glycans are collectively called as congenital disorders of glycosylation (CDG). The CDGs are human diseases due to errors in *N*-glycan metabolism caused by the deficiencies of N-glycosylation enzymes. The vast majority of CDGs are mainly due to the absence of exoglycosidases such as α-mannosidase, β-mannosidase, sialidase, α-fucosidase, α-galactosidase, etc., which are involved in the biosynthesis and/or degradation of *N*-glycans. Many of these enzymes are located in the lysosomes and hence, some of the CDGs are also referred as lysosomal storage diseases (Nagueh 2014).

## N-glycosylation of RTPs

Many RTPs are glycoproteins and contain variable amounts of *N*-glycans (Yang et al. 2015). RTPs that are N-glycosylated include rhDNase, rhEPO, rhtPA, rhTNK, rhFactor VIII, etc. *N*-Glycans present in RTPs are highly heterogeneous and have been shown to impact physicochemical, biological, and pharmacokinetic properties of RTPs (Raju et al. 2001). For example, rhEPO is produced in CHO cells and is N-glycosylated. *N*-Glycans of rhEPO are highly heterogeneous and affects its biological activities along with the pharmacokinetic properties (Macdougall et al. 1991). The heterogeneous *N*-glycans of rhEPO contain sialic acid residues as well as Gal as terminal sugar residues. Because of the presence of terminal Gal residues, serum half-life of rhEPO is not optimal. An engineered version of rhEPO, called novel erythropoiesis stimulating protein (NESP), contains more homogeneously sialylated glycans and hence exhibits improved serum half-life compared to rhEPO (Egrie and Browne 2001). Similarly, rhtPA contain highly heterogeneous *N*-glycans that include complex, high mannose and hybrid structures (Briggs et al. 2009). The high mannose structures in rhtPA are mostly localized to one glycosylation site (Bennett et al. 1991). Due to the presence of high mannose glycans, the serum half-life of rhtPA is not optimal. An engineered version of rhtPA, namely TNK-tPA that does not contain appreciable amounts of high mannose structures exhibit improved serum half-life and biological activity compared to rhtPA (Modi et al. 1998). As mentioned earlier, rhEPO, rhtPA, and TNK-tPA are produced in CHO cells. However, recombinant factor VIII and other RTPs are produced in BHK and other type of cell lines. The glycosylation of RTPs produced in CHO cells are somewhat different than the glycosylation of RTPs produced in BHK cells. Also, different RTPs are produced in different culture conditions and at different production scales. The N-glycosylation pattern of RTPs produced in different cell lines and cell culture conditions may vary. In addition, *N*-glycan heterogeneity may also vary with production scales. Such variation may result in lot-to-lot variation of N-glycosylation. Lot-to-lot variation may impact the product quality, efficacy, and serum half-life of RTPs. Hence, very often *N*-glycans of RTPs may become critical quality attributes (CQAs).

Therapeutic immunoglobulins, such as mAbs, are glycosylated in the $CH_2$ domain of Fc domain. The Fc glycans are highly heterogeneous and may contain sialic acid, Gal, GlcNAc, Man, etc., as terminal sugars. The Fc glycans of mAbs impact antibody functions including antibody effector functions (Raju 2008). The heterogeneity of Fc glycans of mAbs may vary with cell types, culture conditions, manufacturing scale, etc. For example, Fc glycans of Rituximab impact ADCC, ADCP, CDC, apoptosis, trogocytosis, etc. (Boross et al. 2012). Additionally, core fucose of mAbs significantly impacts antibody binding to FcγRIIIa and hence ADCC activity (Raju 2008). Accordingly, Fc glycans may become CQAs depending on the mechanism of action (MOA) of mAbs.

Additionally, some mAbs may contain *N*-glycans in the Fab domains. For example, Cetuximab is a chimeric mAb that contain an N-glycosylation site in the Fab domains (Qian et al. 2007). Cetuximab is produced in mouse myeloma cells and hence, contains α-Gal containing *N*-glycans in the Fab domain. The α-Gal containing *N*-glycans have been attributed to be the reason for anaphylaxis reactions observed in some patients treated with Cetuximab (Chung et al. 2008). Further, mAbs produced in plants may contain α1,2-Xyl and α1,3-Fuc residues in the *N*-glycan core regions (Triguero et al. 2010). Both α1,2-Xyl and α1,3-Fuc epitopes are nonhuman and may be immunogenic to humans (Raju 2008; Altmann 2016).

## Methods to Analyze *N*-Glycans

Analysis of *N*-glycans of glycoproteins used to be very tedious and a highly complex task. However, recent advances in analytical methods and instrumentation have made the task of analyzing *N*-glycans much more user friendly. Historical methods used to analyze the *N*-glycans include metabolic labeling of sugar residues using radiolabeled isotopes, followed by the release of *N*-glycans using enzymes and/or chemicals. The released *N*-glycans were separated by chromatographic methods such as paper chromatography or thin-layer chromatography and identifying the sugar residues by measuring radioactivity of the labeled *N*-glycans against the available standards. Additionally, *N*-glycans were released from unlabeled glycoproteins using either enzymes or chemical and identify them by mass spectrometry, NMR, gas chromatography (GC), gas chromatography–mass spectrometry (GC–MS), etc. The enzymes used to release the *N*-glycans include PNGase F, PNGase A, Endo H, etc. The chemicals used to release *N*-glycans were anhydrous hydrazine, aqueous base, etc. Both enzymatic and chemical release of *N*-glycans have their own shortcomings. For example, enzymes are highly specific and are expensive, whereas chemical treatment may cause peeling reactions that may generate additional heterogeneity to the *N*-glycan repertoire. Hence, many researchers have employed endopeptidases to generate glycopeptides to characterize *N*-glycans. Endopeptidases used to generate glycopeptides from glycoproteins include pronase, proteinase K, trypsin, chymotrypsin, etc. However, methods using endopeptidases to generate glycopeptides may have their own limitations. For examples, trypsin and chymotrypsin may generate large peptides that may not be amenable for further analysis. Hence, still there is no single universal method available to release all the *N*-glycans from glycoproteins. However, releasing the *N*-glycans from glycoproteins using PNGase F and/or generating glycopeptides have been used more than the other methods of releasing the *N*-glycans for further characterization.

Traditionally, the released *N*-glycans were characterized by permethylation followed by GC and GC–MS analyses. In addition, $^{1}$H NMR and $^{13}$C NMR have been employed to characterize *N*-glycans. However, these methods take too long and may require multi-milligram quantities of material. More recently, highly sensitive methods have been employed to characterize sub nanomolar amounts of *N*-glycans. These methods include high-performance anion-exchange chromatography-pulsed amperometry detection/pulsed electric detection (HPAEC-PAD/PED), high performance liquid chromatography (HPLC), capillary electrophoresis (CE), liquid chromatography–mass spectrometry (LC–MS), etc. HPAEC-PAD/PED method requires no derivation of released *N*-glycans. However, other methods such as NP-HPLC, RP-HPLC, IE-HPLC, CE, etc., methods require derivatization of released sugars with UV and/or fluorescent tags at the reducing sugar residue using reductive amination that makes amenable for detection following chromatographic and/or electrophoretic separation. Chemical structure of some of the very common fluorescent tags used to label the released *N*-glycans is shown in Figure 8.3.

The released glycans can also be analyzed by mass spectrometry. For example, matrix assisted laser desorption time-of-flight mass spectrometry (MALDI-TOF-MS) has been extensively used to detect and/or characterize *N*-glycans (Papac et al. 1998). Electro-spray ionization-mass spectrometry

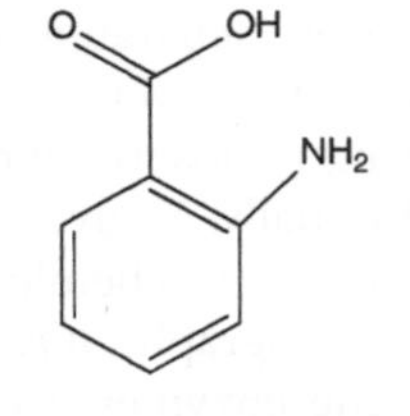

Anthranilic acid (AA)
(2-aminobenzoic acid)

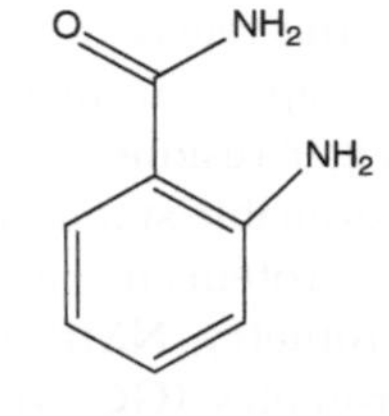

2-Aminobenzamide (2AB)

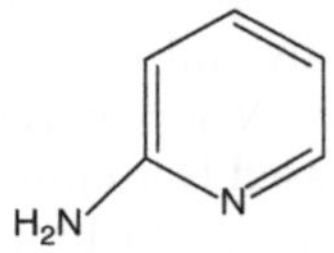

2-Aminopyridine

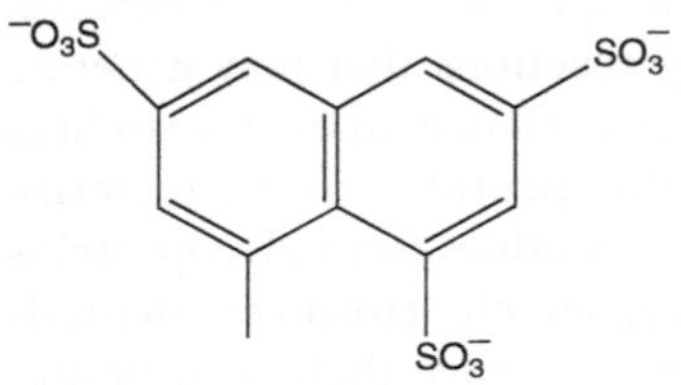

8-Aminonaphthalene-1,3,6-trisulfonic acid
(ANTS)

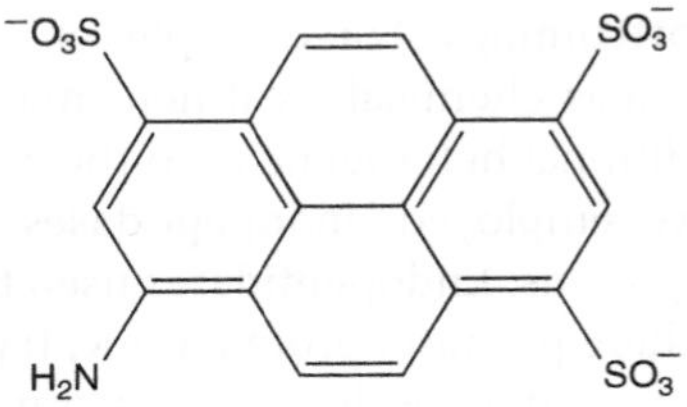

8-Aminopyrene-1,3,6-trisulfonic acid
(APTS)

**Figure 8.3** Chemical structure of some fluorescent tags commonly used to label released *N*-glycans for HPLC and CE analysis.

(ESI-MS) has also been used to characterize the released *N*-glycans. LC–MS analysis has also been used to separate and detect fluorescently labeled glycans. In addition, glycopeptides generated by treating the glycoproteins with endopeptidase have been characterized using LC–MS analysis and also by mass spectrometry alone. Recent advances in mass spectrometry have been enabled to characterize the *N*-glycans of glycoproteins without release or digesting with endopeptidases. The intact mass analysis of mAbs by ESI-MS, MALDI-TOF/TOF, etc., have been enabled to characterize Fc glycans of IgGs without releasing them by enzymes. However, the glycoproteins containing multiple N-glycosylation sites such as rhEPO, rhtPA may not be amenable to such intact mass analysis using high-resolution mass spectrometry to characterize the *N*-glycans.

## References

Aebi, M. (2013). N-linked protein glycosylation in the ER. *Biochim. Biophys. Acta* 1833 (11): 2430–2437. https://doi.org/10.1016/j.bbamcr.2013.04.001.

Altmann, F. (2016). Coping with cross-reactive carbohydrate determinants in allergy diagnosis. *Allergo J. Int.* 25 (4): 98–105.

Ashford, D.A., Alafi, C.D., Gamble, V.M. et al. (1993). Site-specific glycosylation of recombinant rat and human soluble CD4 variants expressed in Chinese hamster ovary cells. *J. Biol. Chem.* 268 (5): 3260–3267.

Barel, M. and Charbit, A. (2017). Role of glycosylation/deglycolysation processes in *Francisella tularensis* pathogenesis. *Front. Cell. Infect. Microbiol.* 7: 71. https://doi.org/10.3389/fcimb.2017.00071.

Basu, M. and Basu, S. (1973). Enzymatic synthesis of a blood group B-related pentaglycosylceramide by an alpha-galactosyltransferase from rabbit bone marrow. *J. Biol. Chem.* 248 (5): 1700–1706.

Becker, D.J. and Lowe, J.B. (2003). Fucose: biosynthesis and biological function in mammals. *Glycobiology* 13 (7): 41R–53R.

Bennett, W.F., Paoni, N.F., Keyt, B.A. et al. (1991). High resolution analysis of functional determinants on human tissue-type plasminogen activator. *J. Biol. Chem.* 266 (8): 5191–5201.

Blasko, E., Brooks, A.R., Ho, E. et al. (2013). Hepatocyte clearance and pharmacokinetics of recombinant factor IX glycosylation variants. *Biochem. Biophys. Res. Commun.* 440 (4): 485–489. https://doi.org/10.1016/j.bbrc.2013.09.001.

Bora de Oliveira, K., Spencer, D., Barton, C., and Agarwal, N. (2017). Site-specific monitoring of N-glycosylation profiles of a CTLA4-Fc-fusion protein from the secretory pathway to the extracellular environment. *Biotechnol. Bioeng.* 114 (7): 1550–1560. https://doi.org/10.1002/bit.26266.

Boross, P., Jansen, J.H., Pastula, A. et al. (2012). Both activating and inhibitory Fc gamma receptors mediate rituximab-induced trogocytosis of CD20 in mice. *Immunol. Lett.* 143 (1): 44–52. https://doi.org/10.1016/j.imlet.2012.01.004.

Briggs, J.B., Keck, R.G., Ma, S. et al. (2009). An analytical system for the characterization of highly heterogeneous mixtures of N-linked oligosaccharides. *Anal. Biochem.* 389 (1, 1): 40–51. https://doi.org/10.1016/j.ab.2009.03.006.

Butler, M. and Spearman, M. (2014). The choice of mammalian cell host and possibilities for glycosylation engineering. *Curr. Opin. Biotechnol.* 30: 107–112. https://doi.org/10.1016/j.copbio.2014.06.010.

Butters, T.D., Hughes, R.C., and Vischer, P. (1981). Steps in the biosynthesis of mosquito cell membrane glycoproteins and the effects of tunicamycin. *Biochim. Biophys. Acta* 640 (3): 672–686.

Byrd-Leotis, L., Cummings, R.D., and Steinhauer, D.A. (2017). The interplay between the host receptor and influenza virus hemagglutinin and neuraminidase. *Int. J. Mol. Sci.* 18 (7): 1541. https://doi.org/10.3390/ijms18071541.

Byrne, G.W., McGregor, C.G.A., and Breimer, M.E. (2015). Recent investigations into pig antigen and anti-pig antibody expression. *Int. J. Surg.* 23 (Pt B): 223–228. https://doi.org/10.1016/j.ijsu.2015.07.724.

Cabanes-Macheteau, M., Fitchette-Lainé, A.C., Loutelier-Bourhis, C. et al. (1999). N-glycosylation of a mouse IgG expressed in transgenic tobacco plants. *Glycobiology* 9 (4): 365–372.

Campbell, C. and Stanley, P. (1984). A dominant mutation to ricin resistance in Chinese hamster ovary cells induces UDP-GlcNAc:glycopeptide beta-4-*N*-acetylglucosaminyltransferase III activity. *J. Biol. Chem.* 259 (21): 13370–13378.

Chalifour, R.J. and Spiro, R.G. (1988). Effect of phospholipids on thyroid oligosaccharyltransferase activity and orientation. Evaluation of structural determinants for stimulation of N-glycosylation. *J. Biol. Chem.* 263 (30): 15673–15680.

Chou, H.H., Takematsu, H., Diaz, S. et al. (1998). A mutation in human CMP-sialic acid hydroxylase occurred after the Homo-Pan divergence. *Proc. Natl. Acad. Sci. U. S. A.* 95 (20): 11751–11756.

Chung, C.H., Mirakhur, B., Chan, E. et al. (2008). Cetuximab-induced anaphylaxis and IgE specific for galactose-alpha-1,3-galactose. *N. Engl. J. Med.* 358 (11): 1109–1117. https://doi.org/10.1056/NEJMoa074943.

Clerc, F., Reiding, K.R., Jansen, B.C. et al. (2016). Human plasma protein N-glycosylation. *Glycoconj. J.* 33 (3): 309–343. https://doi.org/10.1007/s10719-015-9626-2.

Cummings, R.D. and Mattox, S.A. (1988). Retinoic acid-induced differentiation of the mouse teratocarcinoma cell line F9 is accompanied by an increase in the activity of UDP-galactose: beta-D-galactosyl-alpha 1,3-galactosyltransferase. *J. Biol. Chem.* 263 (1): 511–519.

Dam, T.K., Talaga, M.L., Fan, N., and Brewer, C.F. (2016). Measuring multivalent binding interactions by isothermal titration calorimetry. *Methods Enzymol.* 567: 71–95. https://doi.org/10.1016/bs.mie.2015.08.013.

Dorokhov, Y.L., Sheshukova, E.V., Kosobokova, E.N. et al. (2016). Functional role of carbohydrate residues in human immunoglobulin G and therapeutic monoclonal antibodies. *Biochemistry (Mosc)* 81 (8): 835–857. https://doi.org/10.1134/S0006297916080058.

Dutta, D., Mandal, C., and Mandal, C. (2017). Unusual glycosylation of proteins: beyond the universal sequon and other amino acids. *Biochim. Biophys. Acta* https://doi.org/10.1016/j.bbagen.2017.08.025.

Egrie, J.C. and Browne, J.K. (2001). Development and characterization of novel erythropoiesis stimulating protein (NESP). *Br. J. Cancer* 84 (Suppl 1): 3–10.

Elbein, A.D. (1984). Inhibitors of the biosynthesis and processing of N-linked oligosaccharides. *CRC Crit. Rev. Biochem.* 16 (1): 21–49.

Elliott, S., Egrie, J., Browne, J. et al. (2004). Control of rHuEPO biological activity: the role of carbohydrate. *Exp. Hematol.* 32 (12): 1146–1155.

Epp, A., Sullivan, K.C., Herr, A.B., and Strait, R.T. (2016). Immunoglobulin glycosylation effects in allergy and immunity. *Curr Allergy Asthma Rep* 16 (11): 79.

Flint, F.F., Schulte, B.A., and Spicer, S.S. (1986). Glycoconjugate with terminal alpha galactose. A property common to basal cells and a subpopulation of columnar cells of numerous epithelia in mouse and rat. *Histochemistry* 84 (4–6): 387–395.

Friedman, Y. and Higgins, E.A. (1995). A method for monitoring the glycosylation of recombinant glycoproteins from conditioned medium, using fluorophore-assisted carbohydrate electrophoresis. *Anal. Biochem.* 228 (2): 221–225.

Gabius, H.J. (2017). The sugar code: why glycans are so important. *Biosystems* https://doi.org/10.1016/j.biosystems.2017.07.003.

Galili, U. (2015). Significance of the evolutionary α1,3-galactosyltransferase (GGTA1) gene inactivation in preventing extinction of apes and old-world monkeys. *J. Mol. Evol.* 80 (1): 1–9. https://doi.org/10.1007/s00239-014-9652-x.

Gardner, B.M., Pincus, D., Gotthardt, K. et al. (2013). Endoplasmic reticulum stress sensing in the unfolded protein response. *Cold Spring Harb. Perspect. Biol.* 5 (3): https://doi.org/10.1101/cshperspect.a013169.

Geisler, C., Mabashi-Asazuma, H., and Jarvis, D.L. (2015). An overview and history of glyco-engineering in insect expression systems. *Methods Mol. Biol.* 1321: 131–152. https://doi.org/10.1007/978-1-4939-2760-9_10.

Hammond, C., Braakman, I., and Helenius, A. (1994). Role of N-linked oligosaccharide recognition, glucose trimming, and calnexin in glycoprotein folding and quality control. *Proc. Natl. Acad. Sci. U. S. A.* 91 (3): 913–917.

Hanover, J.A. and Lennarz, W.J. (1980). N-linked glycoprotein assembly. Evidence that oligosaccharide attachment occurs within the lumen of the endoplasmic reticulum. *J. Biol. Chem.* 255 (8): 3600–3604.

Hebert, D.N., Simons, J.F., Peterson, J.R., and Helenius, A. (1995). Calnexin, calreticulin, and Bip/Kar2p in protein folding. *Cold Spring Harb. Symp. Quant. Biol.* 60: 405–415.

Henschen-Edman, A.H. (2001). Fibrinogen non-inherited heterogeneity and its relationship to function in health and disease. *Ann. N. Y. Acad. Sci.* 936: 580–593.

Howard, S.C., Wittwer, A.J., and Welply, J.K. (1991). Oligosaccharides at each glycosylation site make structure-dependent contributions to biological properties of human tissue plasminogen activator. *Glycobiology* 1 (4): 411–418.

Huang, R.B., Cheng, D., Liao, S.M. et al. (2017). The intrinsic relationship between structure and function of the sialyltransferase ST8Sia family members. *Curr. Top. Med. Chem.* 17 (21): 2359–2369. https://doi.org/10.2174/1568026617666170414150730.

Hubbard, S.C. and Robbins, P.W. (1980). Synthesis of the N-linked oligosaccharides of glycoproteins. Assembly of the lipid-linked precursor oligosaccharide and its relation to protein synthesis in vivo. *J. Biol. Chem.* 255 (24): 11782–11793.

Hughes, R.C. (1975). The complex carbohydrates of mammalian cell surfaces and their biological roles. *Essays Biochem.* 11: 1–36.

Iwase, H. (1988). Variety and microheterogeneity in the carbohydrate chains of glycoproteins. *Int. J. Biochem.* 20 (5): 479–491.

Jakab, L. (2016). Biological role of heterogeneous glycoprotein structures. *Orv. Hetil.* 157 (30): 1185–1192. https://doi.org/10.1556/650.2016.30436.

Jones, A.J., Papac, D.I., Chin, E.H. et al. (2007). Selective clearance of glycoforms of a complex glycoprotein pharmaceutical caused by terminal *N*-acetylglucosamine is similar in humans and cynomolgus monkeys. *Glycobiology* 17 (5): 529–540.

Kelleher, D.J. and Gilmore, R. (2006). An evolving view of the eukaryotic oligosaccharyltransferase. *Glycobiology* 16 (4): 47R–62R.

Kinoshita, M., Mitsui, Y., Kakoi, N. et al. (2014). Common glycoproteins expressing polylactosamine-type glycans on matched patient primary and metastatic melanoma cells show different glycan profiles. *J. Proteome Res.* 13 (2): 1021–1033. https://doi.org/10.1021/pr401015b.

Kobata, A. (2000). A journey to the world of glycobiology. *Glycoconj. J.* 17 (7-9): 443–464.

Kobata, A. (2013). Exo- and endoglycosidases revisited. *Proc. Jpn. Acad. Ser. B Phys. Biol. Sci.* 89 (3): 97–117.

Kornfeld, R. and Kornfeld, S. (1985). Assembly of asparagine-linked oligosaccharides. *Annu. Rev. Biochem.* 54: 631–664.

Lee, J., Sundaram, S., Shaper, N.L. et al. (2001). Chinese hamster ovary (CHO) cells may express six beta 4-galactosyltransferases (beta 4GalTs). Consequences of the loss of functional beta 4GalT-1, beta 4GalT-6, or both in CHO glycosylation mutants. *J. Biol. Chem.* 276 (17): 13924–13934.

Lowe, J.B. (2005). Fucosylated glycans in innate and adaptive immmunity. *Adv. Exp. Med. Biol.* 564: 127–128.

Ma, B., Simala-Grant, J.L., and Taylor, D.E. (2006). Fucosylation in prokaryotes and eukaryotes. *Glycobiology* 16 (12): 158R–184R.

Macdougall, I.C., Roberts, D.E., Coles, G.A., and Williams, J.D. (1991). Clinical pharmacokinetics of epoetin (recombinant human erythropoietin). *Clin. Pharmacokinet.* 20 (2): 99–113.

Malhotra, R., Wormald, M.R., Rudd, P.M. et al. (1995). Glycosylation changes of IgG associated with rheumatoid arthritis can activate complement via the mannose-binding protein. *Nat. Med.* 1 (3): 237–243.

Mazumder, R., Morampudi, K.S., Motwani, M. et al. (2012). Proteome-wide analysis of single-nucleotide variations in the N-glycosylation sequon of human genes. *PLoS ONE* 7 (5): https://doi.org/10.1371/journal.pone.0036212.

Mellquist, J.L., Kasturi, L., Spitalnik, S.L., and Shakin-Eshleman, S.H. (1998). The amino acid following an Asn-X-Ser/Thr sequon is an important determinant of N-linked core glycosylation efficiency. *Biochemistry* 37 (19): 6833–6837.

Mire, C.E. and Geisbert, T.W. (2017). Neutralizing the threat: pan-ebolavirus antibodies close the loop. *Trends Mol. Med.* 23 (8): 669–671. https://doi.org/10.1016/j.molmed.2017.06.008.

Modi, N.B., Eppler, S., Breed, J. et al. (1998). Pharmacokinetics of a slower clearing tissue plasminogen activator variant, TNK-tPA, in patients with acute myocardial infarction. *Thromb. Haemost.* 79 (1): 134–139.

Nagueh, S.F. (2014). Anderson-Fabry disease and other lysosomal storage disorders. *Circulation* 130 (13): 1081–1090. https://doi.org/10.1161/CIRCULATIONAHA.114.009789.

Okerblom, J. and Varki, A. (2017). Biochemical, cellular, physiological, and pathological consequences of human loss of *N*-glycolylneuraminic acid. *ChemBioChem* 18 (13): 1155–1171. https://doi.org/10.1002/cbic.201700077.

Opdenakker, G., Rudd, P.M., Ponting, C.P., and Dwek, R.A. (1993). Concepts and principles of glycobiology. *FASEB J.* 7 (14): 1330–1337.

Papac, D.I., Briggs, J.B., Chin, E.T., and Jones, A.J. (1998). A high-throughput microscale method to release N-linked oligosaccharides from glycoproteins for matrix-assisted laser desorption/ionization time-of-flight mass spectrometric analysis. *Glycobiology* 8 (5): 445–454.

Plesner, T. and Krejcik, J. (2018). Daratumumab for the treatment of multiple myeloma. *Front. Immunol.* 9: 1228. https://doi.org/10.3389/fimmu.2018.01228.

Qian, J., Liu, T., Yang, L. et al. (2007). Structural characterization of N-linked oligosaccharides on monoclonal antibody cetuximab by the combination of orthogonal matrix-assisted laser desorption/ionization hybrid quadrupole–quadrupole time-of-flight tandem mass spectrometry and sequential enzymatic digestion. *Anal. Biochem.* 364 (1): 8–18.

Quast, I., Peschke, B., and Lünemann, J.D. (2017). Regulation of antibody effector functions through IgG Fc N-glycosylation. *Cell. Mol. Life Sci.* 74 (5): 837–847. https://doi.org/10.1007/s00018-016-2366-z.

Raju, T.S. (2008). Terminal sugars of Fc glycans influence antibody effector functions of IgGs. *Curr. Opin. Immunol.* 20 (4): 471–478. https://doi.org/10.1016/j.coi.2008.06.007.

Raju, T.S. (2017). Antibody mixtures to treat Ebola. *Curr Biotechnol* 6 (1): 5–8.

Raju, T.S. and Davidson, E.A. (1994). New approach towards deglycosylation of sialoglycoproteins and mucins. *Biochem. Mol. Biol. Int.* 34 (5): 943–954.

Raju, T.S. and Jordan, R.E. (2012). Galactosylation variations in marketed therapeutic antibodies. *MAbs* 4 (3): 385–391. https://doi.org/10.4161/mabs.19868.

Raju, T.S. and Scallon, B.J. (2006). Glycosylation in the Fc domain of IgG increases resistance to proteolytic cleavage by papain. *Biochem. Biophys. Res. Commun.* 341 (3): 797–803.

Raju, T.S. and Scallon, B. (2007). Fc glycans terminated with *N*-acetylglucosamine residues increase antibody resistance to papain. *Biotechnol. Prog.* 23 (4): 964–971.

Raju, T.S., Lerner, L., and O'Connor, J.V. (1996). Glycopinion: biological significance and methods for the analysis of complex carbohydrates of recombinant glycoproteins. *Biotechnol. Appl. Biochem.* 24 (Pt 3): 191–194.

Raju, T.S., Briggs, J.B., Borge, S.M., and Jones, A.J. (2000). Species-specific variation in glycosylation of IgG: evidence for the species-specific sialylation and branch-specific galactosylation and importance for engineering recombinant glycoprotein therapeutics. *Glycobiology* 10 (5): 477–486.

Raju, T.S., Briggs, J.B., Chamow, S.M. et al. (2001). Glycoengineering of therapeutic glycoproteins: in vitro galactosylation and sialylation of glycoproteins with terminal *N*-acetylglucosamine and galactose residues. *Biochemistry* 40 (30): 8868–8876.

Rudd, P.M. and Dwek, R.A. (1997). Glycosylation: heterogeneity and the 3D structure of proteins. *Crit. Rev. Biochem. Mol. Biol.* 32 (1): 1–100.

Sanyal, S. and Menon, A.K. (2009). Specific transbilayer translocation of dolichol-linked oligosaccharides by an endoplasmic reticulum flippase. *Proc. Natl. Acad. Sci. U. S. A.* 106 (3): 767–772. https://doi.org/10.1073/pnas.0810225106.

Schachter, H. (2014). Complex *N*-glycans: the story of the "yellow brick road". *Glycoconj. J.* 31 (1): 1–5. https://doi.org/10.1007/s10719-013-9507-5.

Schachter, H. and Jaeken, J. (1999). Carbohydrate-deficient glycoprotein syndrome type II. *Biochim. Biophys. Acta* 1455 (2–3): 179–192.

Sethuraman, N. and Stadheim, T.A. (2006). Challenges in therapeutic glycoprotein production. *Curr. Opin. Biotechnol.* 17 (4): 341–346.

Shiyan, S.D. and Bovin, N.V. (1997). Carbohydrate composition and immunomodulatory activity of different glycoforms of alpha1-acid glycoprotein. *Glycoconj. J.* 14 (5): 631–638.

Shrimal, S., Cherepanova, N.A., and Gilmore, R. (2015). Cotranslational and posttranslocational N-glycosylation of proteins in the endoplasmic reticulum. *Semin. Cell Dev. Biol.* 41: 71–78. https://doi.org/10.1016/j.semcdb.2014.11.005.

Stanley, P. (2002). Biological consequences of overexpressing or eliminating *N*-acetylglucosaminyltransferase-TIII in the mouse. *Biochim. Biophys. Acta* 1573 (3): 363–368.

Stanley, P. and Chaney, W. (1985). Control of carbohydrate processing: the lec1A CHO mutation results in partial loss of *N*-acetylglucosaminyltransferase I activity. *Mol. Cell. Biol.* 5 (6): 1204–1211.

Stanley, P., Sundaram, S., and Sallustio, S. (1991). A subclass of cell surface carbohydrates revealed by a CHO mutant with two glycosylation mutations. *Glycobiology* 1 (3): 307–314.

Stanley, P., Taniguchi, N., and Aebi, M. (2017). *N*-Glycans. In: *Essentials of Glycobiology* [Internet], Chapter 9, 3e (ed. A. Varki, R.D. Cummings, J.D. Esko, et al.). Cold Spring Harbor, NY: Cold Spring Harbor Laboratory Press; 2015–2017.

Staudacher, E., Altmann, F., Wilson, I.B., and März, L. (1999). Fucose in *N*-glycans: from plant to man. *Biochim. Biophys. Acta* 1473 (1): 216–236.

Steer, C.J. and Ashwell, G. (1986). Hepatic membrane receptors for glycoproteins. *Prog. Liver Dis.* 8: 99–123.

Sun, S. and Zhang, H. (2015). Identification and validation of atypical N-glycosylation sites. *Anal. Chem.* 87 (24): 11948–11951. https://doi.org/10.1021/acs.analchem.5b03886.

Suzuki, A. (2006). Genetic basis for the lack of *N*-glycolylneuraminic acid expression in human tissues and its implication to human evolution. *Proc. Jpn. Acad. Ser. B Phys. Biol. Sci.* 82 (3): 93–103.

Terada, M., Khoo, K.H., Inoue, R. et al. (2005). Characterization of oligosaccharide ligands expressed on SW1116 cells recognized by mannan-binding protein. A highly fucosylated polylactosamine type *N*-glycan. *J. Biol. Chem.* 280 (12): 10897–10913.

Triguero, A., Cabrera, G., Royle, L. et al. (2010). Chemical and enzymatic *N*-glycan release comparison for *N*-glycan profiling of monoclonal antibodies expressed in plants. *Anal. Biochem.* 400 (2): 173–183. https://doi.org/10.1016/j.ab.2010.01.027.

Tulp, A., Barnhoorn, M., Bause, E., and Ploegh, H. (1986). Inhibition of N-linked oligosaccharide trimming mannosidases blocks human B cell development. *EMBO J.* 5 (8): 1783–1790.

Varki, A. (1993). Biological roles of oligosaccharides: all of the theories are correct. *Glycobiology* 3 (2): 97–130.

Varki, A. (2017). Biological roles of glycans. *Glycobiology* 27 (1): 3–49. https://doi.org/10.1093/glycob/cww086.

Waechter, C.J. and Lennarz, W.J. (1976). The role of polyprenol-linked sugars in glycoprotein synthesis. *Annu. Rev. Biochem.* 45: 95–112.

Wang, Q., Stuczynski, M., Gao, Y., and Betenbaugh, M.J. (2015). Strategies for engineering protein N-glycosylation pathways in mammalian cells. *Methods Mol. Biol.* 1321: 287–305. https://doi.org/10.1007/978-1-4939-2760-9_20.

Welply, J.K., Shenbagamurthi, P., Lennarz, W.J., and Naider, F. (1983). Substrate recognition by oligosaccharyltransferase. Studies on glycosylation of modified Asn-X-Thr/Ser tripeptides. *J. Biol. Chem.* 258 (19): 11856–11863.

Wilson, I.B. and Altmann, F. (1998). Structural analysis of *N*-glycans from allergenic grass, ragweed and tree pollens: core alpha1,3-linked fucose and xylose present in all pollens examined. *Glycoconj. J.* 15 (11): 1055–1070.

Woodward, H.D., Ringler, N.J., Selvakumar, R. et al. (1987). Deglycosylation studies on tracheal mucin glycoproteins. *Biochemistry* 26 (17): 5315–5322.

Yang, Z., Wang, S., Halim, A. et al. (2015). Engineered CHO cells for production of diverse, homogeneous glycoproteins. *Nat. Biotechnol.* 33 (8): 842–844. https://doi.org/10.1038/nbt.3280.

Yang, N., Goonatilleke, E., Park, D. et al. (2016). Quantitation of site-specific glycosylation in manufactured recombinant monoclonal antibody drugs. *Anal. Chem.* 88 (14): 7091–7100. https://doi.org/10.1021/acs.analchem.6b00963.

Zeck, A., Pohlentz, G., Schlothauer, T. et al. (2011). Cell type-specific and site directed N-glycosylation pattern of FcγRIIIa. *J. Proteome Res.* 10 (7): 3031–3039. https://doi.org/10.1021/pr1012653.

# 9

# O-glycosylation of Proteins

## Introduction

More than 70% of eukaryotic proteins are glycosylated with covalently linked *N*- and/or *O*-glycans (Corfield 2017). The N- and/or O-glycosylation of proteins is a highly heterogeneous and often a very complex co-translational modification (CTM) and/or post-translational modification (PTM) which involves more than 200 different enzymes including glycosyltransferases, glycosidases, nucleotide sugar synthetases, transporters, etc. (Raju et al. 1996). Many of the enzymes involved in the biosynthesis of glycans are located in the endoplasmic reticulum (ER) and Golgi apparatus (Rini and Esko 2017). Hence, the biosynthesis of *N*- and *O*-glycans mainly occurs in the ER and Golgi compartments. However, some of the glycosylating enzymes involved in the biosynthesis and/or degradation of glycans are also located in the cytoplasm, lysosomes, and in the nucleus (Ruba and Yang 2016). Hence, some protein glycosylation may also occur in the nucleus and some degradation of glycans may occur in the acidic compartments of the cells like in lysosomes.

The *O*-glycans are defined as the sugar moieties covalently linked to oxygen atom of hydroxyl side chains of amino acid residues of proteins (Mulagapati et al. 2017). Most of the eukaryotic proteins contain *O*-glycans covalently linked to hydroxyl groups of Ser and/or Thr residues (Staudacher 2015). However, in some proteins, amino acid residues such as Tyr, hydroxylysine, hydroxyproline are found to be modified with *O*-glycans (Zauner et al. 2012). Most common *O*-glycans are mucin type *O*-glycans in which GalNAc is attached to hydroxyl group of Ser and/or Thr residues in α-configuration (Mendonça-Previato et al. 2013; Tran and Ten Hagen 2013). Mucins are a family of high molecular weight heavily glycosylated proteins that contain tandem repeats of Ser and Thr residues (Feldhoff et al. 1979). In most mammals, mucins are produced in epithelial tissues to form gels like layers which are involved in lubrication, cell signaling, cell barriers, etc. (Bhavanandan and Davidson 1976; Raju and

*Co- and Post-Translational Modifications of Therapeutic Antibodies and Proteins*, First Edition. T. Shantha Raju.

Davidson 1994). Mucin glycoproteins are heavily O-glycosylated as they are rich in Ser and/or Thr residues (Verma and Davidson 1994).

Most of the naturally occurring mucin type *O*-glycans may be very simple monosaccharide moieties or they can be disaccharides, trisaccharides, etc. Mucin type *O*-glycans can also be linear polysaccharides such as polylactosamine chains or highly branched complex oligosaccharides or polysaccharides. For example, *O*-glycans found in porcine submaxillary mucin glycoproteins are simple mono-, di-, tri-, and tetrasaccharides. However, the *O*-glycans found in glycoproteins from human and dogs are highly complex oligosaccharides and/or polysaccharides (Feldhoff et al. 1979). Hence, the nature and structural complexity of *O*-glycans may vary from species to species.

In addition to mucin type *O*-glycans, many proteins also contain O-linked GlcNAc, Fuc, Glc, Gal, Xyl, etc. residues (Varki 2017). The O-linked GlcNAc residues are found in many nuclear and cytosolic proteins and is β-linked to Ser/Thr residues (Zachara et al. 2017). The O-linked Fuc residues are mostly found in proteins with epidermal growth factor (EGF) like repeats and are linked to hydroxyls of Ser/Thr residues (Harris et al. 1991). In some proteins, O-Fuc may be elongated with the addition of GlcNAc, Gal, and sialic acid residues, respectively (Harris et al. 1993). The O-linked Xyl is the initiator for the biosynthesis of proteoglycans (Thorsheim et al. 2015) and O-linked Man is found in many eukaryotic and prokaryotic proteins (Sheikh et al. 2017). Also, O-linked Man is very abundant in yeast glycoproteins (Herscovics and Orlean 1993). In addition, O-linked Gal and Glc residues covalently attached to hydroxyl groups of Ser and/or Thr residues are found in many proteins (Haltiwanger et al. 2017). However, in some proteins, such as collagen, Gal and Glc residues are found to be O-linked to hydroxyl groups of hydroxylysine residues (Haltiwanger et al. 2017). Liver and muscle proteins such as glycogenin contain O-linked glucose attached to Tyr residues (de Paula et al. 2005). In plants, sugar residues such as Gal, arabinose (Ara) are found to be O-linked to hydroxyproline residues (Basu et al. 2015).

*O*-Glycans may impact the physicochemical, biological, and pharmacokinetic properties of proteins. O-glycosylation may increase the solubility of proteins and may also increase the viscosity of proteins (Raju and Davidson 1994). For example, mucin glycoproteins are highly viscous in solutions, and deglycosylated mucins are highly insoluble in aqueous solutions (Raju and Davidson 1994). O-glycosylation may also increase the resistance of proteins to proteolysis. *O*-Glycans may function as cell surface receptors for binding of pathogens. For example, sialic acid residues on glycophorin act as receptors for malaria parasite to invade the host red blood cells (Davidson and Gowda 2001). *O*-Glycans may protect the cells and cell membrane by acting as barrier from pathogens. Hence, *O*-glycans not only affect the protein functions, they may also play an important role in cellular functions of organisms.

## Biosynthesis of *O*-Glycans

Biosynthesis of *O*-glycans is described in the following sections as mucin type *O*-glycans, *O*-GlcNAc, *O*-Fuc, *O*-Glc, *O*-Gal, *O*-Xyl, etc.

### Biosynthesis of Mucin Type *O*-Glycans

Biosynthesis of mucin type *O*-glycans begins with the addition of GalNAc to the hydroxyl side chain of Ser/Thr residues of proteins in the α-configuration. The addition of GalNAc to Ser/Thr residues is mediated by α-*N*-acetylgalactosaminyltransferases (α-GalNAcTs) that mediates the transfer of GalNAc from UDP-GalNAc to the hydroxyl side chains of Ser and/or Thr residues. Unlike N-glycosylation, no consensus sequence is required for O-glycosylation of proteins by α-GalNAcT enzymes. The α-GalNAcT enzymes are a group of at least 12 isoenzymes and are located in the trans-Golgi compartment similar to the other glycosyltransferases involved in the biosynthesis of *N*-glycans. Many glycoproteins contain only GalNAc residue as their O-linked glycan repertoire, and these GalNAc epitopes are termed as Tn-antigens. Normally, the Tn antigens are not found in healthy cells but are found in many tumor cells. The Tn-antigen is most often capped by α-2,6-sialic acid which is termed as sialyl Tn-antigen (sTn antigen). The *O*-glycan specific α-2,6-sialyltransferases are involved in the transfer of sialic acid from cytidine monophosphate (CMP)-sialic acid to O-6 position of Tn-antigen. The sialyl Tn-antigens are the major O-linked glycans of many mucin glycoproteins (Carlson 1966).

In a majority of *O*-glycans, the GalNAc is extended by the addition of Gal, GlcNAc, Sia, etc. residues. The β-1,3-galactosyltransferase mediates the transfer of Gal residue from UDP-Gal to O-3 position of GalNAc residue which is O-linked to Ser/Thr residues. The Gal β-1,3-GalNAc disaccharide epitopes that are linked to Ser/Thr residues are called as Thomsen–Friedenreich antigens (TF-antigens or T-antigens). There are eight different mucin type core *O*-glycans and their schematic diagrams are shown in Figure 9.1. In Chinese hamster ovary (CHO) cells, this disaccharide, T-antigen, remains as it is or get capped by the addition of two sialic acid residues one to Gal residue and the other to GalNAc residue. The *O*-glycan-specific α-2,3-sialyltransferases are involved in the transfer of sialic acid residue from CMP-sialic acid to O-3 position of Gal residue of *O*-glycans. Hence, *O*-glycans of CHO cells-derived glycoproteins may contain a mixture of monosaccharide (GalNAc), disaccharides (Gal-GalNAc), trisaccharides (Sia–Gal–GlcNAc), and tetrasaccharides (Sia–Gal–(Sia)–GalNAc) as shown in Figure 9.2.

More complex branching of *O*-glycans into very heterogeneous glycans occurs with the addition of GlcNAc residue to O-6 position of GalNAc residue. This addition of GlcNAc is initiated by *O*-glycan specific

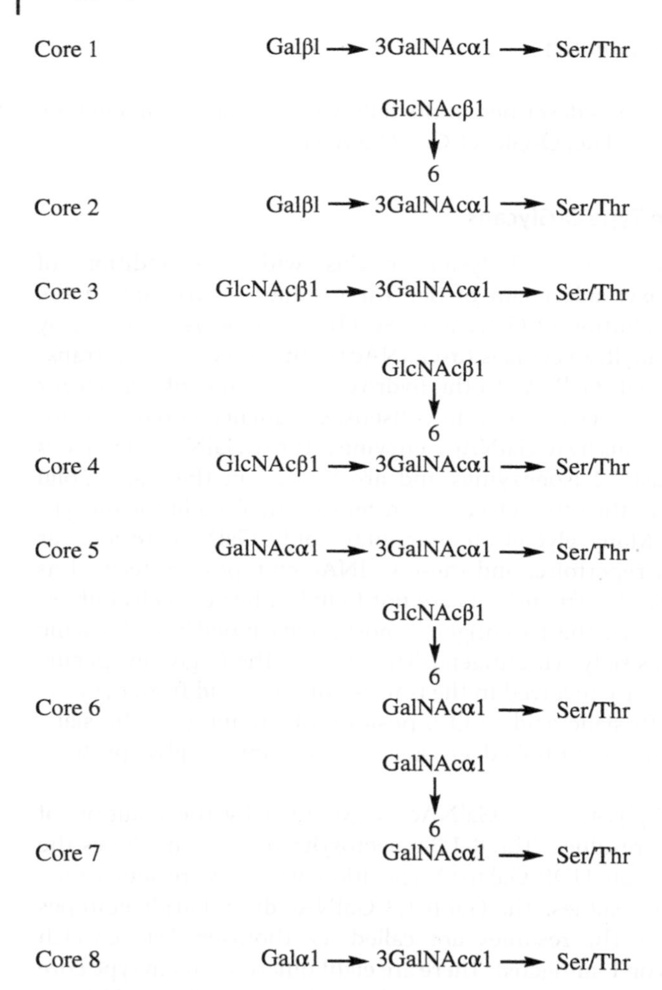


**Figure 9.1** Chemical structure of the core region of *O*-glycans commonly found in mucin glycoproteins.

β-1,6-*N*-acetylglucosaminyltransferase that mediates the transfer of GlcNAc residue from UDP-GlcNAc to O-6 position of GalNAc residue. Chain elongation and further branching occurs with the addition of Gal residue to GlcNAc residue, addition of GlcNAc residue to Gal residue, etc. These processes in turn repeats to form both linear polylactosamine chains as well as branched and more complex *O*-glycan polysaccharide structures. Accordingly, the complexity and heterogeneity of *O*-glycans are often much more complex than the heterogeneous structural complexity found in *N*-glycans. In addition, the

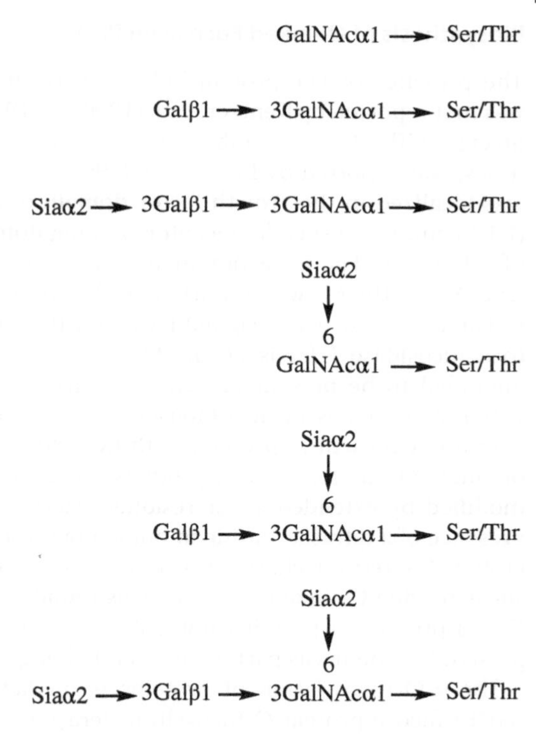


**Figure 9.2** Chemical structure of *O*-glycans found in glycoproteins derived from CHO cells (Sia, *N*-acetylneuraminic acid or sialic acid).

ABO blood group antigenic epitopes are most commonly found on *O*-glycans than on *N*-glycans.

## Biosynthesis of *O*-linked GlcNAc on Proteins

Many nuclear and cytosolic proteins are shown to contain O-linked GlcNAc residues covalently attached to hydroxyl groups of Ser and/or Thr residues (Hart 2016; Zachara et al. 2017). These *O*-GlcNAc modifications are mediated by *O*-β-*N*-acetylglucosaminyltransferases (OGTs) that transfer the GlcNAc from UDP-GlcNAc to the hydroxyl side chain of Ser and/or Thr residues. The systemic name of OGT enzyme is UDP-*N*-acetylglucosamine: protein-OGT and is encoded by *OGT* gene in humans. The *O*-GlcNAc modification is a dynamic modification and is reversible (Hart 2016). This reversible *O*-GlcNAc modification is catalyzed by two enzymes, the transferase and the hydrolase. The hydrolase is called neutral *N*-acetylglucosaminidase. The *O*-GlcNAc modification competes with protein phosphorylation of proteins (Wang et al. 2007). Accordingly, OGT competes with kinases and hence if phosphorylation occurs, then *O*-GlcNAcylation does not occur and vice versa. The *O*-GlcNAc modification is involved in cell signaling including insulin signaling and regulation of RNA transcription.

## Biosynthesis of *O*-linked Fucose on Proteins

The presence of Fuc-protein linkage in recombinant human pro-urokinase was first reported by Kentzer et al. (1990). In 1991, the presence of Fuc-protein linkage at Thr-18 in recombinant urinary-type plasminogen activators (urokinases) was reported by Buko et al. (1991). The Fuc-protein linkage at Thr-18 was localized in the growth factor domain of urinary plasminogen activator (UPA) and is the specific receptor-binding domain of the UPA. The presence of α-linked *O*-Fuc in recombinant human tissue type plasminogen activator (rhtPA) at Thr-61 was reported by Harris et al. (1991). The α-linkage of *O*-Fuc to Thr-61 was assigned based on the susceptibility of the Fuc residue to α-fucosidase (Harris et al. 1991, 1992). Subsequently, the α-*O*-Fuc was identified to be present in EGF domains of proteins (Harris and Spellman 1993). Now it has been established that the *O*-fucose modification is very commonly present in proteins with EGF domain. In addition to the presence of single *O*-Fuc residues in proteins, in some proteins the *O*-Fuc is further modified by extended sugar residues (Harris et al. 1993). For example, the presence of a tetrasaccharide in human factor IX was reported by Harris et al. (1993). The tetrasaccharide was linked to Ser-61 in human factor IX, and the sugar residue O-linked to Ser-61 was found to be α-linked Fuc. Accordingly, Fuc is present in proteins not only as a single monosaccharide but is also present in proteins as part of an extended oligosaccharide side chain.

The α-O-fucosylation of proteins is mediated by guanidine diphosphate (GDP)-fucose protein *O*-fucosyltransferase (*O*-FucT). The enzymatic activity of *O*-FucT was first reported by Wang et al. (1996) and is localized to ER unlike other glycosyltransferases that are localized to Golgi (Wang and Spellman 1998). The enzyme *O*-FucT preferentially adds Fuc to Ser/Thr residues in EGF domains as it does not recognize the Ser/Thr residues in the synthetic peptides (Kao et al. 1999). The enzyme was purified from CHO cells and found to be type II membrane glycoprotein with a molecular weight of ~44 kDa. Although the enzyme does not require $Mn_2^+$ for its activity, its activity substantially increases in the presence of $Mn_2^+$ ions. However, the enzyme requires properly folded EGF domain for its activity to add *O*-Fuc to Ser/Thr residues. The human cDNA encoding the human *O*-FucT was isolated from a screen of a human heart cDNA library using a probe deduced from the N-terminal sequence analysis of purified *O*-FucT from CHO cells. The cDNA of *O*-FucT contains an open reading frame encoding a polypeptide chain that contains 388 amino acid residues (Wang et al. 2001).

The *O*-Fuc residues on proteins are shown to be involved in notch signaling. Very recently, *O*-Fuc residues are shown to be involved in the host–pathogen interactions of malaria parasites (Lopaticki et al. 2017). The presence of *O*-Fuc is presumably involved in increasing the antigenicity of malaria vaccine candidate (Lopaticki et al. 2017). *O*-Fuc is also involved in signal transduction and protein folding.

## Biosynthesis of *O*-linked Glc Residues

Similar to *O*-Fuc, the *O*-Glc residues are added to Ser/Thr residues between the first and second conserved Cys residues of EGF like domains of notch proteins (Harvey and Haltiwanger 2018). The enzyme UDP-glucose: protein-*O*-glucosyltransferase (Poglut) mediates the transfer of Glc residues from UDP-Glc to oxygen atom of hydroxyl functional groups of Ser/Thr. In Drosophila the enzyme Poglut is called as Rumi and is involved in regulating notch signaling (Fernandez-Valdivia et al. 2011). The Rumi knock-out mouse embryos die before 9.5 embryonic day (Fernandez-Valdivia et al. 2011). The Rumi (–/–) mouse embryos exhibit posterior axis truncation and have severe defects in cardiogenesis, neural tube development, somitogenesis, and vascular remodeling. Biochemical studies have indicated that the absence of Rumi activity results in reduced *O*-Glc residues and loss of Notch signaling. *O*-Glc modification is also important for proper folding of Notch protein and secretion of the receptor (Takeuchi and Haltiwanger 2014).

## Biosynthesis of *O*-linked Gal Residues

Collagen and proteins containing collagen-like domains are rich in hydroxylysine residues. Very often the hydroxylysine residues are modified with *O*-Gal residues that in turn are modified by the addition of Glc residues. The Gal residues are added to hydroxyl group of hydroxylysine in β-configuration, whereas the Glc residues are added to the O-2 position of Gal in α-configuration. The two-step glycosylation process is mediated by two different enzymes and is conserved in most animals. The hydroxylysine glycosylation of collagen and proteins with collagen domains occurs in the ER as well as in the Golgi (Terajima et al. 2014). The hydroxylysine glycosylation influences the rate of triple helix formation and the size of collagen fibrils. Hence, glycosylation of hydroxylysine residues is required for the proper functioning of collagen. In addition to collagen, adiponectin, pulmonary surfactants proteins, etc., are reported to contain glycosylation on hydroxylysine residues (Varki 2017). Lack of hydroxylysine glycosylation causes embryonic lethality and deposition of misfolded collagen in the ER in mice.

## Biosynthesis of *O*-linked Man Residues

*O*-mannosylation is observed in both eukaryotic and prokaryotic proteins. The enzyme protein-*O*-mannosyltransferase is involved in transferring mannose from mannose-*p*-dolicol to oxygen atom of hydroxyl side chain of Ser and/or Thr residues (Baker et al. 2018). *O*-Man residues are very common in proteins found in the secretory pathway. The *O*-Man residues are in α-configuration and was first discovered in yeast in 1950s. In yeast and fungi, the *O*-Man residues are elongated to form highly heterogonous and highly complex glycans by a group of mannosyltransferases. In mammals, *O*-Man linked to rat brain proteoglycan was fist observed in 1979, and in some mammalian tissues, it is now estimated about one-third of all *O*-glycans are *O*-Man glycans. Defects in O-mannosylation causes many diseases including Walker–Warburg syndrome, muscle–eye–brain disease, Fukuyama congenital muscular dystrophy, etc. (Diesen et al. 2004).

### *O*-Glycans on Hydroxyproline Residues

In collagen, Pro residues are hydroxylated but are not glycosylated. This is because the hydroxyproline residues are used in the hydrogen bond formation in collagen to form triple helix structure. However, in *Dictyostelium discoideum* a protein called Skp1 is O-glycosylated at hydroxyproline residues with *O*-GlcNAc residues (Taylor et al. 2012). Additionally, in plants hydroxyproline residues are O-glycosylated with O-linked Gal residues and/or with O-linked Ara residues (Ogawa-Ohnishi and Matsubayashi 2015).

## Physicochemical Properties of *O*-Glycosylated Proteins

As discussed in Chapter 8, *O*-glycans also contains monosaccharides as their monomeric units and hence contain hydroxyl groups. In addition, *O*-glycans may also contain carboxylic, sulfate, phosphate, acetyl, methyl, etc., functional groups attached to them. These functional groups exert great influence on the physicochemical properties of O-glycosylated proteins. The role of *O*-glycans on the physicochemical properties of O-glycosylated proteins has been extensively studied using mucin glycoproteins (Ringler et al. 1987). Mucins glycoproteins forms highly viscous gel like solutions. Mucins are highly O-glycosylated glycoproteins and the *O*-glycans significantly contributes to their high viscosity (Raju and Davidson 1994). The *O*-glycans also contributes to their solubility as the deglycosylated mucins are highly insoluble in aqueous solutions (Woodward et al. 1987). Removal of sialic acid residues greatly reduces viscosity of mucin glycoproteins suggesting that sialic acid plays a major role in the physicochemical properties of mucin glycoproteins (Raju and Davidson 1994). *O*-Glycans can also have significant influence on the conformation of proteins.

## Biological Functions of *O*-Glycans

*O*-Glycans provides increased resistance of O-glycosylated proteins to proteases. O-glycosylated mucin domains are difficult to digest with proteases, but removal of *O*-glycans makes the mucin domains susceptible to protease digestions. Mucins secreted at the apical membrane of epithelial cells form hydrogen bonds and absorb water molecules by hydrophilic interactions. Such interactions of mucin glycoproteins with water molecules form mucous membrane on the surface of epithelial cells. Such mucous membrane protects the cells from hostile environment as well as from proteases. Mucous membranes also provides lubrication for swallowing and also protects inner stomach layer from stomach acids. Further, mucous membranes provides protection from infection by pathogens such as bacteria, virus (Verma et al. 1994). Many pathogens express carbohydrate-binding proteins, receptors, etc., which binds to *O*-glycans on the mucous membrane trapping the pathogen thus preventing further entry of pathogens into cells (Janssen et al. 2016). Removal of pathogens such as microbes and particles from mucous membrane is an important physiological process. However, under abnormal conditions and in the absence of mucous membrane, entry of pathogens becomes easier. Additionally, in patients with diseases such as cystic fibrosis patient's abnormal secretion of mucin glycoproteins leads to blocking of air pathways which might ultimately lead to life-threatening conditions (Kumar et al. 2017).

Sialyl Lewis X epitopes present on the *O*-glycans are ligands for selectins that are involved in leukocyte rolling (Impellizzeri and Cuzzocrea 2014). The sialyl Lewis X epitopes are also involved in the selectin-mediated interactions between endothelial cells and leukocytes. The selectin–glycan interactions are also important for leukocyte attachment to capillary endothelium during homing of lymphocytes (Kawashima and Fukuda 2012). The extravasation of leukocytes during inflammatory response is also regulated by *O*-glycan interactions with selectin. The sialyl Lewis X epitopes are also expressed by certain tumors that use glycan–selectin interactions to invade tissues (Hobbs and Nolz 2017).

*O*-Glycans of cell surface receptors are also involved in the regulation of receptor stability and expression levels (Santos et al. 2016). *O*-Glycans are also expressed in reproductive tissues and are shown to be involved in the process of fertilization (Grasa et al. 2012). In several species, certain terminal sugar residues of *O*-glycans are involved in sperm–egg interactions (Pang et al. 2011). *O*-GlcNAc is a dynamic modification that switches with phosphorylation and is involved in cellular signaling mechanism including insulin signaling. The alternative switching of *O*-GlcNAc with phosphorylation is also involved in the regulation of RNA transcription (Zachara et al. 2017). Further, *O*-GlcNAc processing is also involved in cellular signaling events and insulin signaling. *O*-Fuc modification by *O*-fucosyltransferase (OFT) is involved in notch signaling and activation of notch receptors (Haltiwanger et al. 2017). In addition, *O*-Fuc is also involved in ligand-induced receptor signaling. Malaria is caused by parasite *Plasmodium falciparum*. The surface proteins of circumsporozoite protein (CSP) and thrombospondin-related anonymous protein (TRAP) in *Plasmodium* sporozoites are O-fucosylated. The protein

*O*-fucosyltransferase (POFUT2) is responsible for O-fucosylation of CSP and TRAP surface proteins. Genetic disruption of POFUT2 in the malaria parasite results in the lack of malaria infection by the parasite (Lopaticki et al. 2017). Further, the CSP and TRAP proteins without *O*-Fuc residues are less antigenic compared to the fucosylated proteins (Lopaticki et al. 2017). The *O*-glycans also plays important roles in hematopoiesis and inflammation. *O*-Glycans of P-selectin glycoprotein ligand 1 (PSGL-1) contain sialyl Lewis X epitopes that are involved in the leukocyte rolling and recruitment of leukocytes into inflammation sites. *O*-Glycans also carry other human ABO blood group antigens. Lack of O-glycosylation may result in human diseases like congenital disorder of glycosylation (CDG). For example, Schindler's disease is a CDG related to *O*-glycans and is caused by the absence of α-*N*-acetylgalactosaminidase (α-GalNAcase).

## O-glycosylation in RTPs

Many recombinant therapeutic proteins (RTPs) contain *O*-glycans, and their structure vary with the expression system. For example, *O*-glycans of CHO cells-derived RTPs are different than the *O*-glycans of human embryonic kidney (HEK) cells-derived RTPs. *O*-Glycans found in therapeutic glycoproteins produced in CHO cells are shown in Figure 9.1. These are simple monosaccharide, di-, tri-, and tetrasaccharides. However, *O*-glycans found in RTPs produced in mouse myeloma cells, human cells, and other mammalian cells are much more complex than the CHO cells-derived RTPs. Few examples of O-glycosylated RTPs are shown in Table 9.1.

Although intact and full-length mAbs do not appear to contain *O*-glycans, the IgG fusion proteins are found to be O-glycosylated in the hinge region (Houel et al. 2014). Enbrel is an IgG fusion protein that contain multiple O-glycosylation sites in the hinge region (Houel et al. 2014). Although

**Table 9.1** Few examples of O-glycosylated RTPs.

| RTPs | Cells used to produce | Comments |
|---|---|---|
| Enbrel (TNFR-IgG) | CHO cells | *O*-Glycans are present in the hinge region |
| rhtPA | CHO cells | *O*-Fucose is present |
| rhEPO | CHO cells | |
| rhFactor VIII | BHK cells | |
| rhDNase | CHO cells | |

BHK, baby hamster kidney.

O-glycosylation is very common in certain immunoglobulins like IgAs that contain multiple O-glycosylation sites, IgGs are not commonly O-glycosylated. Hence, it appears as though the O-glycosylation of IgG fusion proteins in the hinge region is due to the accessibility of the Ser/Thr residues in the hinge region to *O*-GalNAc transferases. In addition, recently, human $IgG_3$ isotype is reported to contain *O*-glycans in the hinge region (Plomp et al. 2015). The reported O-glycosylation of $IgG_3$ isotype is due to the elongated hinge region in this IgG isotype.

In addition to the common *O*-GalNAc type *O*-glycans, many RTPs are reported to contain unusual *O*-glycans. For example, rhtPA has been reported to contain *O*-Fuc modification (Harris et al. 1991). *O*-Glc and other types of unusual modifications have also been reported in several RTPs. For example, RTPs containing Gly–Ser linkers have been shown to contain *O*-Xyl modifications (Spencer et al. 2013). In a few cases, the Xyl residue was found to be further elongated with the addition of hexose that is further modified by the addition of sulfated hexuronic acid (Spahr et al. 2013; Spencer et al. 2013). The Xyl modification found in RTPs may be mediated by xylosyltransferase involved in the biosynthesis of glycosaminoglycans (GAGs).

## Analysis of *O*-Glycans

Analysis of *O*-glycans is also very complex similar to *N*-glycans. Complexity of the analysis of *O*-glycans is mainly due to sparse availability of methods to release the O-linked glycans from glycoproteins. Hence, in many instances, peptide mapping methods have been adopted to identify and to characterize *O*-glycans from many RTPs (Houel et al. 2014). Peptide mapping with liquid chromatography–mass spectrometry (LC-MS) detection allows the identification of sites that are O-glycosylated. But the method provides very limited information on the structure and quantitation of *O*-glycans. High-resolution mass spectrometry is also not very useful to obtain structural information on *O*-glycans. However, lectin affinity chromatography using immobilized jacalin lectin provides information on the amount of O-glycosylated versus non-O-glycosylated proteins. But the lectin affinity chromatography may not be useful to obtain the detailed structural information of *O*-glycans. Hence, glycan mapping of the released glycans is still being developed and used widely.

Similar to *N*-glycans, *O*-glycans are also being released using either enzymatic methods or chemical methods. Among the chemical methods, there are two widely used chemical methods available to release *O*-glycans. Hydrazinolysis has been used to release *O*-glycans from anhydrous glycoproteins. Hydrazine is a nonselective reagent and hence releases all types of *O*-glycans. The *O*-glycans released using hydrozinolysis contain reducing sugar residues that are amenable for labeling with UV/visible/fluorescent tags. However, absolute anhydrous conditions are required to prevent peeling reactions during hydrazinolysis and also as discussed in Chapter 8 hydrazine is a very dangerous chemical. But the advantage of hydrozinolysis is that the method is nonselective and the availability of reducing sugar moiety for labeling for further analysis using high performance liquid chromatography (HPLC), capillary electrophoresis (CE), etc., methods.

*O*-Glycans undergo β-elimination reaction under alkaline conditions. However, high basic conditions are needed which may lead to high degree of peeling reactions. Such peeling reactions can be minimized by adding reducing agents such as sodium borohydride that converts the reducing sugars into their respective alditols. For example, addition of borohydride during β-elimination reactions converts GalNAc into *N*-acetylgalactosaminitol (GalNAc-ol). The released alditol containing *O*-glycans can be separated using HPLC methods such as high pH anion exchange chromatography (HPAEC) developed by Dionex corporation. The alditol containing *O*-glycans can also be analyzed by matrix assisted laser desorption time-of-flight mass spectrometry (MALDI-TOF-MS) and other mass spectral methods. However, the alditol-containing *O*-glycans are not conducive for labeling with UV/visible/fluorescent tags for analysis using HPLC, CE, etc.

Both the hydrozinolysis and β-elimination reactions to release *O*-glycans have limited applicability in quality control (QC) environment. But there are limited enzymatic methods available to release *O*-glycans. For example, α-*N*-acetylgalactosaminidase releases O-linked GalNAc residues, but the enzyme does not work if the GalNAc is elongated (or capped). For example, the enzyme is not active if the GalNAc is sialylated and/or galactosylated. The enzyme called endo-α-*N*-acetylgalactosaminidase which is also known as *O*-glycosidase releases Gal-β-1,3-GalNAc (T-antigen) disaccharide. Similar to α-*N*-acetylgalactosaminidase, the *O*-glycosidase do not act on the sialylated or GlcNAc containing T-antigen glycans. Hence, the use of α-*N*-acetylgalactosaminidase as well as *O*-glycosidase is very limited for *O*-glycan analysis. However, an enzyme cocktail of these two enzymes with or without sialidase is useful to release *O*-glycans from CHO cells-derived glycoproteins including RTPs. An enzyme cocktail of α-*N*-acetylgalactosaminidase and *O*-glycosidase will release all the nonsialylated GalNAc and Gal-GalNAc *O*-glycans from CHO cells-derived glycoproteins including RTPs. Addition of sialidase will release all the sialylated as well as the nonsialylated GalNAc and Gal–GalNAc *O*-glycans. Hence, using these two enzyme cocktails, it is possible to release both sialylated and nonsialylated *O*-glycans from CHO cells-derived RTPs. Independent analysis of the glycans released from these two enzyme cocktails will give quantitative information on the overall composition of *O*-glycans. Such enzyme cocktail method can also be extended to include β-galactosidase as well as β-*N*-acetylhexosaminidases to release *O*-glycans from glycoproteins containing more complex O-linked oligosaccharides.

The enzymatically released glycans containing reducing sugar residues can be labeled with fluorescent tags using reductive amination reactions as described in Chapter 8. Many investigators routinely used AA tag for labeling of released glycans containing reducing sugars and normal phase-high performance liquid chromatography (NP-HPLC) method to separate the labeled glycans. The NP-HPLC method is a very rugged method with high precision and reproducibility. Similarly, hydrophilic interaction liquid chromatography (HILIC), RP-HPLC, etc., methods are also found to be very useful to analyze *O*-glycans as well as *N*-glycans.

## References

Baker, R., Nakamura, N., Chandel, I. et al. (2018). Protein O-mannosyltransferases affect sensory axon wiring and dynamic chirality of body posture in the Drosophila embryo. *J. Neurosci.* 38 (7): 1850–1865. https://doi.org/10.1523/JNEUROSCI.0346-17.2017.

Basu, D., Wang, W., Ma, S. et al. (2015). Two hydroxyproline galactosyltransferases, GALT5 and GALT2, function in arabinogalactan-protein glycosylation, growth and development in arabidopsis. *PLoS ONE* 10 (5): https://doi.org/10.1371/journal.pone.0125624.

Bhavanandan, V.P. and Davidson, E.A. (1976). Characteristics of a mucin-type sialoglycopeptide produced by B16 mouse melanoma cells. *Biochem. Biophys. Res. Commun.* 70 (1): 139–145.

Buko, A.M., Kentzer, E.J., Petros, A. et al. (1991). Characterization of a posttranslational fucosylation in the growth factor domain of urinary plasminogen activator. *Proc. Natl. Acad. Sci. U. S. A.* 88 (9): 3992–3996.

Carlson, D.M. (1966). Oligosaccharides isolated from pig submaxillary mucin. *J. Biol. Chem.* 241 (12): 2984–2986.

Corfield, A. (2017). Eukaryotic protein glycosylation: a primer for histochemists and cell biologists. *Histochem. Cell Biol.* 147 (2): 119–147. https://doi.org/10.1007/s00418-016-1526-4.

Davidson, E.A. and Gowda, D.C. (2001). Glycobiology of *Plasmodium falciparum*. *Biochimie* 83 (7): 601–604.

de Paula, R.M., Wilson, W.A., Roach, P.J. et al. (2005). Biochemical characterization of Neurospora crassa glycogenin (GNN), the self-glucosylating initiator of glycogen synthesis. *FEBS Lett.* 579 (10): 2208–2214.

Diesen, C., Saarinen, A., Pihko, H. et al. (2004). POMGnT1 mutation and phenotypic spectrum in muscle-eye-brain disease. *J. Med. Genet.* 41 (10): e115.

Feldhoff, P.A., Bhavanandan, V.P., and Davidson, E.A. (1979). Purification, properties, and analysis of human asthmatic bronchial mucin. *Biochemistry* 18 (11): 2430–2436.

Fernandez-Valdivia, R., Takeuchi, H., Samarghandi, A. et al. (2011). Regulation of mammalian Notch signaling and embryonic development by the protein O-glucosyltransferase Rumi. *Development* 138 (10): 1925–1934. https://doi.org/10.1242/dev.060020.

Grasa, P., Kaune, H., and Williams, S.A. (2012). Embryos generated from oocytes lacking complex N- and O-glycans have compromised development and implantation. *Reproduction* 144 (4): 455–465. https://doi.org/10.1530/REP-12-0084.

Haltiwanger, R.S., Wells, L., Freeze, H.H., and Stanley, P. (2017). Other classes of eukaryotic glycans. In: *Essentials of Glycobiology* [Internet], Chapter 13, 3e (ed. A. Varki, R.D. Cummings, J.D. Esko, et al.). Cold Spring Harbor, NY: Cold Spring Harbor Laboratory Press; 2015–2017. http://www.ncbi.nlm.nih.gov/books/NBK453017/.

Harris, R.J., Leonard, C.K., Guzzetta, A.W., and Spellman, M.W. (1991). Tissue plasminogen activator has an O-linked fucose attached to threonine-61 in the epidermal growth factor domain. *Biochemistry* 30 (9): 2311–2314.

Harris, R.J., Ling, V.T., and Spellman, M.W. (1992). O-linked fucose is present in the first epidermal growth factor domain of factor XII but not protein C. *J. Biol. Chem.* 267 (8): 5102–5107.

Harris, R.J. and Spellman, M.W. (1993). O-linked fucose and other post-translational modifications unique to EGF modules. *Glycobiology* 3 (3): 219–224.

Harris, R.J., van Halbeek, H., Glushka, J. et al. (1993). Identification and structural analysis of the tetrasaccharide NeuAc $\alpha(2\rightarrow6)$Gal $\beta(1\rightarrow4)$GlcNAc $\beta(1\rightarrow3)$Fuc $\alpha$ $1\rightarrow$O-linked to serine 61 of human factor IX. *Biochemistry* 32 (26): 6539–6547.

Hart, G.W. (2016). Myriad roles of glycans in biology. *J. Mol. Biol.* 428 (16): 3147–3149. https://doi.org/10.1016/j.jmb.2016.06.010.

Harvey, B.M. and Haltiwanger, R.S. (2018). Regulation of notch function by O-glycosylation. *Adv. Exp. Med. Biol.* 1066: 59–78. https://doi.org/10.1007/978-3-319-89512-3_4.

Herscovics, A. and Orlean, P. (1993). Glycoprotein biosynthesis in yeast. *FASEB J.* 7 (6): 540–550.

Hobbs, S.J. and Nolz, J.C. (2017). Regulation of T cell trafficking by enzymatic synthesis of O-glycans. *Front. Immunol.* 8: 600. https://doi.org/10.3389/fimmu.2017.00600.

Houel, S., Hilliard, M., Yu, Y.Q. et al. (2014). N- and O-glycosylation analysis of etanercept using liquid chromatography and quadrupole time-of-flight mass spectrometry equipped with electron-transfer dissociation functionality. *Anal. Chem.* 86 (1): 576–584. https://doi.org/10.1021/ac402726h.

Impellizzeri, D. and Cuzzocrea, S. (2014). Targeting selectins for the treatment of inflammatory diseases. *Expert Opin. Ther. Targets* 18 (1): 55–67. https://doi.org/10.1517/14728222.2013.841140.

Janssen, W.J., Stefanski, A.L., Bochner, B.S., and Evans, C.M. (2016). Control of lung defence by mucins and macrophages: ancient defence mechanisms with modern functions. *Eur. Respir. J.* 48 (4): 1201–1214. https://doi.org/10.1183/13993003.00120-2015.

Kao, Y.H., Lee, G.F., Wang, Y. et al. (1999). The effect of O-fucosylation on the first EGF-like domain from human blood coagulation factor VII. *Biochemistry* 38 (22): 7097–7110.

Kawashima, H. and Fukuda, M. (2012). Sulfated glycans control lymphocyte homing. *Ann. N. Y. Acad. Sci.* 1253: 112–121. https://doi.org/10.1111/j.1749-6632.2011.06356.x.

Kentzer, E.J., Buko, A., Menon, G., and Sarin, V.K. (1990). Carbohydrate composition and presence of a fucose-protein linkage in recombinant human pro-urokinase. *Biochem. Biophys. Res. Commun.* 171 (1): 401–406.

Kumar, S., Cruz, E., Joshi, S. et al. (2017). Genetic variants of mucins: unexplored conundrum. *Carcinogenesis* 38 (7): 671–679. https://doi.org/10.1093/carcin/bgw120.

Lopaticki, S., Yang, A.S.P., John, A. et al. (2017). Protein O-fucosylation in *Plasmodium falciparum* ensures efficient infection of mosquito and vertebrate hosts. *Nat. Commun.* 8 (1): 561. https://doi.org/10.1038/s41467-017-00571-y.

Mendonça-Previato, L., Penha, L., Garcez, T.C. et al. (2013). Addition of α-O-GlcNAc to threonine residues define the post-translational modification of mucin-like molecules in *Trypanosoma cruzi*. *Glycoconj. J.* 30 (7): 659–666. https://doi.org/10.1007/s10719-013-9469-7.

Mulagapati, S., Koppolu, V., and Raju, T.S. (2017). Decoding of O-linked glycosylation by mass spectrometry. *Biochemistry* 56 (9): 1218–1226. https://doi.org/10.1021/acs.biochem.6b01244.

Ogawa-Ohnishi, M. and Matsubayashi, Y. (2015). Identification of three potent hydroxyproline O-galactosyltransferases in Arabidopsis. *Plant J.* 81 (5): 736–746. https://doi.org/10.1111/tpj.12764.

Pang, P.C., Chiu, P.C., Lee, C.L. et al. (2011). Human sperm binding is mediated by the sialyl-Lewis(*x*) oligosaccharide on the zona pellucida. *Science* 333 (6050): 1761–1764. https://doi.org/10.1126/science.1207438.

Plomp, R., Dekkers, G., Rombouts, Y. et al. (2015). Hinge-region O-glycosylation of human immunoglobulin G3 ($IgG_3$). *Mol. Cell. Proteomics* 14 (5): 1373–1384. https://doi.org/10.1074/mcp.M114.047381.

Raju, T.S. and Davidson, E.A. (1994). Role of sialic acid on the viscosity of canine tracheal mucin glycoprotein. *Biochem. Biophys. Res. Commun.* 205 (1): 402–409.

Raju, T.S., Lerner, L., and O'Connor, J.V. (1996). Glycopinion: biological significance and methods for the analysis of complex carbohydrates of recombinant glycoproteins. *Biotechnol. Appl. Biochem.* 24 (Pt 3): 191–194.

Ringler, N.J., Selvakumar, R., Woodward, H.D. et al. (1987). Structure of canine tracheobronchial mucin glycoprotein. *Biochemistry* 26 (17): 5322–5328.

Rini, J.M. and Esko, J.D. (2017). Glycosyltransferases and glycan-processing enzymes. In: *Essentials of Glycobiology* [Internet], Chapter 6, 3e (ed. A. Varki, R.D. Cummings, J.D. Esko, et al.). Cold Spring Harbor (NY): Cold Spring Harbor Laboratory Press; 2015–2017. http://www.ncbi.nlm.nih.gov/books/NBK453021/.

Ruba, A. and Yang, W. (2016). O-GlcNAc-ylation in the nuclear pore complex. *Cell. Mol. Bioeng.* 9 (2): 227–233. https://doi.org/10.1007/s12195-016-0440-0.

Santos, S.N., Junqueira, M.S., Francisco, G. et al. (2016). O-glycan sialylation alters galectin-3 subcellular localization and decreases chemotherapy sensitivity in gastric cancer. *Oncotarget* 7 (50): 83570–83587. https://doi.org/10.18632/oncotarget.13192.

Sheikh, M.O., Halmo, S.M., and Wells, L. (2017). Recent advancements in understanding mammalian O-mannosylation. *Glycobiology* 27 (9): 806–819. https://doi.org/10.1093/glycob/cwx062.

Spahr, C., Kim, J.J., Deng, S. et al. (2013). Recombinant human lecithin-cholesterol acyltransferase Fc fusion: analysis of N- and O-linked glycans and identification and elimination of a xylose-based O-linked tetrasaccharide core in the linker region. *Protein Sci.* 22 (12): 1739–1753. https://doi.org/10.1002/pro.2373.

Spencer, D., Novarra, S., Zhu, L. et al. (2013). O-xylosylation in a recombinant protein is directed at a common motif oncglycine-serine linkers. *J. Pharm. Sci.* 102 (11): 3920–3924. https://doi.org/10.1002/jps.23733.

Staudacher, E. (2015). Mucin-type O-glycosylation in invertebrates. *Molecules* 20 (6): 10622–10640. https://doi.org/10.3390/molecules200610622.

Takeuchi, H. and Haltiwanger, R.S. (2014). Significance of glycosylation in Notch signaling. *Biochem. Biophys. Res. Commun.* 453 (2): 235–242. https://doi.org/10.1016/j.bbrc.2014.05.115.

Taylor, C.M., Karunaratne, C.V., and Xie, N. (2012). Glycosides of hydroxyproline: some recent, unusual discoveries. *Glycobiology* 22 (6): 757–767. https://doi.org/10.1093/glycob/cwr188.

Terajima, M., Perdivara, I., Sricholpech, M. et al. (2014). Glycosylation and cross-linking in bone type I collagen. *J. Biol. Chem.* 289 (33): 22636–22647. https://doi.org/10.1074/jbc.M113.528513. Epub.

Thorsheim, K., Siegbahn, A., Johnsson, R.E. et al. (2015). Chemistry of xylopyranosides. *Carbohydr. Res.* 418: 65–88. https://doi.org/10.1016/j.carres.2015.10.004.

Tran, D.T. and Ten Hagen, K.G. (2013). Mucin-type O-glycosylation during development. *J. Biol. Chem.* 288 (10): 6921–6929. https://doi.org/10.1074/jbc.R112.418558.

Varki, A. (2017). Biological roles of glycans. *Glycobiology* 27 (1): 3–49. https://doi.org/10.1093/glycob/cww086.

Verma, M., Sanadi, A.R., and Davidson, E.A. (1994). Mucin synthesis in immortalized canine tracheal epithelial cells. *Glycobiology* 4 (6): 825–836.

Verma, M. and Davidson, E.A. (1994). Mucin genes: structure, expression and regulation. *Glycoconj. J.* 11 (3): 172–179.

Wang, Y., Lee, G.F., Kelley, R.F., and Spellman, M.W. (1996). Identification of a GDP-L-fucose: polypeptide fucosyltransferase and enzymatic addition of O-linked fucose to EGF domains. *Glycobiology* 6 (8): 837–842.

Wang, Y., Shao, L., Shi, S. et al. (2001). Modification of epidermal growth factor-like repeats with O-fucose. Molecular cloning and expression of a novel GDP-fucose protein O-fucosyltransferase. *J. Biol. Chem.* 276 (43): 40338–40345.

Wang, Y. and Spellman, M.W. (1998). Purification and characterization of a GDP-fucose: polypeptide fucosyltransferase from Chinese hamster ovary cells. *J. Biol. Chem.* 273 (14): 8112–8118.

Wang, Z., Pandey, A., and Hart, G.W. (2007). Dynamic interplay between O-linked N-acetylglucosaminylation and glycogen synthase kinase-3-dependent phosphorylation. *Mol. Cell. Proteomics* 6 (8): 1365–1379.

Woodward, H.D., Ringler, N.J., Selvakumar, R. et al. (1987). Deglycosylation studies on tracheal mucin glycoproteins. *Biochemistry* 26 (17): 5315–5322.

Zachara, N., Akimoto, Y., and Hart, G.W. (2017). The O-GlcNAc modification. In: *Essentials of Glycobiology* [Internet], Chapter 19, 3e (ed. A. Varki, R.D. Cummings, J.D. Esko, et al.). Cold Spring Harbor, NY: Cold Spring Harbor Laboratory Press; 2015–2017. http://www.ncbi.nlm.nih.gov/books/NBK453063/.

Zauner, G., Kozak, R.P., Gardner, R.A. et al. (2012). Protein O-glycosylation analysis. *Biol. Chem.* 393 (8): 687–708. https://doi.org/10.1515/hsz-2012-0144.

# 10

# Hydroxylation of Proteins

## Introduction

Hydroxylation is a chemical reaction in which a hydroxyl group (–OH) is introduced in to a chemical molecule in place of a covalently linked hydrogen atom to carbon atom of many molecules (Huang and Groves 2017). In organic chemistry, hydroxylation converts a —C—H bond into a —C—OH bond, and the chemical reaction is an oxidative in nature (Yang and Yang 2014). Accordingly, hydroxylation converts an aliphatic hydrocarbon into aliphatic alcohol (Wolf 1998). For example, hydroxylation converts methane into methanol (methyl alcohol), ethane into ethanol (ethyl alcohol), butane into butanol (butyl alcohol), etc. Similarly, upon hydroxylation, organic aromatic hydrocarbons convert into aromatic alcohols. For example, hydroxylation converts benzene into benzenol (phenol), toluene into benzyl alcohol, etc.

In organic chemistry, hydroxylation reactions are mediated by a combination of catalysts and heat (Yoshizawa 2000). Conversion of benzene into phenol is a classic example of hydroxylation reaction. Most often metal ions are used as catalysts for hydroxylation (Periana et al. 1998). In addition, many organometallic compounds as well as organic molecules are also used as catalysts for hydroxylation (Sims and Grover 1974).

Hydroxylation converts hydrophobic molecules into hydrophilic molecules. Hence, hydroxylated molecules become more water-soluble. Hydroxylation is very important in detoxification as it converts the lipophilic compounds into hydrophilic compounds (Simpson et al. 1969). In addition, hydroxylation can activate or deactivate certain drugs like steroids (Marsheck 1971). The hydrophilic compounds can be readily removed from the body by kidneys and liver, and hence they eventually excreted (Cooksley and Powell 1971).

In biochemistry, hydroxylation reactions are mediated by enzymes that are known as hydroxylases (Kaufman and Friedman 1965). In biological systems, cytochrome P-450 is the principal hydroxylation agent that has hundreds of variations (Estabrook et al. 1972; Gunsalus et al. 1972). Flavins, α-ketoglutarate-dependent hydroxylases, diiron hydroxylases, etc., are also

*Co- and Post-Translational Modifications of Therapeutic Antibodies and Proteins,* First Edition. T. Shantha Raju.

involved in hydroxylation of biomolecules including proteins (Boyd 1972; Ullrich 1972). In endogenous proteins, Pro is the most hydroxylated amino acid residue (Kuttan and Radhakrishnan 1973). In addition to Pro and Lys, other amino acid residues such as Phe, Tyr, Arg, and Trp also often get hydroxylated. Hydroxylation of Pro can occur at both C3 and C4 carbons, whereas Lys hydroxylation can occur at C5 carbon. The C3 hydroxylation of Pro is mediated by prolyl 3-hydroxylase and C4 hydroxylation of Pro is mediated by prolyl 4-hydroxylase (Kuttan and Radhakrishnan 1973; Ananthanarayanan 1983). Hydroxylation of Lys at C5 is mediated by lysyl 5-hydroxylase (Zurlo et al. 2016; Gjaltema and Bank 2017). Similarly, hydroxylation of Phe is mediated by phenylalanine hydroxylase and Tyr hydroxylation is mediated by tyrosine hydroxylase (Hole et al. 2016; Lai et al. 2018). The prolyl 3-hydroxlase, prolyl 4-hydroxylase, and lysyl 5-hydroxylase enzymes are very large multisubunit proteins, and they require iron, molecular oxygen along with α-ketoglutarate for the catalysis of hydroxylation reaction (Hayaishi 1969; Sih 1969; Estabrook et al. 1972). This chapter provides an overview of protein hydroxylation at Pro, Lys, Phe, and Try residues. This chapter also covers the impact of hydroxylation on the structure and functions of proteins. However, hydroxylation of Arg, Trp, etc., amino acid residues is not covered in this chapter.

## Mechanism of Hydroxylation

### Mechanism of Hydroxylation in Organic Molecules

Hydroxylation of chemical entities (molecules) is an oxidative process and hence requires atmospheric oxygen molecules (Yang et al. 2012). Since the oxidation involving atmospheric oxygen molecules is a very slow process, catalysts and heat are used to increase the speed of the chemical process of oxidation, i.e. hydroxylation (Liang and Jiao 2017). In organic molecules, the C—H bond is the most stable bond and hence, to replace or to convert the C—H bond requires time, energy, and catalysts (Yang et al. 2012; Liang and Jiao 2017). Accordingly, conversion of C—H bond into C—OH bond is mediated by catalysts and heat in organic chemistry, whereas selective and efficient enzymes are needed in biological systems (Hayaishi 1969; Yang et al. 2012; Liang and Jiao 2017).

Conversion of methane into methanol is often a very difficult reaction that may involve multistep organic synthesis reactions. For example, usually methane is converted into methyl halides and then converted into methanol (Spivey and Egbebi 2007). However, recently single-step conversion of methane into methanol has been developed using catalysts (Khirsariya and Mewada 2013). But, the reaction is often an incomplete conversion of methane into methanol. Similarly, conversion of benzene into phenol is also a complex

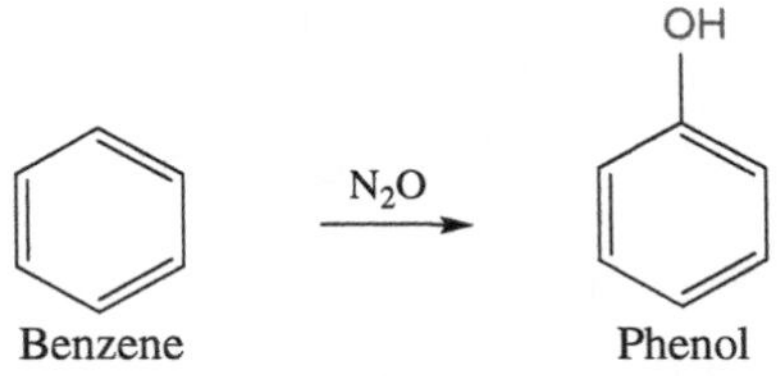

Figure 10.1 Conversion of benzene into phenol using $N_2O$ as catalyst. (*See insert for color representation of this figure.*)



reaction. Most common reaction to obtain phenol is to convert toluene into phenol using multiple reactions. However, recently benzene is converted into phenol in a single step using $N_2O$ as catalyst (see Figure 10.1) (Shiju et al. 2006; Xin et al. 2009; Zhao et al. 2009).

## Mechanism of Hydroxylation in Biomolecules

In the biomolecules, hydroxylation reactions are usually catalyzed by the enzymes that are collectively called as hydroxylases (Bishop and Ratcliffe 2015). For the biomolecules, cytochrome P-450 is the main hydroxylation agent. In living organisms, hundreds of variations of cytochrome P-450 molecules exist (Lee 1998). In addition to cytochrome P-450 and its variants, other hydroxylating agents also exist in biological systems (Sridhar et al. 2012; Liu et al. 2013). The list of other hydroxylating agents includes flavins, α-ketoglutarate-dependent hydroxylases, diiron hydroxylases, etc. Among biomolecules collagen is the most hydroxylated protein (Ramachandran 1963). Collagen is the main component of connective tissues of animal bodies and is the most abundant protein present in mammals (Nyström and Bruckner-Tuderman 2018). The amino acid residues of collagen wound together to form a triple-helix structure that is responsible for the formation of elongated collagen fibrils. Hydroxylation is the principle post-translational modification (PTM) of collagen and hydroxylation of amino acid residues of collagen is responsible for the formation of triple-helix structure (Fidler et al. 2018).

Collagen is very rich in Pro and Lys residues, and hydroxyproline is the principle component of collagen. Both 3-hydroxyproline and 4-hydroxyproline residues are present in collagen in addition to 5-hydroxylysine residues (Gjaltema and Bank 2017; Li and Wu 2018). Chemical structure of 3-hydroxyproline, 4-hydroxyproline, and 5-hydroxylysine is presented in Figure 10.2. Hydroxylation of Pro at C-3 position is mediated by prolyl 3-hydroxylase, whereas hydroxylation of Pro at C-4 position is mediated by prolyl 4-hydroxylase. Similarly, hydroxylation of Lys at C-5 position is mediated by lysyl 5-hydroxylase.

In addition to hydroxyproline and hydroxylysine residues, hydroxyphenylalanine and hydroxytyrosine residues also exist in nature. Hydroxylation of Phe is mediated by phenylalanine hydroxylase that converts Phe residues into Tyr residues (Flydal and Martinez 2013). Conversely, tyrosine hydroxylase

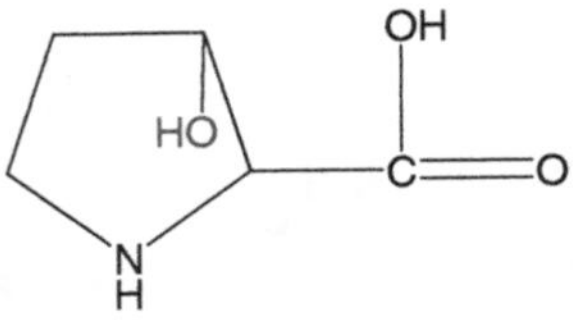

3-Hydroxyproline

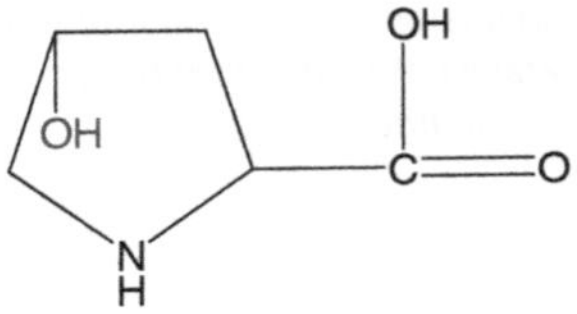

4-Hydroxyproline

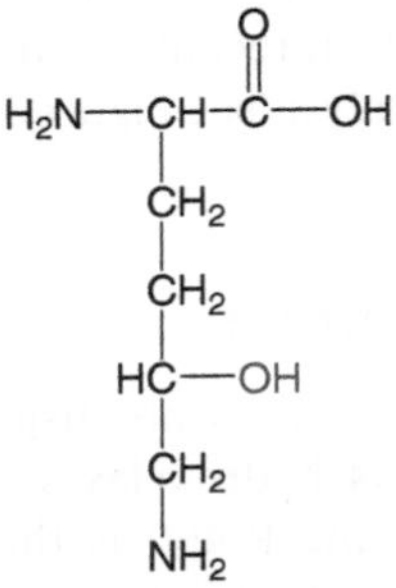

5-Hydroxylysine

Hydroxyl groups are in shown in gray color

**Figure 10.2** Chemical structure of 3-hydroxyproline, 4-hydroxyproline, and 5-hydroxylysine. *(See insert for color representation of this figure.)*

hydroxylates Tyr residues into 3,4-dihydroxy phenylalanine (L-DOPA) (Tekin et al. 2014). Dopamine is derived from L-DOPA and is a precursor for norepinephrine (noradrenaline) and epinephrine (adrenaline). Both adrenaline and noradrenaline are neurotransmitters and catecholamine hormones (Paravati and Warrington 2018).

Hydroxylation of amino acids mostly occurs in the lumen of the endoplasmic reticulum (ER). Most hydroxylases are similar enzymes and almost every one of them requires cosubstrates like iron (Fe), 2-oxoglutarate and molecular oxygen ($O_2$). During hydroxylation, 2-oxyglutarate undergoes decarboxylation, and it is a stoichiometric reaction. One atom of the oxygen molecule ($O_2$) is involved in succinate formation, and the other atom is involved in the formation of hydroxyl (–OH) group for incorporation into amino acids during hydroxylation. The α,α′-dipyridyl is an iron chelator that acts as a very potent inhibitor of hydroxylase mediated hydroxylation reactions (see Figure 10.3 for structure of α,α′-dipyridyl molecule) (Johnston et al. 1954). In the presence of inhibitors of hydroxylation reaction, nonhydroxylated collagen is synthesized

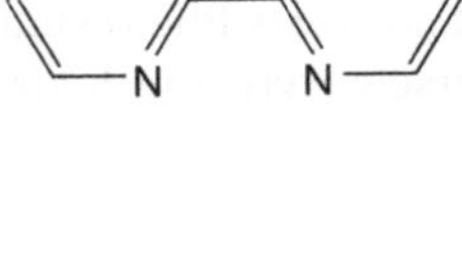

**Figure 10.3** Chemical structure of α,α′-dipyridyl molecule.

(Berg and Prockop 1973). The nonhydroxylated collagen may be secreted as a nonhelical chain protein (Xu et al. 2011).

**Prolyl 4-Hydroxylase**

The enzyme prolyl 4-hydroxylase is also known as protocollagen hydroxylase and is a tetrameric molecule with an apparent molecular weight of about 248 kDa (Vasta and Raines 2018). The enzyme consists of two α-subuints and two β-subunits (Kivirikko et al. 1989). The molecular weight of α-subunit is ~64 kDa and the molecular weight of β-subunit is ~60 kDa. The amino acid sequence of α-subunit protein is highly conserved in mammalians and avians (Smirnova et al. 2012). However, the β-subunit is an unusual multifunctional protein that acts as protein disulfide isomerase in isolation (Koivu et al. 1987). The main catalytic domain of proly 4-hyroxylase is present in α-subunit, although some parts of β-subunits also contributes to the catalytic reaction (Vuori et al. 1992).

**Prolyl 3-Hydroxylase**

Prolyl 3-hydroxylase is also known as procollagen-proline 2-oxoglutarate 3-dioxygenase (Hudson and Eyre 2013). The enzyme requires same cofactors of prolyl 4-hydroxylase but is highly sequence-specific enzymes. The enzyme hydroxylates Pro residues of proteins (mainly collagen) only in the peptide sequence of –Pro–Hyp–Gly– residues (Vranka et al. 2004). Accordingly, prolyl 4-hydroxylase reactions are required for prolyl 3-hydroxylase to act on. The abundance of 3-hydroxyproline varies with the type of collagen. The type IV collagen contains the greatest number of 3-hydroxyproline residues than any other type of collagen molecules (Wu and Crane 2018).

**Lysyl 5-Hydroxylase**

The mechanism of hydroxylation of Lys residues is similar to the hydroxylation of Pro residues in which the enzyme lysyl 5-hydroxylase is involved in catalyzing the enzymatic reaction. The enzyme lysyl 5-hydroxylase is also known as procollagen-lysine 2-oxoglutarate 5-dioxygenase and is a dimer with an apparent molecular weight of ~190 kDa (Myllylä et al. 2007; Chen et al. 2015). The two monomeric polypeptides of lysyl 5-hydroxylase are glycoproteins with different carbohydrate content. The enzyme shares the same cofactors of prolyl hydroxylases and catalyzes the hydroxylation of Lys residues in -X-Lys-Gly- sequence. The hydroxyl group of Lys residues is involved in O-glycosylation and are involved in the formation of intra- and intermolecular collagen crosslinks (Chen et al. 2015). The hydroxylysine content and its nature of O-glycosylation vary with the type of collagen.

**Phenylalanine Hydroxylase**

The enzyme, phenylalanine hydroxylase, is a catalytic enzyme that is involved in hydroxylating the aromatic ring of Phe residues to convert it into Tyr

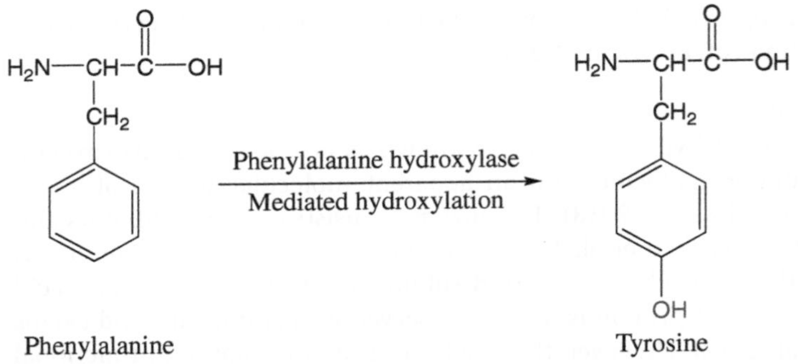

Phenylalanine hydroxylase
Mediated hydroxylation

Phenylalanine

Tyrosine

**Figure 10.4** Conversion of phenylalanine into tyrosine by phenylalanine hydroxylase mediated hydroxylation. The enzymatic reaction requires tetrahydrobiopterin, a non-heme iron and molecular oxygen ($O_2$) as cofactors for the catalytic reaction. (*See insert for color representation of this figure.*)

residues (see Figure 10.4). The enzyme belongs to a family of aromatic amino acid hydroxylases that are biopterin-dependent enzymes (Fitzpatrick 2015). These enzymes need tetrahydrobiopterin, a nonheme iron and molecular oxygen ($O_2$) as cofactors for the catalytic reactions (Pavon and Fitzpatrick 2006). The phenylalanine hydroxylase enzyme is a ~52 kDa protein consisting of three distinct domains, a regulatory N-terminal domain, a catalytic domain, and a C-terminal domain. The enzyme plays an important role in the metabolic pathway and is involved in the degradation of excess amounts of Phe residues.

**Tyrosine Hydroxylase**

The tyrosine hydroxylase is also called as tyrosine 3-monooxygenase that catalyzes the hydroxylation of Tyr residues at C-3 position (Daubner et al. 2011). The enzyme is a homotetramer consisting of four identical subunits. Each subunit of the enzyme contains three domains, a carboxy terminal domain, a central domain, and an N-terminal domain (Zhang et al. 2014). The carboxy terminal domain contains an α helix that is responsible for tetramerization of the enzyme domains. The central domain is the catalytic core consisting of ~300 amino acid residues. The N-terminal domain is the regulatory domain which consists ~150 amino acid residues. The hydroxylation at C-3 position converts Try residues into L-DOPA (see Figure 10.5). The L-DOPA converts into dopamine that is a precursor to catecholamine hormones such as adrenaline and noradrenaline (Weiss and Rossi 1963). Both adrenaline and noradrenaline are neurotransmitters.

## Biological Significance of Hydroxylation

Hydroxylation plays a very important role in the structure and functions of organic molecules as well as biomolecules. Methanol, ethanol, butanol,

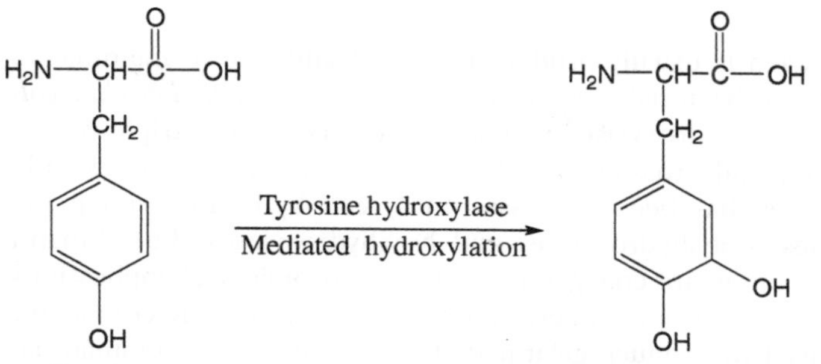

Tyrosine hydroxylase
Mediated hydroxylation

Tyrosine

L-3,4-Dihydroxyphenylalanine

**Figure 10.5** Hydroxylation of tyrosine into L-3,4-dihydroxyphenylalanine (L-DOPA).

etc. are liquids, whereas methane, ethane, butane, etc., are gases at ambient temperature. Although all these molecules are combustible, storage of liquids is easier than the storage of gaseous molecules (Xiao et al. 2014). Ethane is a gas and is highly combustible, whereas ethanol is a liquid and is also a consumable liquid for humans (Xiao et al. 2014). Hydroxylation converts hydrophobic molecules into hydrophilic molecules and hence hydroxylated molecules are more water-soluble. Oxidative degradation of organic molecules involves their hydroxylation as a first step (Cochrane 1991). Hydroxylation is very important for detoxification of organic molecules (Zhang et al. 2018). Hydroxylated molecules are more readily removed by kidneys and liver, and hence hydroxylated molecules are easily excreted compared to their nonhydroxylated counterparts. Certain organic molecules such as steroids are activated or deactivated by hydroxylation (Svoboda and Weirich 1995).

Hydroxylation converts Phe residues into Try residues and hence plays an important role in controlling the excessive amounts of Phe residues in living organisms. Hydroxylation also plays an important role in converting Try into L-DOPA, which is very important for the biosynthesis of dopamine. Neurotransmitters such as adrenaline and noradrenaline are catecholamine hormones that are derived from dopamine. Hence, hydroxylation is important to produce neurotransmitters such as adrenaline and noradrenaline hormones (Tank and Lee Wong 2015).

Hydroxyproline and hydroxylysine residues are very important for the triple helical structure of collagen (Song and Mechref 2013). Hydroxyproline is a major component of collagen and it constitutes about 13.5% of collagen derived from mammalian sources. Hydroxyproline residues are involved in the hydrogen bond formation, and hydroxylysine residues are involved in cross-linking of collagen molecules. Hydroxylysine residues are also responsible for the presence of O-glycan residues in collagen (see Chapter 7) (Bann and Bächinger 2000). The glycan residues linked to hydroxylysine are involved in stabilizing the triple helical structure of collagen.

### Hydroxylation in RTPs

Recombinant collagen is usually produced in yeast and also in mammalian cells. This is because bacterial expression systems such as *Escherichia coli* lack the enzymes like hydroxylases that are necessary for the triple helical structure formation and crosslinking of collagen fibrils. However, recently recombinant collagen has been expressed in *E. coli* by cotransfecting the genes that encodes prolylhydroxylases and lysylhydroxylases (Rutschmann et al. 2014). The recombinant collagen is used in many biological applications including tissue engineering, surgeries, drug delivery, etc. Recombinant collagen is also used to produce gelatin that has many uses as biomaterial in biomedical field. Hydroxylation plays a major role in the structure and functions of recombinant collagen. Production of recombinant collagen in yeast and mammalian expression systems is also necessary to obtain properly glycosylated collagen. Otherwise, it is necessary to co-express glycosylating enzymes along with the hydroxylating enzymes in bacterial expression systems to produce properly glycosylated collagen (Rutschmann et al. 2014).

### Analysis of Hydroxylation

Hydroxylation adds an oxygen atom that is about 16 Da in mass. Accordingly, hydroxylation is basically an oxygenation reaction involving catalysts or enzymes. The increase in 16 Da in molecular weight of hydroxylated molecules is easily detectable by high-resolution mass spectrometry. Hydroxylation can also be easily analyzed by NMR (both $^{1}$H and $^{13}$C), high performance liquid chromatography (HPLC), ultra-performance liquid chromatography (UPLC), and capillary electrophoresis (CE) methods (Sroga and Vashishth 2011). The $^{1}$H and $^{13}$C NMR are very useful to detect hydroxylation in organic molecules as well as in biomolecules. Hydroxylation impacts the resonance of protons attached to the carbon atom directly linked to the hydroxyl group as well as the protons attached to the neighboring carbon atom in the hydroxylated molecule. Hence, $^{1}$H NMR will be able to detect and quantitate hydroxylation in molecules. Similarly, hydroxylation impacts the resonance of the carbon atom that is directly linked to hydroxyl group. Also, hydroxylation impacts the resonance of neighboring carbon atoms, and hence, it is possible to identify hydroxylation by $^{13}$C NMR also. Since, hydroxyl groups can form hydrogen bonds with water and other functional groups such as amines, amides, carboxyl groups, NMR and chromatographic properties of such groups are also impacted.

Hydroxylation increases the molecular polarity and hence the water solubility of the hydroxylated molecules. Accordingly, hydroxylated molecules can be separated by chromatographic methods such as ion-exchange chromatography (IEC), hydrophobic interaction chromatography (HIC), and reversed phase chromatography (RPC) methods (Qiu et al. 2014). Hydroxylated molecules can also be separated by electrophoretic methods such as CE, capillary iso-electric focusing (cIEF), and gel-electrophoresis methods. Hydroxylated small organic molecules can also be separated and quantitated by thin-layer chromatography (TLC), paper chromatography (PC), etc. methods.

## References

Ananthanarayanan, V.S. (1983). Structural aspects of hydroxyproline-containing proteins. *J. Biomol. Struct. Dyn.* 1 (3): 843–855.

Bann, J.G. and Bächinger, H.P. (2000). Glycosylation/hydroxylation-induced stabilization of the collagen triple helix. 4-trans-hydroxyproline in the Xaa position can stabilize the triple helix. *J. Biol. Chem.* 275 (32): 24466–24469.

Berg, R.A. and Prockop, D.J. (1973). The thermal transition of a non-hydroxylated form of collagen. Evidence for a role for hydroxyproline in stabilizing the triple-helix of collagen. *Biochem. Biophys. Res. Commun.* 52 (1, 1): 115–120.

Bishop, T. and Ratcliffe, P.J. (2015). HIF hydroxylase pathways in cardiovascular physiology and medicine. *Circ. Res.* 117 (1): 65–79. https://doi.org/10.1161/CIRCRESAHA.117.305109.

Boyd, G.S. (1972). Biological hydroxylation reactions. *Biochem. Soc. Symp.* 34: 1–9.

Chen, Y., Terajima, M., Yang, Y. et al. (2015). Lysyl hydroxylase 2 induces a collagen cross-link switch in tumor stroma. *J. Clin. Invest.* 125 (3): 1147–1162. https://doi.org/10.1172/JCI74725.

Cochrane, C.G. (1991). Cellular injury by oxidants. *Am. J. Med.* 91 (3C): 23S–30S.

Cooksley, W.G. and Powell, L.W. (1971). Drug metabolism and interaction with particular reference to the liver. *Drugs* 2 (3): 177–189.

Daubner, S.C., Le, T., and Wang, S. (2011). Tyrosine hydroxylase and regulation of dopamine synthesis. *Arch. Biochem. Biophys.* 508 (1): 1–12. https://doi.org/10.1016/j.abb.2010.12.017.

Estabrook, R.W., Baron, J., Peterson, J., and Ishimura, Y. (1972). Oxygenated cytochrome P-450 as an intermediate in hydroxylation reactions. *Biochem. Soc. Symp.* 34: 159–185.

Fidler, A.L., Boudko, S.P., Rokas, A., and Hudson, B.G. (2018). The triple helix of collagens – an ancient protein structure that enabled animal multicellularity and tissue evolution. *J. Cell Sci.* 131 (7): https://doi.org/10.1242/jcs.203950.

Fitzpatrick, P.F. (2015). Structural insights into the regulation of aromatic amino acid hydroxylation. *Curr. Opin. Struct. Biol.* 35: 1–6. https://doi.org/10.1016/j.sbi.2015.07.004.

Flydal, M.I. and Martinez, A. (2013). Phenylalanine hydroxylase: function, structure, and regulation. *IUBMB Life* 65 (4): 341–349. https://doi.org/10.1002/iub.1150.

Gjaltema, R.A. and Bank, R.A. (2017). Molecular insights into prolyl and lysyl hydroxylation of fibrillar collagens in health and disease. *Crit. Rev. Biochem. Mol. Biol.* 52 (1): 74–95. https://doi.org/10.1080/10409238.2016.1269716.

Gunsalus, I.C., Lipscomb, J.D., Marshall, V. et al. (1972). Structure and reactions of oxygenase active centers: cytochrome P-450 and iron sulfur proteins. *Biochem. Soc. Symp.* 34: 135–157.

Hayaishi, O. (1969). Enzymic hydroxylation. *Annu. Rev. Biochem.* 38: 21–44.

Hole, M., Jorge-Finnigan, A., Underhaug, J. et al. (2016). Pharmacological chaperones that protect tetrahydrobiopterin dependent aromatic amino acid hydroxylases through different mechanisms. *Curr. Drug Targets* 17 (13): 1515–1526.

Huang, X. and Groves, J.T. (2017). Beyond ferryl-mediated hydroxylation: 40 years of the rebound mechanism and C–H activation. *J. Biol. Inorg. Chem.* 22 (2–3): 185–207. https://doi.org/10.1007/s00775-016-1414-3.

Hudson, D.M. and Eyre, D.R. (2013). Collagen prolyl 3-hydroxylation: a major role for a minor post-translational modification? *Connect. Tissue Res.* 54 (4–5): 245–251. https://doi.org/10.3109/03008207.2013.800867.

Johnston, F.A., Ingalls, R.L., and Muka, B.O. (1954). The use of AA′-dipyridyl for determining the amount of ferrous iron formed in the digestive tract of women before and after the addition of beef to the diet. *J. Nutr.* 53 (1): 83–91.

Kaufman, S. and Friedman, S. (1965). Dopamine-beta-hydroxylase. *Pharmacol. Rev.* 17: 71–100.

Khirsariya, P. and Mewada, R.K. (2013). Single step oxidation of methane to methanol–towards better understanding. *Procedia Eng.* 51: 409–415. https://doi.org/10.1016/j.proeng.2013.01.057.

Kivirikko, K.I., Myllylä, R., and Pihlajaniemi, T. (1989). Protein hydroxylation: prolyl 4-hydroxylase, an enzyme with four cosubstrates and a multifunctional subunit. *FASEB J.* 3 (5): 1609–1617.

Koivu, J., Myllylä, R., Helaakoski, T. et al. (1987). A single polypeptide acts both as the beta subunit of prolyl 4-hydroxylase and as a protein disulfide-isomerase. *J. Biol. Chem.* 262 (14): 6447–6449.

Kuttan, R. and Radhakrishnan, A.N. (1973). Biochemistry of the hydroxyprolines. *Adv. Enzymol. Relat. Areas Mol. Biol.* 37: 273–347.

Lai, X., Wichers, H.J., Soler-Lopez, M., and Dijkstra, B.W. (2018). Structure and function of human tyrosinase and tyrosinase-related proteins. *Chemistry* 24 (1): 47–55. https://doi.org/10.1002/chem.201704410.

Lee, R.F. (1998). Annelid cytochrome P-450. *Comp. Biochem. Physiol. C Pharmacol. Toxicol. Endocrinol.* 121 (1-3): 173–179.

Li, P. and Wu, G. (2018). Roles of dietary glycine, proline, and hydroxyproline in collagen synthesis and animal growth. *Amino Acids* 50 (1): 29–38. https://doi.org/10.1007/s00726-017-2490-6.

Liang, Y.F. and Jiao, N. (2017). Oxygenation via C–H/C–C bond activation with molecular oxygen. *Acc. Chem. Res.* 50 (7): 1640–1653. https://doi.org/10.1021/acs.accounts.7b00108.

Liu, J., Sridhar, J., and Foroozesh, M. (2013). Cytochrome P450 family 1 inhibitors and structure-activity relationships. *Molecules* 18 (12): 14470–14495. https://doi.org/10.3390/molecules181214470.

Marsheck, W.J. Jr., (1971). Current trends in the microbiological transformation of steroids. *Prog. Ind. Microbiol.* 10: 49–103.

Myllylä, R., Wang, C., Heikkinen, J. et al. (2007). Expanding the lysyl hydroxylase toolbox: new insights into the localization and activities of lysyl hydroxylase 3 (LH3). *J. Cell. Physiol.* 212 (2): 323–329.

Nyström, A. and Bruckner-Tuderman, L. (2018). Matrix molecules and skin biology. *Semin. Cell Dev. Biol.* https://doi.org/10.1016/j.semcdb.2018.07.025.

Paravati, S. and Warrington, S.J. (2018. StatPearls [Internet]). *Catecholamines.* Treasure Island, FL: StatPearls Publishing http://www.ncbi.nlm.nih.gov/books/NBK507716/.

Pavon, J.A. and Fitzpatrick, P.F. (2006). Insights into the catalytic mechanisms of phenylalanine and tryptophan hydroxylase from kinetic isotope effects on aromatic hydroxylation. *Biochemistry* 45 (36): 11030–11037.

Periana, R.A., Taube, D.J., Gamble, S. et al. (1998). Platinum catalysts for the high-yield oxidation of methane to a methanol derivative. *Science* 280 (5363): 560–564.

Qiu, B., Wei, F., Sun, X. et al. (2014). Measurement of hydroxyproline in collagen with three different methods. *Mol. Med. Rep.* 10 (2): 1157–1163. https://doi.org/10.3892/mmr.2014.2267.

Ramachandran, G.N. (1963). Molecular structure of collagen. *Int. Rev. Connect. Tissue Res.* 1: 127–182.

Rutschmann, C., Baumann, S., Cabalzar, J. et al. (2014). Recombinant expression of hydroxylated human collagen in *Escherichia coli*. *Appl. Microbiol. Biotechnol.* 98 (10): 4445–4455. https://doi.org/10.1007/s00253-013-5447-z.

Shiju, N.R., Fiddy, S., Sonntag, O. et al. (2006). Selective oxidation of benzene to phenol over FeAlPO catalysts using nitrous oxide as oxidant. *Chem. Commun. (Camb.)* (47): 4955–4957.

Sih, C.J. (1969). Enzymatic mechanism of steroid hydroxylation. *Science* 163 (3873): 1297–1300.

Simpson, E.R., Cooper, D.Y., and Estabrook, R.W. (1969). Metabolic events associated with steroid hydroxylation by the adrenal cortex. *Recent Prog. Horm. Res.* 25: 523–562.

Sims, P. and Grover, P.L. (1974). Epoxides in polycyclic aromatic hydrocarbon metabolism and carcinogenesis. *Adv. Cancer Res.* 20: 165–274.

Smirnova, N.A., Hushpulian, D.M., Speer, R.E. et al. (2012). Catalytic mechanism and substrate specificity of HIF prolyl hydroxylases. *Biochemistry (Mosc)* 77 (10): 1108–1119. https://doi.org/10.1134/S0006297912100033.

Song, E. and Mechref, Y. (2013). LC-MS/MS identification of the O-glycosylation and hydroxylation of amino acid residues of collagen α-1 (II) chain from bovine cartilage. *J. Proteome Res.* 12 (8): 3599–3609. https://doi.org/10.1021/pr400101t.

Spivey, J.J. and Egbebi, A. (2007). Heterogeneous catalytic synthesis of ethanol from biomass-derived syngas. *Chem. Soc. Rev.* 36 (9): 1514–1528.

Sridhar, J., Liu, J., Foroozesh, M., and Stevens, C.L. (2012). Insights on cytochrome p450 enzymes and inhibitors obtained through QSAR studies. *Molecules* 17 (8): 9283–9305. https://doi.org/10.3390/molecules17089283.

Sroga, G.E. and Vashishth, D. (2011). UPLC methodology for identification and quantitation of naturally fluorescent crosslinks in proteins: a study of bone collagen. *J. Chromatogr. B Analyt. Technol. Biomed. Life Sci.* 879 (5–6): 379–385. https://doi.org/10.1016/j.jchromb.2010.12.024.

Svoboda, J.A. and Weirich, G.F. (1995). Sterol metabolism in the tobacco hornworm, Manduca sexta – a review. *Lipids* 30 (3): 263–267.

Tank, A.W. and Lee Wong, D. (2015). Peripheral and central effects of circulating catecholamines. *Compr. Physiol.* 5 (1): 1–15. https://doi.org/10.1002/cphy.c140007.

Tekin, I., Roskoski, R. Jr., Carkaci-Salli, N., and Vrana, K.E. (2014). Complex molecular regulation of tyrosine hydroxylase. *J. Neural Transm. (Vienna)* 121 (12): 1451–1481. https://doi.org/10.1007/s00702-014-1238-7.

Ullrich, V. (1972). Enzymatic hydroxylations with molecular oxygen. *Angew. Chem. Int. Ed. Engl.* 11 (8): 701–712.

Vasta, J.D. and Raines, R.T. (2018). Collagen prolyl 4-hydroxylase as a therapeutic target. *J. Med. Chem.* https://doi.org/10.1021/acs.jmedchem.8b00822.

Vranka, J.A., Sakai, L.Y., and Bächinger, H.P. (2004). Prolyl 3-hydroxylase 1, enzyme characterization and identification of a novel family of enzymes. *J. Biol. Chem.* 279 (22): 23615–23621.

Vuori, K., Pihlajaniemi, T., Marttila, M., and Kivirikko, K.I. (1992). Characterization of the human prolyl 4-hydroxylase tetramer and its multifunctional protein disulfide-isomerase subunit synthesized in a baculovirus expression system. *Proc. Natl. Acad. Sci. U. S. A.* 89 (16): 7467–7470.

Weiss, B. and Rossi, G.V. (1963). Catecholamines, Biosynthesis and inhibitors of formation. *Am. J. Pharm. Sci. Support. Public Health* 135: 206–218.

Wolf, D. (1998). High yields of methanol from methane by C–H bond activation at low temperatures. *Angew. Chem. Int. Ed. Engl.* 37 (24): 3351–3353. https://doi.org/10.1002/(SICI)1521-3773.

Wu, M. and Crane, J.S. (2018. StatPearls [Internet]). *Biochemistry, Collagen Synthesis*. Treasure Island, FL: StatPearls Publishing http://www.ncbi.nlm.nih.gov/books/NBK507709/.

Xiao, D.J., Bloch, E.D., Mason, J.A. et al. (2014). Oxidation of ethane to ethanol by $N_2O$ in a metal-organic framework with coordinatively unsaturated iron(II) sites. *Nat. Chem.* 6 (7): 590–595. https://doi.org/10.1038/nchem.1956.

Xin, H., Koekkoek, A., Yang, Q. et al. (2009). A hierarchical Fe/ZSM-5 zeolite with superior catalytic performance for benzene hydroxylation to phenol. *Chem. Commun. (Camb.)* (48): 7590–7592. https://doi.org/10.1039/b917038c.

Xu, X., Gan, Q., Clough, R.C. et al. (2011). Hydroxylation of recombinant human collagen type I alpha 1 in transgenic maize co-expressed with a recombinant human prolyl 4-hydroxylase. *BMC Biotechnol.* 11: 69. https://doi.org/10.1186/1472-6750-11-69.

Yang, P. and Yang, W. (2014). Hydroxylation of organic polymer surface: method and application. *ACS Appl. Mater. Interfaces* 6 (6): 3759–3770. https://doi.org/10.1021/am405857m.

Yang, Y., Moinodeen, F., Chin, W. et al. (2012). Pentanidium-catalyzed enantioselective α-hydroxylation of oxindoles using molecular oxygen. *Org. Lett.* 14 (18): 4762–4765.

Yoshizawa, K. (2000). Two-step concerted mechanism for methane hydroxylation on the diiron active site of soluble methane monooxygenase. *J. Inorg. Biochem.* 78 (1): 23–34.

Zhang, S., Huang, T., Ilangovan, U. et al. (2014). The solution structure of the regulatory domain of tyrosine hydroxylase. *J. Mol. Biol.* 426 (7): 1483–1497. https://doi.org/10.1016/j.jmb.2013.12.015.

Zhang, D., Tang, Z., and Liu, W. (2018). Biosynthesis of lincosamide antibiotics: reactions associated with degradation and detoxification pathways play a constructive role. *Acc. Chem. Res.* 51 (6): 1496–1506. https://doi.org/10.1021/acs.accounts.8b00135.

Zhao, L., Liu, Z., Guo, W. et al. (2009). Theoretical investigation of the gas-phase Mn(+)- and Co(+)-catalyzed oxidation of benzene by N(2)O. *Phys. Chem. Chem. Phys.* 11 (21): 4219–4229. https://doi.org/10.1039/b901019j.

Zurlo, G., Guo, J., Takada, M. et al. (2016). New insights into protein hydroxylation and its important role in human diseases. *Biochim. Biophys. Acta* 1866 (2): 208–220. https://doi.org/10.1016/j.bbcan.2016.09.004.

# 11

# Methylation of Proteins

## Introduction

Methylation of organic compounds is a chemical reaction in which a hydrogen atom is replaced by a methyl group ($CH_3$–) in hydroxyl, amine, and carboxylic acid functional groups (Bobst et al. 1969). Methylation of hydroxyl groups converts alcohols into ethers, methylation of amine groups converts them into methylamines, and methylation of carboxylic acid groups converts the acids into esters (Holman and Wiegand 1948; Korhonen 1967; Bobst et al. 1969). Methylation of such functional groups involves electrophilic methyl donors such as iodomethane, dimethyl sulfate, dimethyl carbonate, tertramethylammonium chloride, etc. (Korhonen 1967). Conversion of phenol into anisole is a very common methylation reaction in organic chemistry (see Figure 11.1). Methylation analysis is a very common analytical method used in the linkage determination of glycosidic bonds of oligosaccharides, polysaccharides, and other glycoconjugates.

Methylation of biomolecules such as DNA and proteins is mediated by methyltransferases that catalyzes the transfer of methyl groups from *S*-adenosyl methionine (AdoMet) (Creveling and Daly 1971). The methionine adenosyltransferase is involved in the biosynthesis of AdoMet from adenosine triphosphate (ATP) and methionine (Tabor and Tabor 1984). As a cosubstrate AdoMet is also involved in trans-sulfuration and aminopropylation reactions (Giulidori et al. 1984). In a biological process, methylation is observed to be very common, and hence, AdoMet is believed to be the second most used cosubstrate in enzymatic reactions after ATP (Bottiglieri 2002).

The most common DNA methylation occurs at cytosine–phosphate–guanine sites (CpG) in which cytosine residue is directly linked to phospate followed by guanine residue in DNA sequence (Datta and Datta 1969). The DNA methylation reaction is mediated by DNA methyltransferase that converts cytosine into 5-methylcytosine (MeC) (Lilischkis et al. 2001). About 90% of CpG sites in human DNA can get methylated (Zhang et al. 2017). However, some GC-rich sequences termed as CpG islands do not get methylated

*Co- and Post-Translational Modifications of Therapeutic Antibodies and Proteins*, First Edition. T. Shantha Raju.

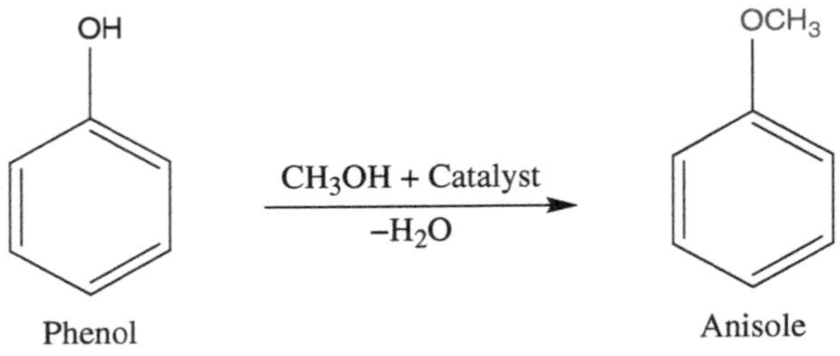

Phenol ($OH$) $\xrightarrow[-H_2O]{CH_3OH + \text{Catalyst}}$ Anisole ($OCH_3$)

**Figure 11.1** Methylation of phenol into anisole. *(See insert for color representation of this figure.)*

(Bird 1986). Approximately, 2% of human genome contains CpG clusters, and methylation of these clusters is involved in the regulation of transcriptional activity (Bird 1986).

In proteins amine groups of Arg and Lys residues may be N-methylated by methyltransferases (Paik and Kim 1975). The methyltransferases that mediates the transfer of methyl group to arginine residues are called protein arginine methyltransferases (PRMTs) (Cheng et al. 2005). Arg residues can be monomethylated or dimethylated and in the dimethylated Arg residues, the two methyl groups can be on one nitrogen atom or on both nitrogen groups (McBride and Silver 2001). If the two methyl groups are on one nitrogen atom, such dimethyl arginine residues are termed as asymmetric dimethylarginine. If the two methyl groups are on two nitrogen atoms such dimethyl arginine residues are called symmetric dimethylarginine. The methyltransferases that mediates the transfer of methyl groups to Lys residues are termed as lysine methyltransferases (Lachner and Jenuwein 2002). The Lys residues can be methylated by one or two or three methyl groups (Comb et al. 1966).

In some proteins, carboxylic acid functional groups at the C-terminus are methylated (Vorburger et al. 1989). Such proteins most commonly contain CAAX motifs in which C is Cys, A is any aliphatic amino acid except Ala, and X is the terminal uncharged amino acid at the C-termini (Leung et al. 2007). Methylation of the C-termini of CAAX motif occurs in three steps (Young et al. 2006). In the first step, the cysteine residue forms a thioester linkage with a prenyl lipid anchor. In the second step, the AAX residues were removed by endoproteolysis and hence exposing the α-COOH group of prenylcysteine residue. Methylation of the newly exposed α-COOH occurs in the third and final step. C-Terminal methylation of CAAX motif has been shown to play a role during mid-gestation stage in mouse (Nishimura and Linder 2013).

C-Terminus of Ser/Thr phosphatases is modified by the reversible methylation reaction (Dudiki et al. 2015). Accordingly, Ser/Thr phosphatases may contain leucine carboxy methyl ester in their C-termini (Janssens and Goris 2001). The three-step processing as in CAAX motif is not required for the C-terminus methylation of phosphatases (Kloeker et al. 1997). The C-terminus methylation of phosphatases is mediated by protein phosphatase methyltransferase (PPMT). This is a reversible methylation, and the demethylation process is mediated by protein phosphatase methylesterase (PPME). Methylation and demethylation of phosphatases is a dynamic process that is in response to certain type of stimulations (Hwang et al. 2016).

## Mechanism of Protein Methylation

### Chemical Methylation Reactions

Figure 11.1 illustrates the chemical methylation reaction of phenol into anisole. This is a $SN_2$ nucleophilic substitution reaction in which methyl iodide is a donor of methyl group, and $Li_2CO_3$ is a base to convert –OH group of phenol into oxygen ion ($-O^-$ ion) (Nakagome et al. 1966). In addition to methyl iodide (iodomethane), dimethyl sulfate, dimethyl carbonate, tetramethylammonium chloride, methyl triflate, diazomethane, methyl flurosulfonate (magic methyl), etc., are also used as donors of methyl groups for methylation reactions during organic synthesis (Sorvari and Stoward 1970). In addition to hydroxyl groups, methylation of organic amines, carboxylic acids, sulfhydryls, phosphates, etc., are also very common in organic synthetic chemistry.

Many phenolic compounds undergo O-methylation in plants to form anisole derivatives. Such O-methylation processes in plants are catalyzed by Caffeoyl-CoA *O*-methyltransferases (Lam et al. 2007). These are key enzymatic reactions involved in the biosynthesis of lignols in plants (Provenzano et al. 2014). The lignols are the precursors of lignin, a polyphenolic organic compound found in plants and is one of the major structural components of plants (Sandberg et al. 2012).

In eukaryotic proteins, methylation is mainly through N-methylation of amine residues of Arg and Lys residues (Migliori et al. 2010). Methylation of carboxylic acid residues at the C-termini occurs in specific cases, and O-methylation is very rare in proteins (Kim et al. 2009). However, O-methylated sugar residues have been observed in specific cases, but the literature data is very sparse (Schauer 1988). Mechanism of some of these methylation reactions are discussed briefly in the following sections.

## Biological Methylation Reactions

### Methylation of Arg Residues

Arg methylation occurs in the nucleus of cells, and most of the arginine-methylated proteins are RNA-binding proteins (Nishida et al. 2017). The Arg methylation is mediated by PRMTs. There are at least two types of PRMTs that are involved in the methylation of Arg residues of proteins (Wesche et al. 2017).

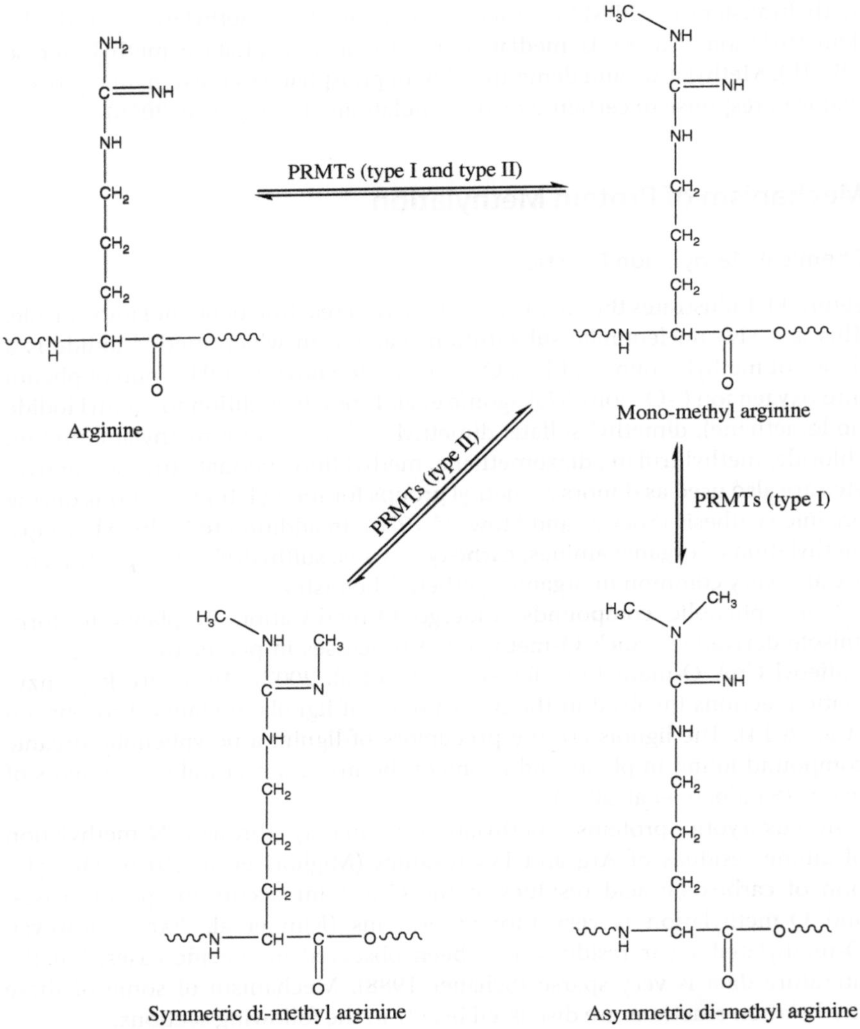

PRMTs (type I and type II)

Arginine

Mono-methyl arginine

PRMTs (type II)

PRMTs (type I)

Symmetric di-methyl arginine

Asymmetric di-methyl arginine

**Figure 11.2** Symmetric and asymmetric methylation of Arg residues.

They are Type I and Type II enzymes in which Type I enzymes mediates the formation of asymmetric dimethyl arginine residues, whereas the Type II enzymes mediate the formation of symmetric dimethyl arginine residues (Kaniskan and Jin 2017). Both Type I and Type II PRMTs mediates the formation of monomethyl arginine residues as intermediates as shown in Figure 11.2.

### Methylation of Lys Residues

Methylation of Lys residues is mediated by lysine methyltransferases that can transfer upto three methyl groups to Lys residues (Milite et al. 2016). Many of these Lys methyltransferases contain an evolutionarily conserved SET domain which is named after three Drosophila proteins (Katoh 2016). These three Drosophila proteins harbor Su (var), Enhancer-of-Zeste and Trithorax and hence named as SET domain. The SET domains contain *S*-adenosyl methionine (SAM)-dependent methyltransferase activities and are structurally different than the other SAM-binding proteins. Lysine methyltransferases with different SET domain possess different substrate specificities (Spellmon et al. 2015). For instance, SET1, SET7, and MLL mediate the methylation of Lysine 4 (K4) of histone H3. In contrast, Suv39h1, ESET, and G9a specifically mediates the methylation of lysine 9 (residue # 9) of histone H3. Lysine 4 and lysine 9 methylation of histone H3 are site-specific methylations. In addition, other lysine residues on histone H3 and histone H4 are also methylated by specific SET domain-containing methyltransferases. Methylation of histones at lysine 4 and lysine 9 are mutually exclusive and are diametrically opposed to each other. Histones are the prime targets for lysine methyltransferases. In addition to histones, other cellular proteins such as elongation factor 1A, calcium-sensing protein calmodulin may also carry N-methylated lysine residues (Binda 2013).

### Methylation of Prenylcysteine Residues

Methylation of carboxyl residue at the C-termini of eukaryotic proteins containing CAAX motif occurs after a series of post-translational modifications (Vorburger et al. 1989). The processing of CAAX-tail is a three-step reaction process in which first step is the attachment of prenyl lipid anchor to the Cys residue through a thioester bond formation. In the second step, the CAAX tail containing prenyl lipid anchor undergoes endoproteolysis to remove the last three amino acid residues (AAX) that exposes the prenylcysteine α-COOH group. In the third and final step, the newly exposed prenylcysteine α-COOH group undergoes methylation that is mediated by isoprenylcysteine carboxyl methyltransferases (ICMTs) (Hancock et al. 1991). This is a reversible methylation reaction in which the demethylation reaction is catalyzed by isoprenylcysteine carboxyl methylesterases (Dunten et al. 1995).

### Methylation of Protein Phosphatase 2A

Phosphatases are the enzymes that catalyze the hydrolysis of phosphate groups from tyrosine phosphate, serine phosphate, and threonine phosphate residues in phosphoproteins of eukaryotic cells (Walker et al. 1954). The C-termini of some major Ser/Thr phosphatases (protein phosphatases, PP2A) is covalently modified by the leucine carboxyl methyltransferase (PPMT) to form leucine carboxyl methyl ester (Xie and Clarke 1993; Onofrio et al. 2006). This is a reversible modification that requires no C-terminal processing for methylation unlike the methylation of CAAX motif (Ota and Clarke 1989). Specific protein phosphatase methylesterase (PPME) mediates the conversion of leucine carboxy methyl ester into leucine (Li and Stock 2009; Yabe et al. 2015). Thus, both PPMT and PPME enzymes are involved in the catalysis of opposing enzymatic reactions, i.e. methylation and demethylation. Accordingly, methylation and demethylation reactions of Ser/Thr phosphatases are dynamic processes.

### Methylation of Isoaspartyl Residues

In eukaryotic proteins, asparagine residues undergo deamidation to form either aspartic acid and/or isoaspartic acid residues (Hao and Sze 2014). The isoaspartic acid is an unnatural amino acid residue and accumulation of proteins with isoaspartic acid residues is harmful to cells (Fossati et al. 2013). Hence, eukaryotic cells contain a pathway to convert the isoaspartyl residues to aspartyl residues (McFadden and Clarke 1987). Deamidation of asparagine residues is a stepwise process that involves the conversion of asparagine residue into succinimidyl intermediate as discussed in deamidation of proteins chapter (see Chapter 5). The succinimidyl intermediate undergoes spontaneous hydrolysis to form aspartyl or isoaspartyl residues. The protein isoaspartyl *O*-methyltransferase (PIMT) catalyzes the formation of methyl ester of isoaspartyl residues which in turn converts back to succinimidyl intermediate and hence preventing the accumulation of deleterious isoaspartyl containing proteins (Kagan et al. 1997).

### O-Methylation of Sugar Residues

In addition to methylation of amino acid residues of proteins through amine, hydroxyl, and carboxyl functional groups, O-methylation of sugar residues of oligosaccharides, polysaccharides, and other glycoconjugates have been reported (Hakomori and Saito 1969; Ueno et al. 1975). For example, glycoproteins purified from Helix pomatia have been shown to contain *O*-methylgalactose and *O*-methylmannose residues (Hall et al. 1977). Presence of *O*-methylmannose, *O*-methylxylose, and *O*-methyltalose has been reported in lipopolysaccharides from *Rhodopseudomonas palustris* (Weckesser et al. 1973a; Weckesser et al. 1973b; Mayer et al. 1974). In addition, 4-*O*-methyl and 8-*O*-methyl sialic acid residues have been reported to be present in various glycoconjugates (Kamerling and Gerwig 2006). Biosynthesis of 8-*O*-methyl sialic acid residues has been shown to be mediated by *S*-adenosyl-L-methionine: sialiate 8-*O*-methyltransferase that uses SAM as a methyl group donor (Kelm et al. 1998). Similarly, methylation of Man, Xyl and talose is mediated by their respective methyltransferases.

## Physicochemical and Biological Significance of Methylation of Proteins

Methyl group is a hydrophobic group, and hence, methylation increases hydrophobicity of proteins. N-Methylation of amine groups may affect the overall positive charge of proteins. Similarly, methylation of carboxylic acid groups to form esters may affect the overall negative charge of proteins. O-Methylation of hydroxyl groups may affect the hydrophilic interactions of proteins. Additionally, methylation increases chemical repertoire and heterogeneity of a given protein. The addition of one carbon and three hydrogen containing methyl group to amines, hydroxyls, carboxylic acid groups impact the proteins charge heterogeneity. Hence, methylation of proteins may affect their physicochemical and biological properties.

Methylation of Arg residues are mainly found in RNA-binding proteins and hence, affect their binding affinity to RNA (Nishida et al. 2017). Arg methylation affects the protein–protein interactions and may also affect many cellular processes including regulation of transcription, signal transduction, protein trafficking, etc. (Poulard et al. 2016). Methylation of Lys residues affects the protein–protein interactions of histone proteins (Rona et al. 2016). In addition, methylation of Lys residues is also associated with the transcriptional repression and heterochromatin formation (Beaver and Waters 2016). In the case of prenylcysteine methylation, the absence of ICMT, the enzyme involved in the biosynthesis of prenylcysteine methylation results in mid-gestation lethality in mice (Svensson et al. 2006). Also, prenylcysteine methylation plays a role in facilitating the transfer of CAAX motif containing proteins to membrane surfaces within the cells (Rodríguez-Concepción et al. 2000). Examples of proteins that contain methylated prenylcysteine residues include Ras, GTP binding proteins, nuclear lamins, certain protein kinases, etc. These proteins are involved in cell signaling and are functional when they are on the cytosolic surface of the plasma membrane (Nishimura and Linder 2013). The C-terminal methylation of phosphatases regulates the recruitment of regulatory proteins into PP2A complexes through protein–protein interactions and hence indirectly controls the PP2A complex activity (Kaur and Westermarck 2016). The methylation and demethylation of phosphatases is a dynamic process that fluctuates during cell cycle and also in response to cAMP stimuli (Liu et al. 2013). Presence of isoaspartyl residues causes instability of proteins, loss of biological activity, and stimulates autoimmune responses (Patananan et al. 2014). Methylation prevents the accumulation of proteins with isoaspartyl residues and hence prevents the instability, loss of biological activity, stimulation of autoimmune responses, etc. Mice lacking PIMT have been shown to die at young age, whereas the overexpression of PIMT in flies are shown to have improved life span of more than 30% (Bennett et al. 2003; Dimitrijevic et al. 2014). In addition, many associations between methylation of proteins and cancer are consistent because the enzymes involved in methylation and demethylation play important regulatory role in the cells. At this point, the biological role of O-methylated sugars is not very clear. However, they may play a role in protein–carbohydrate interactions as well as carbohydrate–carbohydrate interactions (Varki 2017).

### Methylation in RTPs

Methylation of protein backbone in recombinant therapeutic proteins (RTPs) has not yet been reported. However, O-methylation of sialic acid residues has been observed in a therapeutic recombinant immunoglobulin G (IgG) fusion protein that was produced using Chinese hamster ovary (CHO) cells as host expression system (unpublished data). The methylation of sialic acid residues was observed during cell culture manipulations to increase the product titer. Since, the IgG fusion protein product development was stopped prior to the submission of marketing application, the methylation data was not published. This unpublished data suggests that the methylation of RTPs may occur in certain cell culture conditions that might affect their biological functions (Varki 2017).

### Methods to Analyze Methylation in Proteins and Glycoproteins

Methylation in proteins may be detected by using mass spectrometry, liquid chromatography, capillary electrophoresis, enzyme-linked immunosorbent assay (ELISA) assays, etc. (Wesche et al. 2017). The mass difference between methyl group ($CH_3$–) and hydrogen atom is large enough (14 amu) to detect the methylation of proteins using high-resolution mass spectrometry (Wang et al. 2017). Peptide mapping methods can be used to identify the sites of methylation (Wang et al. 2017). Since methylation impacts the hydrophobicity and the overall charge, hydrophobic interaction chromatography, ion-exchange chromatography, reversed-phase chromatography, etc., methods can be used to separate the methylated proteins from their nonmethylated counterparts (Chen et al. 2017). ELISA assays can also be used to detect methylated proteins in cells, tissues, crude extracts, purified proteins, etc. (Pichler et al. 2012). Additionally, NMR methods can also be used to detect methylation of proteins (Garay et al. 2016).

Permethylation analysis is a very useful tool in carbohydrate chemistry to identify the glycosidic linkages of oligosaccharides, polysaccharides, and other glycoconjugates (Price 2008; Shubhakar et al. 2016). Using $^{13}C$ labeled iodomethane as a methylating reagent, the presence and the location of *O*-methyl groups can be easily detected (Shajahan et al. 2017). Since *O*-methyl groups are relatively stable than *O*-acetyl groups, monosaccharide composition analysis can also be used to detect O-methylated sugar residues. In addition, NMR is a very useful tool to detect O-methylated sugar residues. The *O*-methyl groups give very distinct proton and carbon resonance in both $^{1}H$ and $^{13}C13$ NMR analyses, respectively, which are useful to detect and quantitate methyl groups in biomolecules (Möbius et al. 2013; Garay et al. 2016). In addition, ELISA methods are also available to detect and quantitate protein methylation (Kremer et al. 2012). However, ELISA methods show very high variability compared to chromatographic methods.

## References


Beaver, J.E. and Waters, M.L. (2016). Molecular recognition of Lys and Arg methylation. *ACS Chem. Biol.* 11 (3): 643–653. https://doi.org/10.1021/acschembio.5b00996.

Bennett, E.J., Bjerregaard, J., Knapp, J.E. et al. (2003). Catalytic implications from the *Drosophila* protein L-isoaspartyl methyltransferase structure and site-directed mutagenesis. *Biochemistry* 42 (44): 12844–12853.

Binda, O. (2013). On your histone mark, SET, methylate! *Epigenetics* 8 (5): 457–463. https://doi.org/10.4161/epi.24451.

Bird, A.P. (1986). CpG-rich islands and the function of DNA methylation. *Nature* 321 (6067): 209–213.

Bobst, A.M., Rottman, F., and Cerutti, P.A. (1969). Effect of the methylation of the 2′-hydroxyl groups in polyadenylic acid on its structure in weakly acidic and neutral solutions and on its capability to form ordered complexes with polyuridylic acid. *J. Mol. Biol.* 46 (2): 221–234.

Bottiglieri, T. (2002). *S*-Adenosyl-L-methionine (SAMe): from the bench to the bedside – molecular basis of a pleiotrophic molecule. *Am. J. Clin. Nutr.* 76 (5): 1151S–1157S.

Chen, M., Zhang, M., Zhai, L. et al. (2017). Tryptic peptides bearing C-terminal dimethyllysine need to be considered during the analysis of lysine dimethylation in proteomic study. *J. Proteome Res.* 16 (9): 3460–3469. https://doi.org/10.1021/acs.jproteome.7b00373.

Cheng, X., Collins, R.E., and Zhang, X. (2005). Structural and sequence motifs of protein (histone) methylation enzymes. *Annu. Rev. Biophys. Biomol. Struct.* 34: 267–294.

Comb, D.G., Sarkar, N., and Pinzino, C.J. (1966). The methylation of lysine residues in protein. *J. Biol. Chem.* 241 (8): 1857–1862.

Creveling, C.R. and Daly, J.W. (1971). Assay of enzymes of catecholamine biosynthesis and metabolism. *Methods Biochem. Anal.* 153–182.

Datta, R.K. and Datta, B. (1969). Role of the methylated nucleic acids in carcinogenesis. *Exp. Mol. Pathol.* 10 (2): 129–140.

Dimitrijevic, A., Qin, Z., and Aswad, D.W. (2014). Isoaspartyl formation in creatine kinase B is associated with loss of enzymatic activity; implications for the linkage of isoaspartate accumulation and neurological dysfunction in the

PIMT knockout mouse. *PLoS ONE* 9 (6): e100622. https://doi.org/10.1371/journal.pone.0100622.
Dudiki, T., Kadunganattil, S., Ferrara, J.K. et al. (2015). Changes in carboxy methylation and tyrosine phosphorylation of protein phosphatase PP2A are associated with epididymal sperm maturation and motility. *PLoS ONE* 10 (11): e0141961. https://doi.org/10.1371/journal.pone.0141961.
Dunten, R.L., Wait, S.J., and Backlund, P.S. Jr. (1995). Fractionation and characterization of protein C-terminal prenyl-cysteine methylesterase activities from rabbit brain. *Biochem. Biophys. Res. Commun.* 208 (1): 174–182.
Fossati, S., Todd, K., Sotolongo, K. et al. (2013). Differential contribution of isoaspartate post-translational modifications to the fibrillization and toxic properties of amyloid $\beta$ and the $Asn^{23}$ Iowa mutation. *Biochem. J* 456 (3): 347–360. https://doi.org/10.1042/BJ20130652.
Garay, P.G., Martin, O.A., Scheraga, H.A., and Vila, J.A. (2016). Detection of methylation, acetylation and glycosylation of protein residues by monitoring $^{13}C$ chemical-shift changes: a quantum-chemical study. *PeerJ* 4: e2253. https://doi.org/10.7717/peerj.2253.
Giulidori, P., Galli-Kienle, M., Catto, E., and Stramentinoli, G. (1984). Transmethylation, transsulfuration, and aminopropylation reactions of *S*-adenosyl-L-methionine in vivo. *J. Biol. Chem.* 259 (7): 4205–4211.
Hakomori, S. and Saito, T. (1969). Isolation and characterization of a glycosphingolipid having a new sialic acid. *Biochemistry* 8 (12): 5082–5088.
Hall, R.L., Wood, E.J., Kamberling, J.P. et al. (1977). 3-*O*-methyl sugars as constituents of glycoproteins. Identification of 3-*O*-methylgalactose and 3-*O*-methylmannose in pulmonate gastropod haemocyanins. *Biochem. J* 165 (1): 173–176.
Hancock, J.F., Cadwallader, K., and Marshall, C.J. (1991). Methylation and proteolysis are essential for efficient membrane binding of prenylated p21K-ras(B). *EMBO J.* 10 (3): 641–646.
Hao, P. and Sze, S.K. (2014). Proteomic analysis of protein deamidation. *Curr. Protoc. Protein Sci.* 78: 24.5.1–24.5.14. https://doi.org/10.1002/0471140864.ps2405s78.
Holman, W.I. and Wiegand, C. (1948). The chemical conversion of nicotinic acid and nicotinamide to derivatives of N-methyl-2-pyridone by methylation and oxidation. *Biochem. J.* 43 (3): 423–426.
Hwang, J., Lee, J.A., and Pallas, D.C. (2016). Leucine carboxyl methyltransferase 1 (LCMT-1) methylates protein phosphatase 4 (PP4) and protein phosphatase 6 (PP6) and differentially regulates the stable formation of different PP4 holoenzymes. *J. Biol. Chem.* 291 (40): 21008–21019.
Janssens, V. and Goris, J. (2001). Protein phosphatase 2A: a highly regulated family of serine/threonine phosphatases implicated in cell growth and signalling. *Biochem. J.* 353 (Pt. 3): 417–439.

Kagan, R.M., Niewmierzycka, A., and Clarke, S. (1997). Targeted gene disruption of the *Caenorhabditis elegans* L-isoaspartyl protein repair methyltransferase impairs survival of dauer stage nematodes. *Arch. Biochem. Biophys.* 348 (2): 320–328.

Kamerling, J.P. and Gerwig, G.J. (2006). Structural analysis of naturally occurring sialic acids. *Methods Mol. Biol.* 347: 69–91.

Kaniskan, H.Ü. and Jin, J. (2017). Recent progress in developing selective inhibitors of protein methyltransferases. *Curr. Opin. Chem. Biol.* 39: 100–108. https://doi.org/10.1016/j.cbpa.2017.06.013.

Katoh, M. (2016). Mutation spectra of histone methyltransferases with canonical SET domains and EZH2-targeted therapy. *Epigenomics* 8 (2): 285–305. https://doi.org/10.2217/epi.15.89.

Kaur, A. and Westermarck, J. (2016). Regulation of protein phosphatase 2A (PP2A) tumor suppressor function by PME-1. *Biochem. Soc. Trans.* 44 (6): 1683–1693.

Kelm, A., Shaw, L., Schauer, R., and Reuter, G. (1998). The biosynthesis of 8-*O*-methylated sialic acids in the starfish *Asterias rubens*--isolation and characterisation of *S*-adenosyl-L-methionine:sialate-8-*O*-methyltransferase. *Eur. J. Biochem.* 251 (3): 874–884.

Kim, J.K., Samaranayake, M., and Pradhan, S. (2009). Epigenetic mechanisms in mammals. *Cell. Mol. Life Sci.* 66 (4): 596–612. https://doi.org/10.1007/s00018-008-8432-4.

Kloeker, S., Bryant, J.C., Strack, S. et al. (1997). Carboxymethylation of nuclear protein serine/threonine phosphatase X. *Biochem. J.* 327 (Pt. 2): 481–486.

Korhonen, L.K. (1967). Specific methylation of carboxyl groups by thionyl chloride in methanol. *Acta Histochem.* 26 (1): 80–86.

Kremer, D., Metzger, S., Kolb-Bachofen, V., and Kremer, D. (2012). Quantitative measurement of genome-wide DNA methylation by a reliable and cost-efficient enzyme-linked immunosorbent assay technique. *Anal. Biochem.* 422 (2): 74–78. https://doi.org/10.1016/j.ab.2011.11.033.

Lachner, M. and Jenuwein, T. (2002). The many faces of histone lysine methylation. *Curr. Opin. Cell Biol.* 14 (3): 286–298.

Lam, K.C., Ibrahim, R.K., Behdad, B., and Dayanandan, S. (2007). Structure, function, and evolution of plant *O*-methyltransferases. *Genome* 50 (11): 1001–1013.

Leung, K.F., Baron, R., Ali, B.R. et al. (2007). Rab GTPases containing a CAAX motif are processed post-geranylgeranylation by proteolysis and methylation. *J. Biol. Chem.* 282 (2): 1487–1497.

Li, Z. and Stock, J.B. (2009). Protein carboxyl methylation and the biochemistry of memory. *Biol. Chem.* 390 (11): 1087–1096. https://doi.org/10.1515/BC.2009.133.

Lilischkis, R., Kneitz, H., and Kreipe, H. (2001). Methylation analysis of CpG islands. *Methods Mol. Med.* 57: 271–283. https://doi.org/10.1385/1-59259-136-1:271.

Liu, Y., Zheng, P., Liu, Y. et al. (2013). An epigenetic role for PRL-3 as a regulator of H3K9 methylation in colorectal cancer. *Gut* 62 (4): 571–581. https://doi.org/10.1136/gutjnl-2011-301059.

Mayer, H., Framberg, K., and Weckesser, J. (1974). 6-*O*-methyl-D-glucosamine in lipopolysaccharides of *Rhodopseudomonas palustris* strains. *Eur. J. Biochem.* 44 (1): 181–187.

McBride, A.E. and Silver, P.A. (2001). State of the Arg: protein methylation at arginine comes of age. *Cell* 106 (1): 5–8.

McFadden, P.N. and Clarke, S. (1987). Conversion of isoaspartyl peptides to normal peptides: implications for the cellular repair of damaged proteins. *Proc. Natl. Acad. Sci. U.S.A.* 84 (9): 2595–2599.

Migliori, V., Phalke, S., Bezzi, M., and Guccione, E. (2010). Arginine/lysine-methyl/methyl switches: biochemical role of histone arginine methylation in transcriptional regulation. *Epigenomics* 2 (1): 119–137. https://doi.org/10.2217/epi.09.39.

Milite, C., Feoli, A., Viviano, M. et al. (2016). The emerging role of lysine methyltransferase SETD8 in human diseases. *Clin. Epigenetics* 8: 102.

Möbius, K., Nordsieck, K., Pichert, A. et al. (2013). Investigation of lysine side chain interactions of interleukin-8 with heparin and other glycosaminoglycans studied by a methylation-NMR approach. *Glycobiology* 23 (11): 1260–1269. https://doi.org/10.1093/glycob/cwt062.

Nakagome, T., Misaki, A., and Murano, A. (1966). Synthesis of pyridazine derivatives. XIV. On the methylation of 4-amino-3(2H)pyridazinone derivatives. *Chem. Pharm. Bull. (Tokyo)* 14 (10): 1090–1096.

Nishida, K., Kuwano, Y., Nishikawa, T. et al. (2017). RNA binding proteins and genome integrity. *Int. J. Mol. Sci.* 18 (7): https://doi.org/10.3390/ijms18071341.

Nishimura, A. and Linder, M.E. (2013). Identification of a novel prenyl and palmitoyl modification at the CaaX motif of Cdc42 that regulates RhoGDI binding. *Mol. Cell. Biol.* 33 (7): 1417–1429. https://doi.org/10.1128/MCB.01398-12.

Onofrio, A.B., Jäger, E., Brandão, T.A. et al. (2006). *N*-(2-carboxybenzoyl)-L-leucine methyl ester. *Acta Crystallogr. C* 62 (Pt. 5): o237–o239.

Ota, I.M. and Clarke, S. (1989). Enzymatic methylation of 23-29-kDa bovine retinal rod outer segment membrane proteins. Evidence for methyl ester formation at carboxyl-terminal cysteinyl residues. *J. Biol. Chem.* 264 (22): 12879–12884.

Paik, W.K. and Kim, S. (1975). Protein methylation: chemical, enzymological, and biological significance. *Adv. Enzymol. Relat. Areas Mol. Biol.* 42: 227–286.

Patananan, A.N., Capri, J., Whitelegge, J.P., and Clarke, S.G. (2014). Non-repair pathways for minimizing protein isoaspartyl damage in the yeast *Saccharomyces cerevisiae*. *J. Biol. Chem.* 289 (24): 16936–16953. https://doi.org/10.1074/jbc.M114.564385.

Pichler, G., Jack, A., Wolf, P., and Hake, S.B. (2012). Versatile toolbox for high throughput biochemical and functional studies with fluorescent fusion proteins. *PLoS ONE* 7 (5): e36967. https://doi.org/10.1371/journal.pone.0036967.

Poulard, C., Corbo, L., and Le Romancer, M. (2016). Protein arginine methylation/demethylation and cancer. *Oncotarget* 7 (41): 67532–67550. https://doi.org/10.18632/oncotarget.11376.

Price, N.P. (2008). Permethylation linkage analysis techniques for residual carbohydrates. *Appl. Biochem. Biotechnol.* 148 (1–3): 271–276. https://doi.org/10.1007/s12010-007-8044-8.

Provenzano, S., Spelt, C., Hosokawa, S. et al. (2014). Genetic control and evolution of anthocyanin methylation. *Plant Physiol.* 165 (3): 962–977.

Rodríguez-Concepción, M., Toledo-Ortiz, G., Yalovsky, S. et al. (2000). Carboxyl-methylation of prenylated calmodulin CaM53 is required for efficient plasma membrane targeting of the protein. *Plant J.* 24 (6): 775–784.

Rona, G.B., Eleutherio, E.C.A., and Pinheiro, A.S. (2016). PWWP domains and their modes of sensing DNA and histone methylated lysines. *Biophys. Rev.* 8 (1): 63–74. https://doi.org/10.1007/s12551-015-0190-6.

Sandberg, T., Eklund, P., and Hotokka, M. (2012). Conformational solvation studies of LIGNOLs with molecular dynamics and conductor-like screening model. *Int. J. Mol. Sci.* 13 (8): 9845–9863. https://doi.org/10.3390/ijms13089845.

Schauer, R. (1988). Sialic acids as antigenic determinants of complex carbohydrates. *Adv. Exp. Med. Biol.* 228: 47–72.

Shajahan, A., Supekar, N.T., Heiss, C. et al. (2017). Tool for rapid analysis of glycopeptide by permethylation (TRAP) via one-pot site mapping and glycan analysis. *Anal. Chem.* https://doi.org/10.1021/acs.analchem.7b01730.

Shubhakar, A., Kozak, R.P., Reiding, K.R. et al. (2016). Automated high-throughput permethylation for glycosylation analysis of biologics using MALDI-TOF-MS. *Anal. Chem.* 88 (17): 8562–8569. https://doi.org/10.1021/acs.analchem.6b01639.

Sorvari, T.E. and Stoward, P.J. (1970). Some investigations of the mechanism of the so-called "methylation" reactions used in mucosubstance histochemistry. I. "Methylation" with methyl iodide, diazomethane, and various organic solvents containing either hydrogen chloride or thionyl chloride. *Histochemie* 24 (2): 106–113.

Spellmon, N., Holcomb, J., Trescott, L. et al. (2015). Structure and function of SET and MYND domain-containing proteins. *Int. J. Mol. Sci.* 16 (1): 1406–1428. https://doi.org/10.3390/ijms16011406.

Svensson, A.W., Casey, P.J., Young, S.G., and Bergo, M.O. (2006). Genetic and pharmacologic analyses of the role of Icmt in Ras membrane association and function. *Methods Enzymol.* 407: 144–159.

Tabor, C.W. and Tabor, H. (1984). Methionine adenosyltransferase (*S*-adenosylmethionine synthetase) and *S*-adenosylmethionine decarboxylase. *Adv. Enzymol. Relat. Areas Mol. Biol.* 56: 251–282.

Ueno, K., Ishizuka, I., and Yamakawa, T. (1975). Glycolipids of the fish testis. *J. Biochem.* 77 (6): 1223–1232.

Varki, A. (2017). Biological roles of glycans. *Glycobiology* 27 (1): 3–49. https://doi.org/10.1093/glycob/cww086.

Vorburger, K., Kitten, G.T., and Nigg, E.A. (1989). Modification of nuclear lamin proteins by a mevalonic acid derivative occurs in reticulocyte lysates and requires the cysteine residue of the C-terminal CXXM motif. *EMBO J.* 8 (13): 4007–4013.

Walker, B.S., Lemon, H.M., Davison, M.M., and Schwartz, M.K. (1954). Acid phosphatases; a review. *Am. J. Clin. Pathol.* 24 (7): 807–837.

Wang, Q., Wang, K., and Ye, M. (2017). Strategies for large-scale analysis of non-histone protein methylation by LC-MS/MS. *Analyst* https://doi.org/10.1039/c7an00954b.

Weckesser, J., Drews, G., Fromme, I., and Mayer, H. (1973a). Isolation and chemical composition of the lipopolysaccharides of *Rhodopseudomonas palustris* strains. *Arch. Mikrobiol.* 92 (2): 123–138.

Weckesser, J., Mayer, H., and Fromme, I. (1973b). *O*-methyl sugars in lipopolysaccharides of Rhodospirillaceae. Identification of 3-*O*-methyl-D-mannose in *Rhodopseudomonas viridis* and of 4-*O*-methyl-D-xylose and 3-*O*-methyl-6-deoxy-D-talose in *Rhodopseudomonas palustris* respectively. *Biochem. J.* 135 (2): 293–297.

Wesche, J., Kühn, S., Kessler, B.M. et al. (2017). Protein arginine methylation: a prominent modification and its demethylation. *Cell. Mol. Life Sci.* 74 (18): 3305–3315. https://doi.org/10.1007/s00018-017-2515-z.

Xie, H. and Clarke, S. (1993). Methyl esterification of C-terminal leucine residues in cytosolic 36-kDa polypeptides of bovine brain. A novel eucaryotic protein carboxyl methylation reaction. *J. Biol. Chem.* 268 (18): 13364–13371.

Yabe, R., Miura, A., Usui, T. et al. (2015). Protein phosphatase methyl-esterase PME-1 protects protein phosphatase 2A from ubiquitin/proteasome degradation. *PLoS ONE* 10 (12): e0145226. https://doi.org/10.1371/journal.pone.0145226.

Young, S.G., Clarke, S.G., Bergoc, M.O. et al. (2006). 10 Genetic approaches to understanding the physiologic importance of the carboxyl methylation of isoprenylated proteins. *Enzymes* 24: 273–301. https://doi.org/10.1016/S1874-6047(06)80012-0.

Zhang, P., Rausch, C., Hastert, F.D. et al. (2017). Methyl-CpG binding domain protein 1 regulates localization and activity of Tet1 in a CXXC3 domain-dependent manner. *Nucleic Acids Res.* 45 (12): 7118–7136. https://doi.org/10.1093/nar/gkx281.

# 12

# Oxidation of Proteins

## Introduction

In chemistry, oxidation and reduction reactions are very common chemical reactions in which a chemical entity such as an ion, atom, or a molecule undergoes a change in their oxidative or reductive state, respectively (Massey and Veeger 1963). For example, during oxidation ions lose electrons, while during reduction ions gain electrons. In general, at the molecular level, oxidizing molecules gains oxygen atoms, whereas reducing molecules lose oxygen atoms. Alternatively, oxidizing molecules lose hydrogen atoms and reducing molecules gain hydrogen atoms. Oxidation and reduction reactions are catalytic reactions in which catalysts acts as oxidizing or reducing agents (Morton 1965). During oxidation, a catalyst which acts as an oxidizing agent undergoes reduction and in the process oxidizes the oxidizing molecule. The most common oxidation reaction of organic molecules is the conversion of ethyl alcohol into acetic acid. During this oxidation reaction, the alcohol functionality in ethyl alcohol converts into carboxylic acid functionality to yield acetic acid as shown in Figure 12.1. For oxidation of alcohols into acids, the most commonly used catalysts are potassium permanganate ($KMnO_4$), ruthenium tetraoxide ($RuO_4$), pyridinium dichromate (PDC) in dimethyl formate (DMF), etc. (Crimmins and DeBaillie 2006).

In biology also, oxidation reactions occur in the presence of oxidizing agents. The most commonly available oxidizing agent in biological systems is hydrogen peroxide (Moody 1964). Other oxidizing agents available in the biological systems are mostly the derivatives of peroxides. Chemically, peroxides are highly unstable that easily converts into oxygen radicals. These oxygen radicals are highly reactive and are often referred as reactive oxygen species (ROS) (Ilas and Surgenor 1946).

In biological systems, water is a very abundant component that normally contains dissolved oxygen (DO) molecules. Aqueous proteins solutions also contain water as a main constituent that normally carries some DO molecules. Whenever the DO molecules are exposed to certain metal ions, ultraviolet light,

*Co- and Post-Translational Modifications of Therapeutic Antibodies and Proteins*, First Edition. T. Shantha Raju.

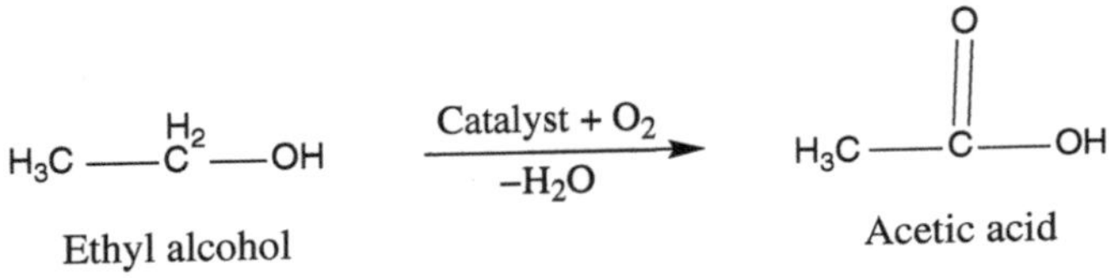

Figure 12.1 Oxidation of ethyl alcohol into acetic acid.

X-ray, γ-ray, etc. peroxides are formed that eventually converts into ROS forms. Ozone is an oxygen molecule containing oxygen–oxygen–oxygen (O—O—O) bonds that are highly unstable and hence easily forms ROS. Additionally, in life, stress is very common, and stressful conditions often lead to ROS formation. Aging is a natural phenomenon in life which also leads to the formation of ROS. In addition, certain human diseases such as Alzheimer, Parkinson, diabetes, and autoimmune diseases also leads to formation of ROS.

As discussed in the earlier chapters, functional groups present in amino acids are highly reactive and are subject to derivatizations/modifications. ROS generated in biological systems, due to various reasons, catalyzes the oxidation of functional groups present in amino acid residues of proteins. Peptides and proteins containing amino acid residues such as Cys, Met, Trp, and Tyr readily undergoes oxidation. For example, Met residue undergoes oxidation in the presence of ROS in a stepwise reaction in which Met residue first converts into methionine sulfoxide and then into methionine sulfone as shown in Figure 12.2. Similarly, Cys, Trp, and Tyr amino acid residues present in proteins are also oxidized by ROS (Glazer 1970).

Oxidation impacts the physicochemical properties of proteins (Aitken 2017). Since oxidation results in the addition of oxygen atom/s, hydrophobicity and hydrophilicity of proteins are impacted. For instance, oxidation results in the overall decrease in hydrophobicity and increase in hydrophilicity of proteins. In addition, overall charge and mass of proteins may also be impacted by

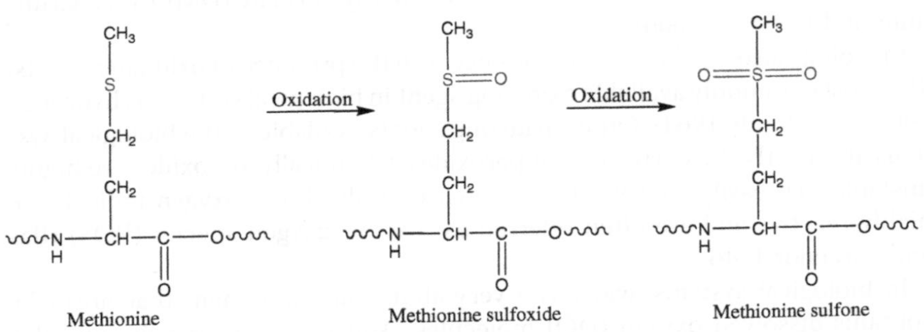

Figure 12.2 Oxidation of methionine into methionine sulfoxide and then into methionine sulfone. (*See insert for color representation of this figure.*)

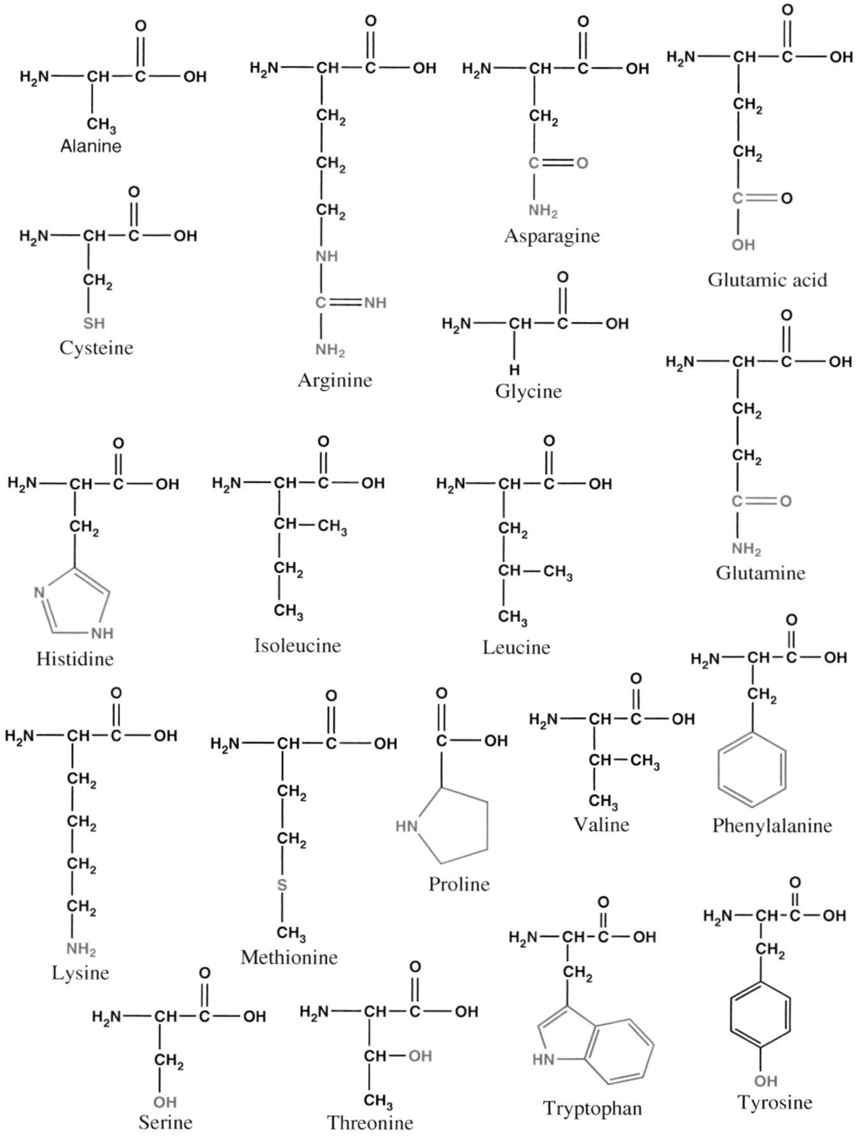

Alanine

Cysteine

Arginine

Asparagine

Glycine

Glutamic acid

Histidine

Isoleucine

Leucine

Glutamine

Lysine

Methionine

Proline

Valine

Phenylalanine

Serine

Threonine

Tryptophan

Tyrosine

**Figure 1.1** Chemical structure of 20 different amino acid residues commonly found in human proteins (reactive functional groups amenable for CTMs and PTMs are shown in red).

*Co- and Post-Translational Modifications of Therapeutic Antibodies and Proteins*, First Edition. T. Shantha Raju.

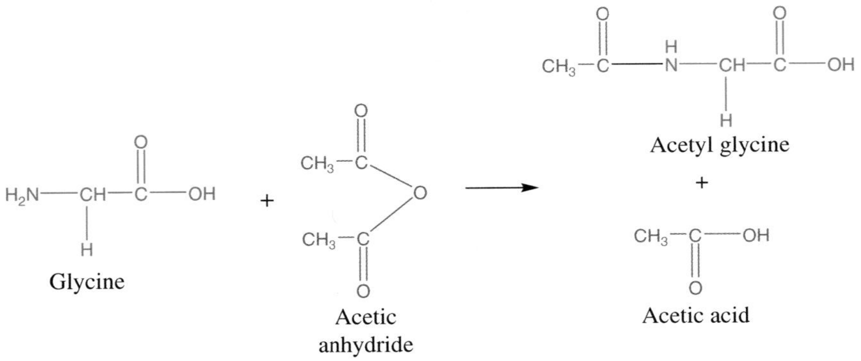

Glycine + Acetic anhydride → Acetyl glycine + Acetic acid

**Figure 2.1** N-acetylation of glycine to form *N*-acetylglycine using acetic anhydride as an acetylating reagent. The final products of the reaction are *N*-acetyl glycine and acetic acid.

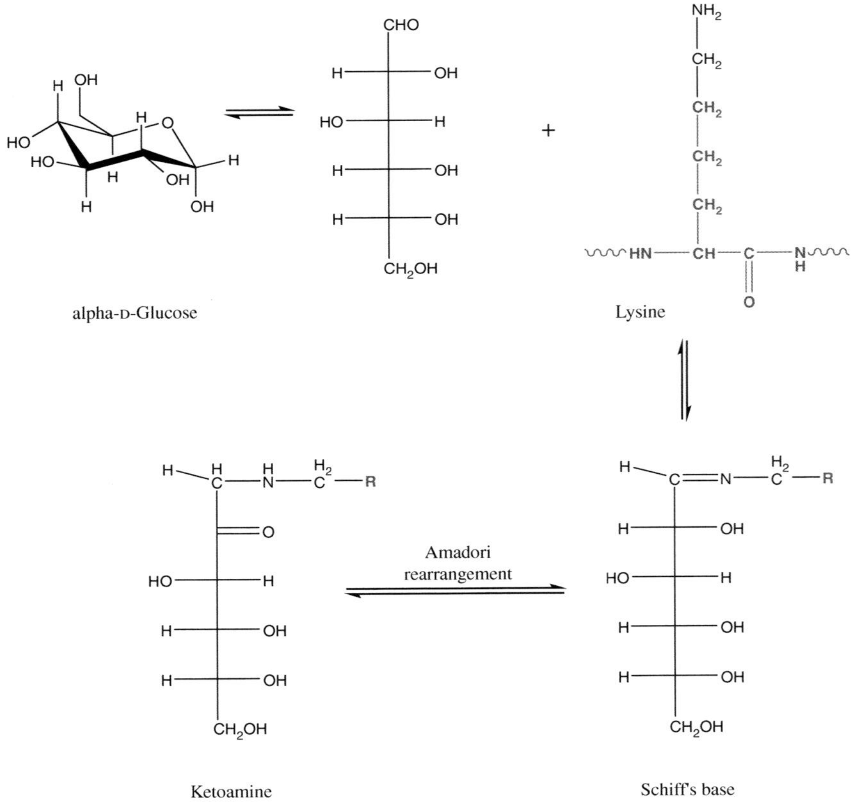

alpha-D-Glucose + Lysine ⇌ Schiff's base ⇌ (Amadori rearrangement) Ketoamine

**Figure 6.1** Mechanism of protein glycation.

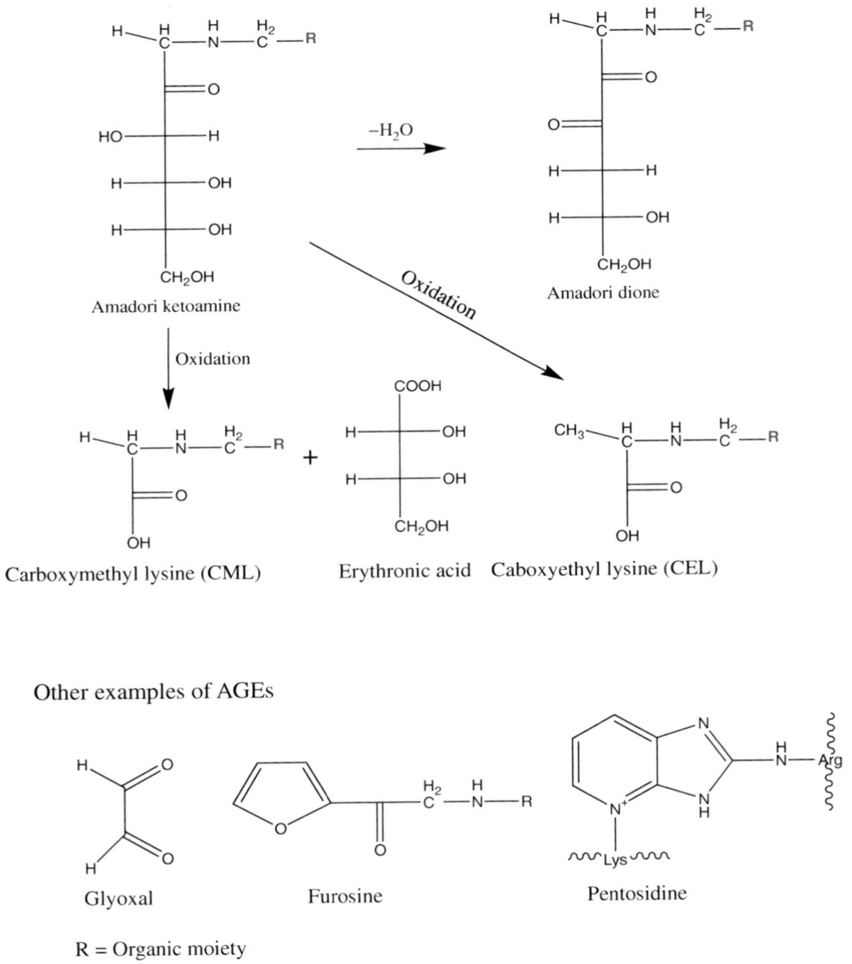

H H H $H_2$
C N C R
O
HO H
H OH
H OH
$CH_2OH$
Amadori ketoamine

$-H_2O$

H H H $H_2$
C N C R
O
O
H H
H OH
$CH_2OH$
Amadori dione

Oxidation

Oxidation

H H H $H_2$
C N C R
O
OH
Carboxymethyl lysine (CML)

+

COOH
H OH
H OH
$CH_2OH$
Erythronic acid

$CH_3$ H H $H_2$
C N C R
O
OH
Caboxyethyl lysine (CEL)

Other examples of AGEs

H O
H O
Glyoxal

O $H_2$ H
C N R
O
Furosine

N
H
N Arg
N
H
$N^+$
Lys
Pentosidine

R = Organic moiety

**Figure 6.2** Mechanism and chemical structure of few examples of advanced glycation end products (AGEs).

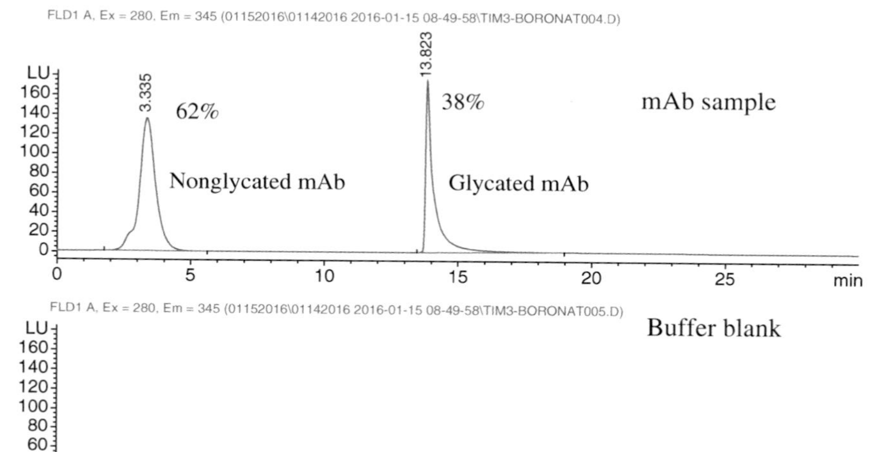


**Figure 6.3** Separation of glycated and non-glycated mAb using boronate affinity chromatography.

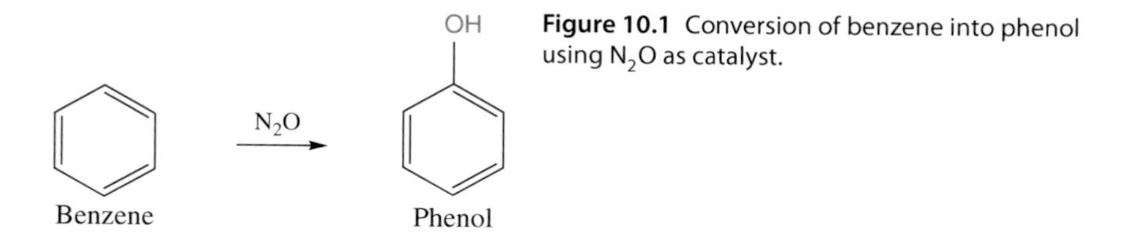


**Figure 10.1** Conversion of benzene into phenol using $N_2O$ as catalyst.

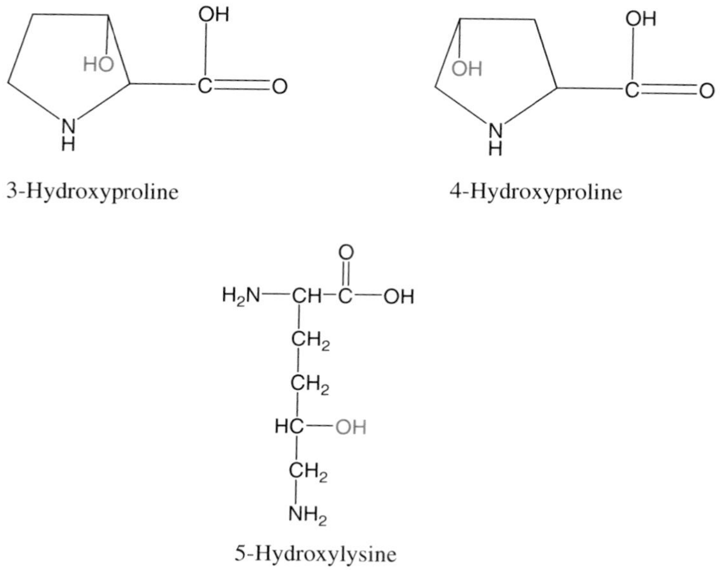

3-Hydroxyproline

4-Hydroxyproline

5-Hydroxylysine

Hydroxyl groups are in shown in red color

**Figure 10.2** Chemical structure of 3-hydroxyproline, 4-hydroxyproline, and 5-hydroxylysine.

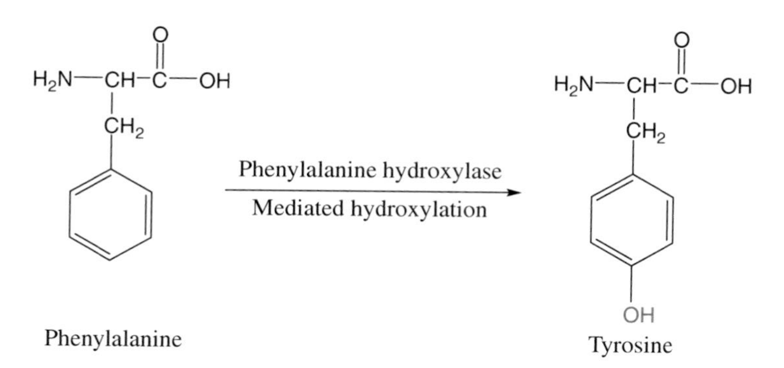


**Figure 10.4** Conversion of phenylalanine into tyrosine by phenylalanine hydroxylase mediated hydroxylation. The enzymatic reaction requires tetrahydrobiopterin, a non-heme iron and molecular oxygen ($O_2$) as cofactors for the catalytic reaction.

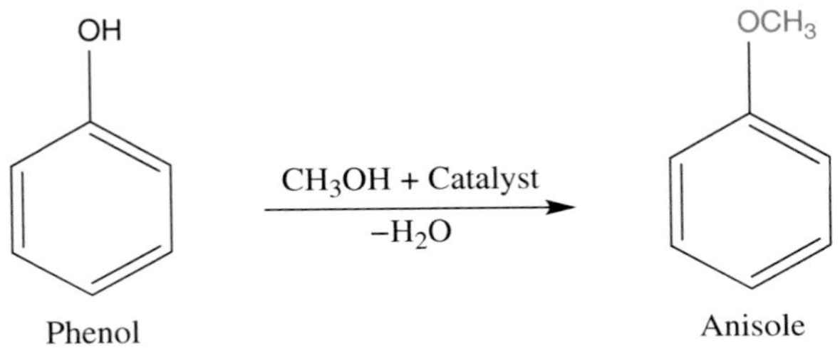

**Figure 11.1** Methylation of phenol into anisole.

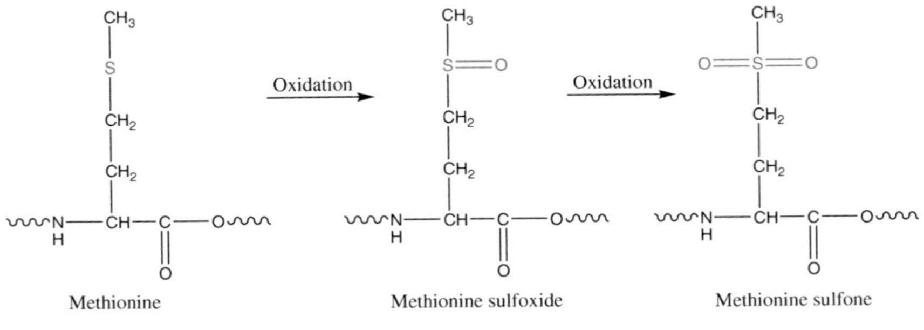

**Figure 12.2** Oxidation of methionine into methionine sulfoxide and then into methionine sulfone.

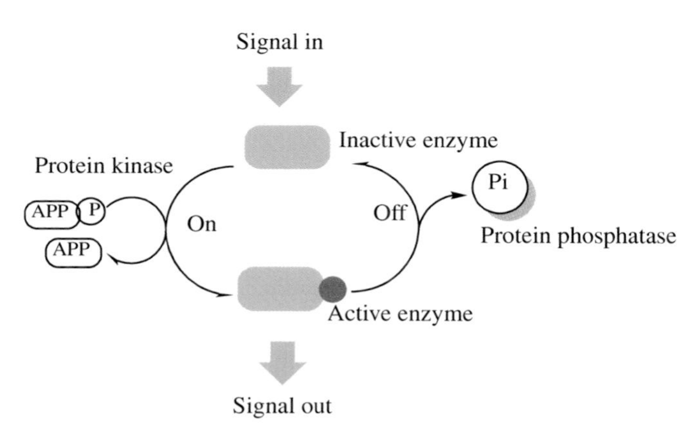


**Figure 13.1** Phosphorylation and dephosphyorylation as on and off switch for proteins.

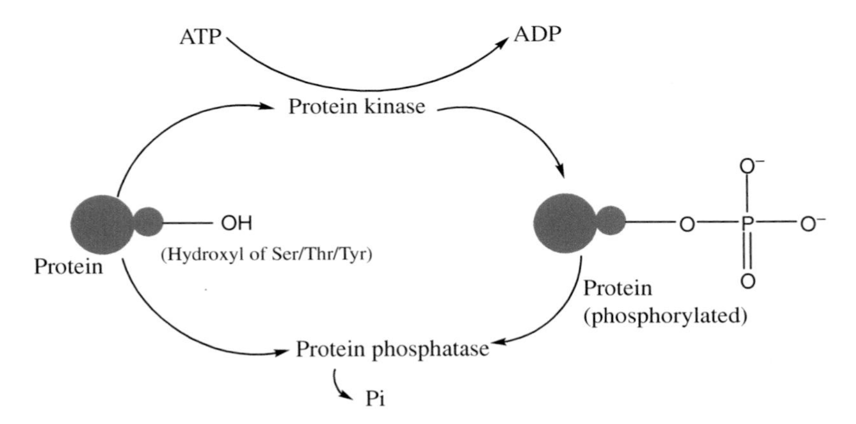


**Figure 13.3** Mechanism of protein phosphorylation mediated by kinases (Ser phosphorylation as an example).

oxidation. Oxidation may impact the hydrogen bond formation and hence may affect the higher-order structure of proteins.

Biological functions of proteins may also be impacted by oxidation. Enzymatic activity of enzymes may be affected by oxidation. Oxidation may also affect protein–protein interactions, receptor binding, antibody–antigen binding, etc. Pharmacokinetics, pharmacodynamics, immunogenicity, and toxicity of proteins may also be affected by oxidation (Dice 1987). Hence, protein oxidation may become a critical quality attribute (CQA) for recombinant therapeutic proteins (RTPs). Protein oxidation may result in increased sensitivity to proteases. This is because increased hydrophobicity may also result in the increased access of impacted peptide bonds to proteases.

In human life, increase in stress conditions, age, disease status, etc., may result in the increased levels of ROS (Gào and Schöttker 2017). The increase in ROS will enhance the rate of protein oxidation. The increase in oxidation of biological macromolecules such as proteins, and nucleic acids may affect the quality of life. Increased protein degradation as a result of oxidation may eventually lead to death. Hence, protein oxidation significantly affects the biological life cycle of living organisms. This chapter discusses these aspects as well as using antioxidants to control and manage protein oxidation in human life.

## Mechanism of Protein Oxidation

### Methionine (Met) Oxidation

Met is a sulfur containing amino acid residue found in many naturally occurring proteins. Met is an essential amino acid and act as an initiator of protein synthesis (Nakamoto 2009). Met contains *S*-methyl thioether functional group covalently linked to 2-aminobutanoic acid and the IUPAC name of Met is 2-amno-4-(methylthio)-butanoic acid. Thioethers can undergo chemical oxidation into sulfoxides that in turn can also be reduced by thioethers. Hence, single oxidation of Met into Met sulfoxide (MetO) is a reversible reaction. Further oxidation of MetO results in the formation of Met sulfone (Schöneich 2005). Similar to Cys, another sulfur containing amino acid, Met is involved in the antioxidant defense mechanism and regulation of cellular functions by reversible oxidation–reduction cycle (Hoshi and Heinemann 2001). Oxidation of Met residues occurs over a wide range of pH and is not necessarily a temperature-dependent reaction. The reduction of methionine sulfoxide into Met is mediated by methionine sulfoxide reductases (Boschi-Muller and Branlant 2014). Most organisms from bacteria to mammals contain thioredoxin-linked methionine sulfoxide reductases (Achilli et al. 2015). Some reductases reduce only the free amino acid, while some enzymes reduce both the free amino acid as well as the peptide bond linked internal amino acids. Oxidation of methionine into methionine sulfoxide and the subsequent reduction back to methionine of surface exposed residues affect the protein function. It has been established that in the case of α-antitrypsin which upon oxidation losses anti-elastase activity but regains the activity after reduction by reductases (Vogt 1995). Similar to α-antitrypsin many proteins containing methionine residues undergo oxidation and reduction mechanism involving the off–on switch on the biological activities (Vogt 1995).

Met also acts as an antioxidant to prevent oxidation of critical amino acid residues required for biological activities of proteins. For example, in the case of α-2-macroglobulin, surface-exposed Met residues undergo oxidation very easily and hence act as scavengers to prevent the oxidation of Trp residue that is critical for its biological activity (Kim et al. 2014). Met oxidation may be regulated by age as in the case of Calmodulin, which contains nine Met residues, isolated from the old rat brain that had six Met residues that were oxidized, whereas the same Calmodulin isolated from young rats had none of the Met residues oxidized (Song 2013). It has been discovered that the methionine sulfoxide reductase in young rats is more active than the old rats. In Alzheimer patients, it was determined that the activity of reductase is drastically reduced compared to normal human (Adams et al. 2017). Hence, Met oxidation is regulated by many factors including stress, age, disease status, etc. (Kim et al. 2014).

### Cysteine (Cys) Oxidation

Cys is a thiol containing amino acid and is similar to Ser in which the hydroxyl group of Ser is replaced by thiol group in Cys. Additionally, replacing the sulfur atom in Cys residues by selenium atom would result in seleniumcysteine (or selenocysteine) residues. The IUPAC name of Cys is 2-amino-3-sulfhydrylpropanoic acid and is classified as a nonessential amino acid. However, in the case of infants, elderly, and individuals with certain rare metabolic diseases and/or individuals suffering from malabsorption syndrome, Cys is an essential amino acid (Elshorbagy et al. 2012). The thiol functional group in Cys acts as a nucleophile and participates in many enzymatic reactions. Oxidation of thiol group converts Cys into cystine, a dimeric molecule containing disulfide bond between the two Cys residues. Oxidation of Cys by ROS is a three-step reaction as shown in the Figure 12.3.

In the first step, Cys converts into sulfinic acid, which in turn converts into sulfenic acid and then into sulfonic acid. Detecting sulfinic acid is difficult as it quickly undergoes either disulfide bond formation or further oxidizes into sulfenic acid and then into sulfonic acid. Recently, using Dimedone as a trapping reagent, conversion of cysteine into cysteine sulfinic acid was studied (Klomsiri et al. 2010; Ratnayake et al. 2013). These studies suggest that the cysteine oxidation by ROS may play important biological as well as pathological roles in living organisms. It has been speculated that the oxidative stress leads to cysteine oxidation by ROS that eventually leads to many acute and chronic diseases

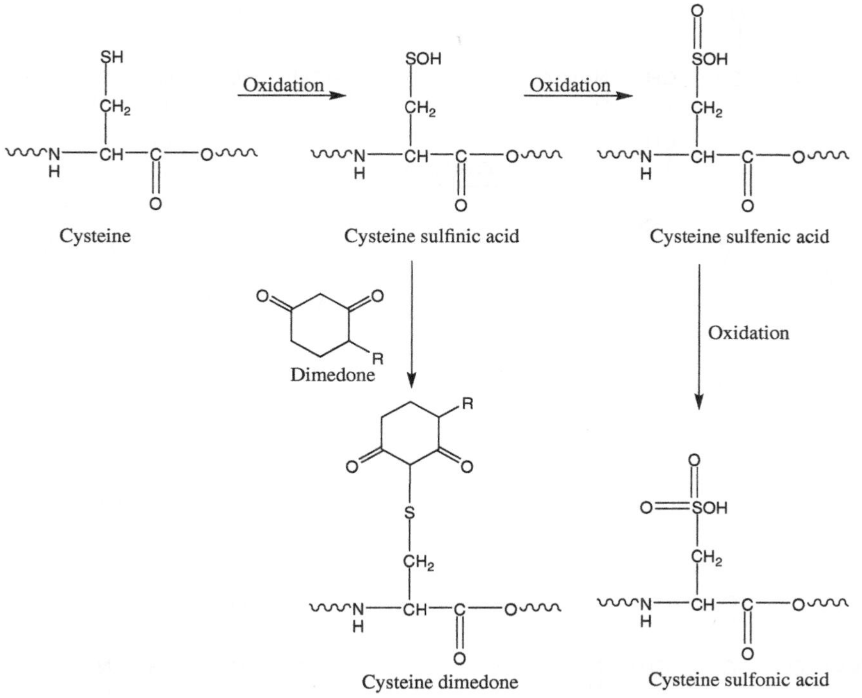

**Figure 12.3** Oxidation of cysteine residues.

(Crump et al. 2012). In addition, cysteine oxidation by ROS is involved in the normal aging process.

## Tryptophan (Trp) Oxidation

Trp is an essential amino acid containing indole side chain as a functional group. It has an α-amino group as well as an α-carboxyl group along with an indole side chain. The IUPAC name of tryptophan is (2*S*)-2-amino-3-(1*H*-indol-3-yl)propanoic acid and is abbreviated as Trp or W. Indole is a bicyclic aromatic compound with a heterocyclic ring structure. In indole, the aromatic six-membered benzene ring is fused to a heterocyclic nitrogen containing five-membered pyrrole ring. Because of its indole side chain, Trp is classified as a nonpolar aromatic amino acid. Trp is a precursor to serotonin and melatonin (Slominski et al. 2002). Serotonin is a neurotransmitter molecule and melatonin is a hormone. In addition, the indole moiety in Trp is an important naturally intrinsic fluorescent probe that can be used to study many biological properties of proteins (Zhang et al. 2017).

The pyrrole ring in Trp is prone to oxidation by ROS and other oxidizing agents (Ehrenshaft et al. 2015). Oxidation of Trp into *N*-formylkynurenine

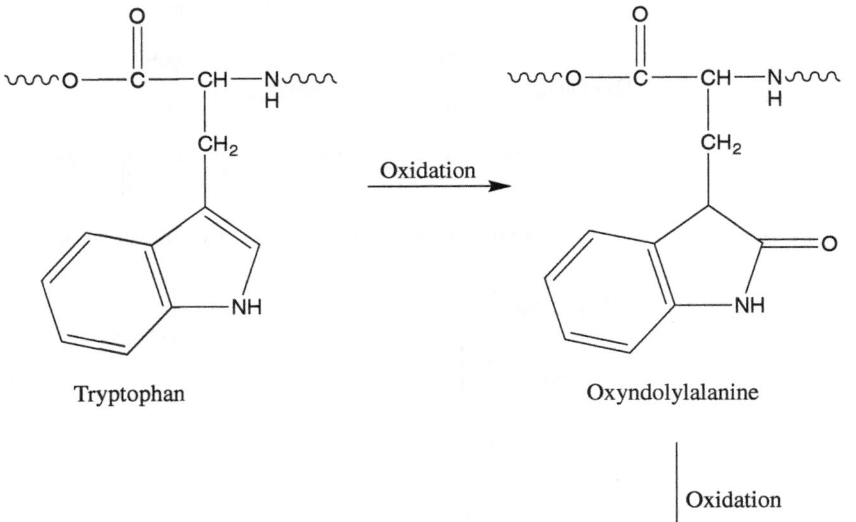

Figure 12.4 Oxidation of Trp residue into *N*-formylkynurenine which undergoes rearrangement into kynurenine.

is a multistep reaction as shown in the Figure 12.4. Initially, Trp oxidation converts it into oxindolylalanine that upon further oxidation converts into *N*-formylkynurenine. The *N*-formylkynurenine undergoes rearrangement to form kynurenine or quinolinic acid (Simat and Steinhart 1998).

Oxidation of Trp in lysozyme impairs its enzymatic activity (Holzman et al. 1968). Similarly, other proteins have been shown to be losing their biological functions due to oxidation of Trp residues (Zhang et al. 2003). A subset of proteins from cardiac mitochondria have been shown to be susceptible for oxidation of Trp residues representing potential hot spots for oxidative modification in respiring mitochondria (Rahim et al. 2016).

### Tyrosine (Tyr) Oxidation

Tyr is a hydrophilic amino acid containing 4-hydroxyphenyl functional group and hence is also called as 4-hydroxyphenylalanine. The IUPAC name of Tyr is (*S*)-tyrosine and is also abbreviated as Y. Tyr is a nonessential amino acid and contains a polar hydroxyl group covalently linked to an aromatic benzene ring. Phenylalanine is a precursor to Tyr and due to the presence of hydroxyl group, Tyr is more water soluble than its precursor amino acid. Also, because of the hydroxyl group, one Tyr molecule can form hydrogen bond with the carboxyl group of another Tyr molecule. In plants, tyrosine residues are involved in the photosynthesis and act as an electron donor in the oxidation of chlorophyll (Narzi et al. 2014).

Oxidation of Tyr residues by ROS is a multistep process that involves a complex mechanism as shown in Figure 12.5. Tyr also undergoes enzymatic oxidation by tyrosinase (Zhang and Zhou 2013). The oxidation of Tyr by ROS and tyrosinase involves different mechanism that leads to different oxidation products. In the case of ROS, Tyr converts into oxygen-free radical that is also referred as tyrosine phenoxyl radical. The free radical oxygen undergoes rearrangement followed by the formation of dityrosine as a major oxidation product (Giulivi et al. 2003). The tyrosine phenoxyl radical can also form other oxidation products such as 3-nitrosotyrosine, iminoxyl radical, and 3-nitrotyrosine as shown in Figure 12.5.

Tyrosinase is a unique bifunctional enzyme present in melanocytes and is responsible for formation of melanin (Niu and Aisa 2017). Tyrosinase is involved in the hydroxylation of Tyr into 3,4-dihydroxyphenylalanine (DOPA) and the subsequent oxidation of DOPA into DOPAquinone (Asanuma et al. 2003). The DOPAquinone is highly reactive that reacts with thiol groups to form thioether derivatives of DOPA. For example, DOPAquinone reacts with Cys to form 5-*S*-cysteinylDOPA and its 2-*S*-isomer. However, in the absence of thiol groups, DOPAquinone undergoes intramolecular nucleophilic addition reaction with amine group to form black melanin.

Dityrosine is very commonly found in proteins that underwent irradiation with UV and $\gamma$-rays. Tyr oxidation is also shown to be associated with human aging and various disease status as well as stress (Cacabelos 2017). Tyr oxidation is used as a biomarker to identify oxidatively damaged proteins. Such proteins also undergo selective and rapid proteolysis. Tyr oxidation may also play a critical role in oxidative damage during inflammation, atherosclerosis, and other autoimmune diseases (Mazzucchi et al. 2015).

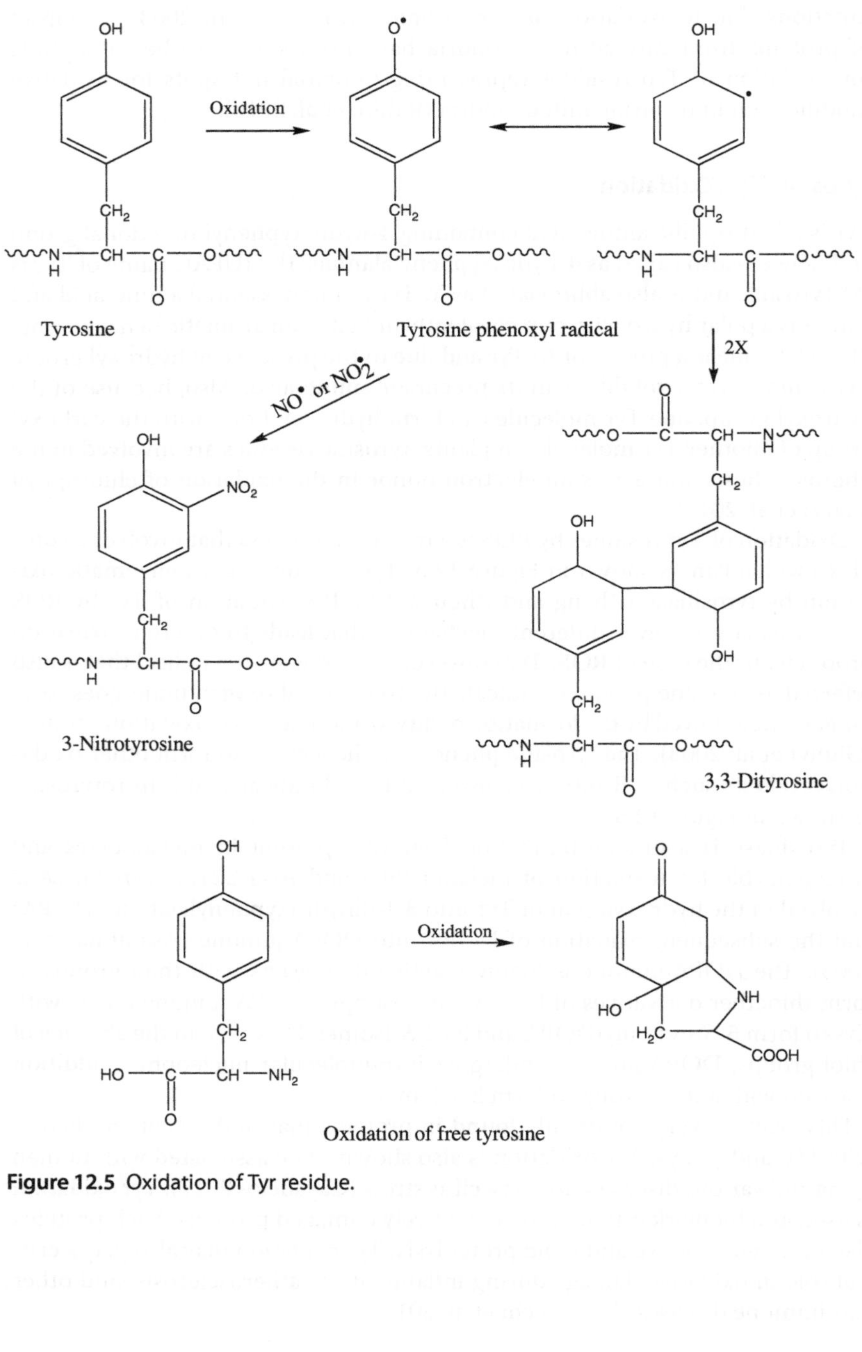

Oxidation

Tyrosine

Tyrosine phenoxyl radical

2X

$NO^{\bullet}$ or $NO_2^{\bullet}$

3-Nitrotyrosine

3,3-Dityrosine

Oxidation

Oxidation of free tyrosine

**Figure 12.5** Oxidation of Tyr residue.

### Oxidation of Other Amino Acid Residues and Protein Backbone

In addition to the amino acids described above, Arg, His, etc. residues may also undergo oxidation in the presence of ROS (Amici et al. 1989). Presence of the excessive amounts of ROS-generating oxidizing agents may also results in the oxidation of Ser, and Thr amino acid residues and eventually may also result in the oxidative degradation of the protein backbone (Davies 2016). Such a process leads to rapid destruction of biological systems that may ultimately lead to death of the living organisms.

### Oxidation in RTPs

RTPs may undergo oxidation during cell culture, purification, formulation, storage, etc. DO present in the cell culture medium may form ROS during manufacturing processes that may lead to oxidation of Met, Cys, Trp, Tyr residues present in RTPs (Lam et al. 1997). The DO may also lead to the formation of ROS during purification processes. For example, when the DO comes in contact with certain metal ions ROS is generated, which in turn oxidizes the susceptible amino acid residues in RTPs (Harris 2005). During and after formulation, the trace metal ions present in the buffer excipients may lead to the formation of ROS, which will result in oxidation of RTPs (Cleland et al. 1993). Oxidation of RTPs may also occur during storage either due to metal containers or contaminating metal ions present in storage containers. Formulation excipients such as Tween-20 may contain trace amounts of peroxides that would generate ROS and hence results in protein oxidation. Oxidation of RTPs may impact their biological functions, hence affecting their potency. In such a scenario, oxidation will impact the safety and efficacy of the drug substances and drug products. In such cases, oxidation of certain amino acid residues of RTPs may become CQAs. Few examples of oxidation affect the biological functions of RTPs and hence may become CQAs are discussed below. In addition, some examples, where oxidation may not impact the biological functions of RTPs and hence may not be CQAs of RTPs are also discussed below.

Recombinant human erythropoietin (rhEPO) and its biosimilar versions are being marketed by many biopharmaceutical companies to treat red blood cell anemia and related diseases (Bommer et al. 1990). rhEPO is similar to endogenous epogen (EPO) that signals bone marrow to produce more red blood cells (Adamson and Eschbach 1990). rhEPO is being produced in in vitro cell culture systems using recombinant DNA technology with mammalian cells as hosts (Elliott et al. 2008). The impact of oxidation of Met residues at position 54 on the structure, stability, and biological functions has been studied by Labrenz et al. (2008). These researchers carried out oxidation of Met-54 by treating rhEPO with tertiary butylhydroperoxide (TBHP) or hydrogen peroxide (HP). As a result of oxidation, a shift in fluorescence spectral wavelength was observed that was linear in response with the amount of oxidation of Met-54 residue. This change in fluorescence spectra was due to change in local change in environment near the Trp residue at position 51. However, circular dichroism (CD) studies and thermal melting temperature (Tm) studies showed no observable change in the secondary structure of rhEPO upon oxidation. But decrease in biological activity of oxidized rhEPO was observed. This decrease in biological activity correlates with the amount of Met-54 oxidation (Labrenz et al. 2008). Accordingly, oxidation of Met-54 residues may become CQA for rhEPO.

Recombinant human tissue-type plasminogen activator (rhtPA) is produced using Chinese hamster ovary (CHO) cells and is being marketed to treat acute ischemic stroke that is caused by blood clot in the brain blood vessels (Chloupek et al. 1989). rhtPA contains five Met residues of which three residues are surface-exposed (Paborsky and Harris 1990). The surface-exposed Met residues were oxidized by treating rhtPA with TBHP to Met-sulfoxide (Keck 1996). The oxidation of surface-exposed Met residues did not affect the biological functions of rhtPA (O'Connor et al. 1994). Hence, in the case of rhtPA, the Met oxidation is not necessarily a CQA.

Recombinant human interpheron γ-1b (rhIFN-γ) is used to treat chronic granulomatous disease (CGD). The GCD is a genetic disorder and is usually diagnosed in childhood. GCD affects the immune system and body's ability to fight diseases (Short et al. 1995). rhIFN-γ contains 5 Met residues of which two are surface-exposed (Keck 1996). Oxidation of surface-exposed Met residues with TBHP converted them into Met-sulfoxide (Keck 1996). The oxidation of the two surface-exposed Met residues did not alter the structural characteristics. The biological properties of rhIFN-γ also did not change upon oxidation suggesting that the Met oxidation is not necessarily a CQA for rhIFN-γ.

Antithrombin (AT) is a serine protease inhibitor and plasma-derived AT contains low levels of oxidation (Van Patten et al. 1999). AT contains four Met residues, two adjacent Met residues, Met 314 and Met 315, near the reactive site loop and two exposed Met residues, Met 17 and Met 20, at the border of heparin-binding region. To study the impact of Met oxidation on its biological activity, recombinant form of AT was treated with hydrogen peroxide. The adjacent Met residues at 314 and 315 were susceptible to chemical oxidation. However, oxidation of Met 314 and Met 315 did not impact the biological activity of AT (Van Patten et al. 1999). At higher concentrations of peroxides, Met 17 and Met 20 were also oxidized that affected the heparin-binding affinity of AT. Detailed structural analysis showed that the oxidation of Met 17 affects the heparin-induced conformation change of AT. Since, Met 17 oxidation may not happen in normal AT preparations, Met oxidation in AT may not be a CQA.

### Met Oxidation in mAbs

Majority of currently marketed monoclonal antibodies (mAbs) are $IgG_1k$ isotype (Ecker et al. 2015). Each heavy chain in the Fc region of $IgG_1$ isotypes of mAbs contain three conserved Met residues at positions 252, 358, and 428. The Met 252 is in $CH_2$ domain, whereas Met 358 and Met 428 are located in the $CH_3$ domain. However, both Met 252 and 428 are positioned at the junction of $CH_2$ and $CH_3$ domains and are involved in binding of $IgG_1$ isotype to FcRn. Hence, both Met 252 and 428 residues play key roles in the serum half-life of antibody molecules (Saxena and Wu 2016). However, Met-358 is not involved in the FcRn binding and hence do not play any role in the antibody serum half-life. Oxidation of Met-252 and 428 reduces antibody binding to FcRn which also reduces the antibody serum half-life of mAbs (Hui et al. 2015). Site-directed mutagenesis studies in conjunction with oxidation studies suggest that the oxidation of Met-252 impacts the antibody binding to FcRn more than the oxidation of Met 428 (Gao et al. 2015). Further, oxidation of Met 428 impacts the thermal stability of antibody more than the oxidation of Met 252. However, oxidation of Met 358 does not impact the antibody binding to FcRn as well as the thermal stability (Gao et al. 2015). Accordingly, Met 252 and Met 428 may be CQA's for $IgG_1$ isotype based mAb molecules.

## Analytical Methods to Measure Protein Oxidation

Oxidation of Met, Cys, Trp, and Tyr residues of proteins is a covalent modification that impact the hydrophobicity and hydrophilicity. Additionally, oxidation may also impact the solubility and aggregation properties of proteins including mAbs. Oxidation results in addition of an oxygen atom thus increasing the overall molecular weight as well as the heterogeneity of the protein samples. Hence, protein oxidation can be analyzed using many analytical methods such as high performance liquid chromatography (HPLC), capillary electrophoresis (CE), mass spectrometry (MS), etc. methods. Both normal phase and reversed phase HPLC methods have been used to separate and quantitate oxidized proteins. Since, oxidation results in the overall surface hydrophobicity, hydrophobic interaction chromatography (HIC) methods can also be used to separate and quantitate oxidized proteins (Boyd et al. 2011). In addition, peptide mapping methods have also been successfully used to identify the sites of oxidation in RTPs (Harris 2005; Keck 1996). Liquid chromatography–mass spectrometry (LC–MS) methods have also been employed to detect and quantitate the sites of oxidation in protein samples (Rykær et al. 2017). Both CE and capillary electrophoresis–mass spectrometry (CE-MS) methods were also used to measure oxidation levels in proteins (Lew et al. 2015). In addition, enzyme-linked immunosorbent assay (ELISA) methods have also been employed to detect and quantitate oxidation levels in proteins (Mirarabshahi et al. 2012). However, ELISA methods show high variability during the sample analysis.

Oxidation of Met residues in the Fc domains affect the FcRn binding and hence the pharmacokinetic properties of mAbs. In addition, if Met, Trp, and Try residues are present in the complementarity-determining region (CDR) regions, oxidation of these residues might impact antibody binding to antigens. Hence, highly sensitive methods have been developed to detect and to quantitate levels of oxidation in mAbs. In addition to LC-MS, HPLC, CE methods, biophysical methods such as CD, near and far UV methods have also been employed to measure the change in protein conformation as a result of oxidation (Islam et al. 2017). Oxidation in mAbs can be detected by measuring intact molecular weight using high-resolution MS techniques. However, identification of sites and/or domain of oxidation is not possible by intact molecular weight measurement using MS. Digestion of mAbs with specific proteases such as papain, pepsin have been adopted to generate Fab and Fc fragments (Raju and Scallon 2007). MS analysis of the Fab and Fc fragments would help to determine the domains of oxidation. Recently, IdeS (IgG degrading enzyme from *Streptococcus pyogenes*) has been employed to digest mAbs to generate $F(ab')_2$ and truncated Fc fragments (An et al. 2014). Reduction of the enzyme digests results in the separation of heavy chain and light chain which can be analyzed by HPLC followed by MS analysis. Such methods are also employed as identity tests for mAb products (An et al. 2014).

## References

Achilli, C., Ciana, A., and Minetti, G. (2015). The discovery of methionine sulfoxide reductase enzymes: an historical account and future perspectives. *Biofactors* 41 (3): 135–152. https://doi.org/10.1002/biof.1214.

Adams, S.L., Benayoun, L., Tilton, K. et al. (2017). Methionine sulfoxide reductase-B3 (MsrB3) protein associates with synaptic vesicles and its expression changes in the hippocampi of Alzheimer's disease patients. *J. Alzheimers Dis.* 60 (1): 43–56. https://doi.org/10.3233/JAD-170459.

Adamson, J.W. and Eschbach, J.W. (1990). The use of recombinant human erythropoietin in humans. *Ciba Found. Symp.* 148: 186–195; discussion 195–200.

Aitken, R.J. (2017). Reactive oxygen species as mediators of sperm capacitation and pathological damage. *Mol. Reprod. Dev.* https://doi.org/10.1002/mrd.22871.

Amici, A., Levine, R.L., Tsai, L., and Stadtman, E.R. (1989). Conversion of amino acid residues in proteins and amino acid homopolymers to carbonyl derivatives by metal-catalyzed oxidation reactions. *J. Biol. Chem.* 264 (6): 3341–3346.

An, Y., Zhang, Y., Mueller, H.M. et al. (2014). A new tool for monoclonal antibody analysis: application of IdeS proteolysis in IgG domain-specific characterization. *MAbs* 6 (4): 879–893. https://doi.org/10.4161/mabs.28762.

Asanuma, M., Miyazaki, I., and Ogawa, N. (2003). Dopamine- or L-DOPA-induced neurotoxicity: the role of dopamine quinone formation and tyrosinase in a model of Parkinson's disease. *Neurotox. Res.* 5 (3): 165–176.

Bommer, J., Barth, H.P., and Schwöbel, B. (1990). rhEPO treatment of anemia in uremic patients. *Contrib. Nephrol.* 87: 59–67.

Boschi-Muller, S. and Branlant, G. (2014). Methionine sulfoxide reductase: chemistry, substrate binding recycling process and oxidase activity. *Bioorg. Chem.* 57: 222–230. https://doi.org/10.1016/j.bioorg.2014.07.002.

Boyd, D., Kaschak, T., and Yan, B. (2011). HIC resolution of an $IgG_1$ with an oxidized Trp in a complementarity determining region. *J. Chromatogr. B Anal. Technol. Biomed. Life Sci.* 879 (13–14): 955–960. https://doi.org/10.1016/j.jchromb.2011.03.006.

Cacabelos, R. (2017). Parkinson's disease: from pathogenesis to pharmacogenomics. *Int. J. Mol. Sci.* 18 (3): https://doi.org/10.3390/ijms18030551.

Chloupek, R.C., Harris, R.J., Leonard, C.K. et al. (1989). Study of the primary structure of recombinant tissue plasminogen activator by reversed-phase high-performance liquid chromatographic tryptic mapping. *J. Chromatogr.* 463 (2): 375–396.

Cleland, J.L., Powell, M.F., and Shire, S.J. (1993). The development of stable protein formulations: a close look at protein aggregation, deamidation, and oxidation. *Crit. Rev. Ther. Drug Carrier Syst.* 10 (4): 307–377.

Crimmins, M.T. and DeBaillie, A.C. (2006). Enantioselective total synthesis of bistramide A. *J. Am. Chem. Soc.* 128 (15): 4936–4937.

Crump, K.E., Juneau, D.G., Poole, L.B. et al. (2012). The reversible formation of cysteine sulfenic acid promotes B-cell activation and proliferation. *Eur. J. Immunol.* 42 (8): 2152–2164. https://doi.org/10.1002/eji.201142289.

Davies, M.J. (2016). Protein oxidation and peroxidation. *Biochem. J.* 473 (7): 805–825. https://doi.org/10.1042/BJ20151227.

Dice, J.F. (1987). Molecular determinants of protein half-lives in eukaryotic cells. *FASEB J.* 1 (5): 349–357; Erratum in: (1988). *FASEB J*. 2 (7): 2262.

Ecker, D.M., Jones, S.D., and Levine, H.L. (2015). The therapeutic monoclonal antibody market. *MAbs* 7 (1): 9–14. https://doi.org/10.4161/19420862.2015.989042.

Ehrenshaft, M., Deterding, L.J., and Mason, R.P. (2015). Tripping up Trp: modification of protein tryptophan residues by reactive oxygen species, modes of detection and biological consequences. *Free Radical Biol. Med.* 89: 220–228. https://doi.org/10.1016/j.freeradbiomed.2015.08.003.

Elliott, S., Pham, E., and Macdougall, I.C. (2008). Erythropoietins: a common mechanism of action. *Exp. Hematol.* 36 (12): 1573–1584. https://doi.org/10.1016/j.exphem.2008.08.003.

Elshorbagy, A.K., Smith, A.D., Kozich, V., and Refsum, H. (2012). Cysteine and obesity. *Obesity (Silver Spring)* 20 (3): 473–481. https://doi.org/10.1038/oby.2011.93.

Gào, X. and Schöttker, B. (2017). Reduction–oxidation pathways involved in cancer development: a systematic review of literature reviews. *Oncotarget* 8 (31): 51888–51906. https://doi.org/10.18632/oncotarget.17128.

Gao, X., Ji, J.A., Veeravalli, K. et al. (2015). Effect of individual Fc methionine oxidation on FcRn binding: Met252 oxidation impairs FcRn binding more profoundly than Met428 oxidation. *J. Pharm. Sci.* 104 (2): 368–377. https://doi.org/10.1002/jps.24136.

Giulivi, C., Traaseth, N.J., and Davies, K.J. (2003). Tyrosine oxidation products: analysis and biological relevance. *Amino Acids* 25 (3–4): 227–232.

Glazer, A.N. (1970). Specific chemical modification of proteins. *Annu. Rev. Biochem.* 39: 101–130.

Harris, R.J. (2005). Heterogeneity of recombinant antibodies: linking structure to function. *Dev. Biol. (Basel)* 122: 117–127.

Holzman, R.S., Gardner, D.E., and Coffin, D.L. (1968). In vivo inactivation of lysozyme by ozone. *J. Bacteriol.* 96 (5): 1562–1566.

Hoshi, T. and Heinemann, S. (2001). Regulation of cell function by methionine oxidation and reduction. *J. Physiol.* 531 (Pt. 1): 1–11.

Hui, A., Lam, X.M., Kuehl, C. et al. (2015). Kinetic modeling of methionine oxidation in monoclonal antibodies from hydrogen peroxide spiking studies. *PDA J. Pharm. Sci. Technol.* 69 (4): 511–525. https://doi.org/10.5731/pdajpst.2015.01059.

Ilas, N.A. and Surgenor, D.M. (1946). Studies in organic peroxides; t-butyl peresters. *J. Am. Chem. Soc.* 68: 642.

Islam, S., Moinuddin, Mir, A.R. et al. (2017). Glycation, oxidation and glycoxidation of IgG: a biophysical, biochemical, immunological and hematological study. *J. Biomol. Struct. Dyn.* 12: 1–17. https://doi.org/10.1080/07391102.2017.1365770.

Keck, R.G. (1996). The use of t-butyl hydroperoxide as a probe for methionine oxidation in proteins. *Anal. Biochem.* 236 (1): 56–62.

Kim, G., Weiss, S.J., and Levine, R.L. (2014). Methionine oxidation and reduction in proteins. *Biochim. Biophys. Acta* 1840 (2): 901–905. https://doi.org/10.1016/j.bbagen.2013.04.038.

Klomsiri, C., Nelson, K.J., Bechtold, E. et al. (2010). Use of dimedone-based chemical probes for sulfenic acid detection evaluation of conditions affecting probe incorporation into redox-sensitive proteins. *Methods Enzymol.* 473: 77–94. https://doi.org/10.1016/S0076-6879(10)73003-2.

Labrenz, S.R., Calmann, M.A., Heavner, G.A., and Tolman, G. (2008). The oxidation of methionine-54 of epoetinum alfa does not affect molecular structure or stability but does decrease biological activity. *PDA J. Pharm. Sci. Technol.* 62 (3): 211–223.

Lam, X.M., Yang, J.Y., and Cleland, J.L. (1997). Antioxidants for prevention of methionine oxidation in recombinant monoclonal antibody HER2. *J. Pharm. Sci.* 86 (11): 1250–1255.

Lew, C., Gallegos-Perez, J.L., Fonslow, B. et al. (2015 Mar). Rapid level-3 characterization of therapeutic antibodies by capillary electrophoresis electrospray ionization mass spectrometry. *J. Chromatogr. Sci.* 53 (3): 443–449. https://doi.org/10.1093/chromsci/bmu229.

Massey, V. and Veeger, C. (1963). Biological oxidations. *Annu. Rev. Biochem.* 32: 579–638.

Mazzucchi, S., Frosini, D., Bonuccelli, U., and Ceravolo, R. (2015). Current treatment and future prospects of dopa-induced dyskinesias. *Drugs Today (Barc.)* 51 (5): 315–329. https://doi.org/10.1358/dot.2015.51.5.2313726.

Mirarabshahi, P., Abdelatti, M., and Krilis, S. (2012). Post-translational oxidative modification of β2-glycoprotein I and its role in the pathophysiology of the antiphospholipid syndrome. *Autoimmun. Rev.* 11 (11): 779–780. https://doi.org/10.1016/j.autrev.2011.12.007.

Moody, G.J. (1964). The action of hydrogen peroxide on carbohydrates and related compounds. *Adv. Carbohydr. Chem.* 19: 149–179.

Morton, R.A. (1965). Quinones as a biological catalysts. *Endeavour* 24: 81–86.

Nakamoto, T. (2009). Evolution and the universality of the mechanism of initiation of protein synthesis. *Gene* 432 (1–2): 1–6. https://doi.org/10.1016/j.gene.2008.11.001.

Narzi, D., Bovi, D., and Guidoni, L. (2014). Pathway for Mn-cluster oxidation by tyrosine-Z in the S2 state of photosystem II. *Proc. Natl. Acad. Sci. U.S.A.* 111 (24): 8723–8728. https://doi.org/10.1073/pnas.1401719111.

Niu, C. and Aisa, H.A. (2017). Upregulation of melanogenesis and tyrosinase activity: potential agents for vitiligo. *Molecules* 22 (8): https://doi.org/10.3390/molecules22081303.

O'Connor, J.V., Keck, R.G., Harris, R.J., and Field, M.J. (1994). New techniques in protein chemistry. *Dev. Biol. Stand.* 83: 165–173.

Paborsky, L.R. and Harris, R.J. (1990). Post-translational modifications of recombinant human tissue factor. *Thromb. Res.* 60 (5): 367–376.

Rahim, R.S., Chen, M., Nourse, C.C. et al. (2016). Mitochondrial changes and oxidative stress in a mouse model of Zellweger syndrome neuropathogenesis. *Neuroscience* 334: 201–213. https://doi.org/10.1016/j.neuroscience.2016.08.001.

Raju, T.S. and Scallon, B. (2007). Fc glycans terminated with *N*-acetylglucosamine residues increase antibody resistance to papain. *Biotechnol. Prog.* 23 (4): 964–971.

Ratnayake, S., Dias, I.H., Lattman, E., and Griffiths, H.R. (2013). Stabilising cysteinyl thiol oxidation and nitrosation for proteomic analysis. *J. Proteomics* 92: 160–170. https://doi.org/10.1016/j.jprot.2013.06.019.

Rykær, M., Svensson, B., Davies, M.J., and Hägglund, P. (2017). Unrestricted mass spectrometric data analysis for identification, localization, and quantification of oxidative protein modifications. *J. Proteome Res.* https://doi.org/10.1021/acs.jproteome.7b00330.

Saxena, A. and Wu, D. (2016). Advances in therapeutic Fc engineering – modulation of IgG-associated effector functions and serum half-life. *Front. Immunol.* 7: 580. https://doi.org/10.3389/fimmu.2016.00580.

Schöneich, C. (2005). Methionine oxidation by reactive oxygen species: reaction mechanisms and relevance to Alzheimer's disease. *Biochim. Biophys. Acta* 1703 (2): 111–119.

Short, S.M., Rubas, W., Paasch, B.D., and Mrsny, R.J. (1995). Transport of biologically active interferon-gamma across human skin in vitro. *Pharm. Res.* 12 (8): 1140–1145.

Simat, T.J. and Steinhart, H. (1998). Oxidation of free tryptophan and tryptophan residues in peptides and proteins. *J. Agric. Food Chem.* 46 (2): 490–498.

Slominski, A., Semak, I., Pisarchik, A. et al. (2002). Conversion of L-tryptophan to serotonin and melatonin in human melanoma cells. *FEBS Lett.* 511 (1-3): 102–106.

Song, Y.H. (2013). A memory molecule, Ca(2+)/calmodulin-dependent protein kinase II and redox stress; key factors for arrhythmias in a diseased heart. *Korean Circ. J.* 43 (3): 145–151. https://doi.org/10.4070/kcj.2013.43.3.145.

Van Patten, S.M., Hanson, E., Bernasconi, R. et al. (1999). Oxidation of methionine residues in antithrombin. Effects on biological activity and heparin binding. *J. Biol. Chem.* 274 (15): 10268–10276.

Vogt, W. (1995). Oxidation of methionyl residues in proteins: tools, targets, and reversal. *Free Radical Biol. Med.* 18 (1): 93–105.

Zhang, H. and Zhou, Q. (2013). Tyrosinase inhibitory effects and antioxidative activities of saponins from *Xanthoceras Sorbifolia* nutshell. *PLoS ONE* 8 (8): e70090. https://doi.org/10.1371/journal.pone.0070090.

Zhang, H., Andrekopoulos, C., Joseph, J. et al. (2003). Bicarbonate-dependent peroxidase activity of human Cu, Zn-superoxide dismutase induces covalent aggregation of protein: intermediacy of tryptophan-derived oxidation products. *J. Biol. Chem.* 278 (26): 24078–24089.

Zhang, H., Chen, H., Pan, S. et al. (2017). Development of an optical sensor for chlortetracycline detection based on the fluorescence quenching of L-tryptophan. *Luminescence* https://doi.org/10.1002/bio.3393.

# 13

# Phosphorylation of Proteins

## Introduction

A molecular entity with a chemical formula $-PO_3^{-2}$ is called phosphoryl group, which may exist in different ionic state in nature (Morrison and Heyde 1972). A phosphoryl group should not be confused with a phosphate group ($-PO_4^{-3}$). A phosphoryl group is usually attached to other atoms or molecules. The biochemical reaction involving the transfer of phosphoryl group from a phosphate containing molecule to another molecule is called phosphorylation. In a biological system, phosphorylation usually occurs between a phosphate containing molecule and a hydroxyl group containing molecule and is a post-translational modification (PTM). In biochemical reactions, phosphorylation is also mediated by enzymes called kinases and dephosphorylation is mediated by phosphatases (Oliva and Hassan 2017).

Protein phosphorylation was first discovered in 1906 by Phoebus Levene at Rockefeller Institute for Medical Research (Levene and Alsberg 1906). Dr. Levene identified the existence of phosphate group in vitellin protein that is also called as phovitin (Levene and Alsberg 1906). During the year 1932, Levene, in collaboration with Fritz Lipmann, discovered phosphoserine in casein, a milk protein (Lipmann and Levene 1932). In 1940, Carl Cori, Gerti Cori, and Arda Green identified that the enzyme "phosphorylase" exists in active and inactive forms (Keller and Cori 1953). They named the active form as phosphorylase A and the inactive form as phosphorylase B (Cowgill and Cori 1955). It was later discovered that the phosphorylase A is active in the absence of 5′AMP, whereas phosphorylase B is active in the presence of 5′AMP (Helmreich et al. 1967). However, later it was discovered that these two are interconvertible enzymes and their interconversion is mediated by another enzyme which is now called as phosphatase (Sasai 1965). In 1954, George Burnett and Eugene P. Kennedy identified the enzymatic process of protein phosphorylation by kinases (Burnett and Kennedy 1954). During 1955–1956, Krebs and Fisher, and also Sutherland and Wosilait identified that the interconversion of phosphorylase A to phosphorylase B involves

*Co- and Post-Translational Modifications of Therapeutic Antibodies and Proteins*, First Edition. T. Shantha Raju.

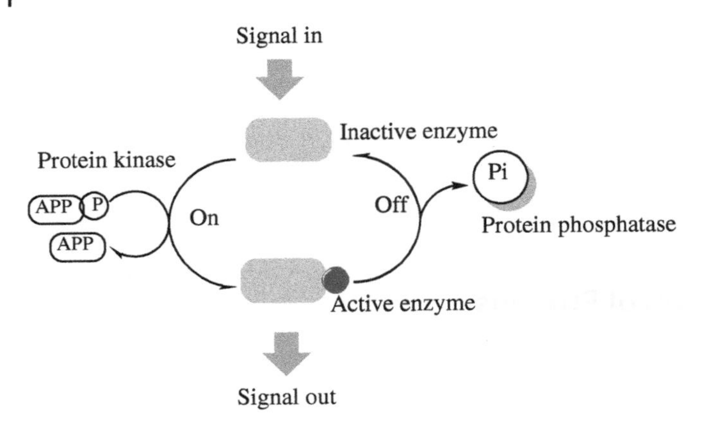


**Figure 13.1** Phosphorylation and dephosphorylation as on and off switch for proteins. (*See insert for color representation of this figure.*)

the phosphorylation and dephosphorylation of proteins (Fischer and Krebs 1955; Sutherland and Wosilait 1955). Now it is well established that the phosphorylation and dephosphorylation reactions are mediated by kinases and phosphatases, respectively.

In essence, protein phosphorylation is a reversible process and is common among all living organisms including animals, plants, fungi, bacteria, archaea, etc. By some estimation approximately 30% of all proteins from eukaryotes are phosphorylated. Phosphorylation has been described as an on or off switch for biological functions for many proteins. For example, biological activity of many enzymes and receptors are switched on by phosphorylation and are switched off by dephosphorylation as shown in Figure 13.1 (Krebs and Beavo 1979).

## Mechanism of Protein Phosphorylation in Living Cells

Protein phosphorylation is a very common PTM that is involved in regulating protein functions. However, protein phosphorylation in eukaryotes only occurs on the side chains of amino acids containing free hydroxyl groups. Amino acids such as Ser, Thr, and Tyr residues of proteins carry free hydroxyl groups. Hence, phosphorylation is very commonly found in eukaryotic proteins containing Ser, Thr, and Tyr residues. Very rarely in some eukaryotic proteins and most commonly in prokaryotic proteins, phosphorylation is also found on His, Lys, and aspartate residues. Protein phosphorylation may follow two types of mechanisms, one is mediated by magnesium ion ($Mg^{2+}$) and the other is mediated by protein kinases. For both mechanisms, universal phosphoryl donor is adenosine triphosphate (ATP).

### $Mg^{2+}$ Mediated Mechanism of Protein Phosphorylation

The hydroxyl functional group of Ser, Thr, and Tyr amino acid residues acts as a nucleophilic group that reacts with the terminal phosphate group of ATP, a universal phosphoryl donor in biological systems. The reaction between the hydroxyl groups and the terminal phosphate groups results in the transfer of phosphate group from ATP to amino acid residues and is mediated by magnesium as shown in Figure 13.2. Magnesium ($Mg^{2+}$) chelates the g and b phosphoryl group on ATP, thus lowering the energy threshold for the transfer of phosphate group to amino acid residues replacing the hydroxyl groups.

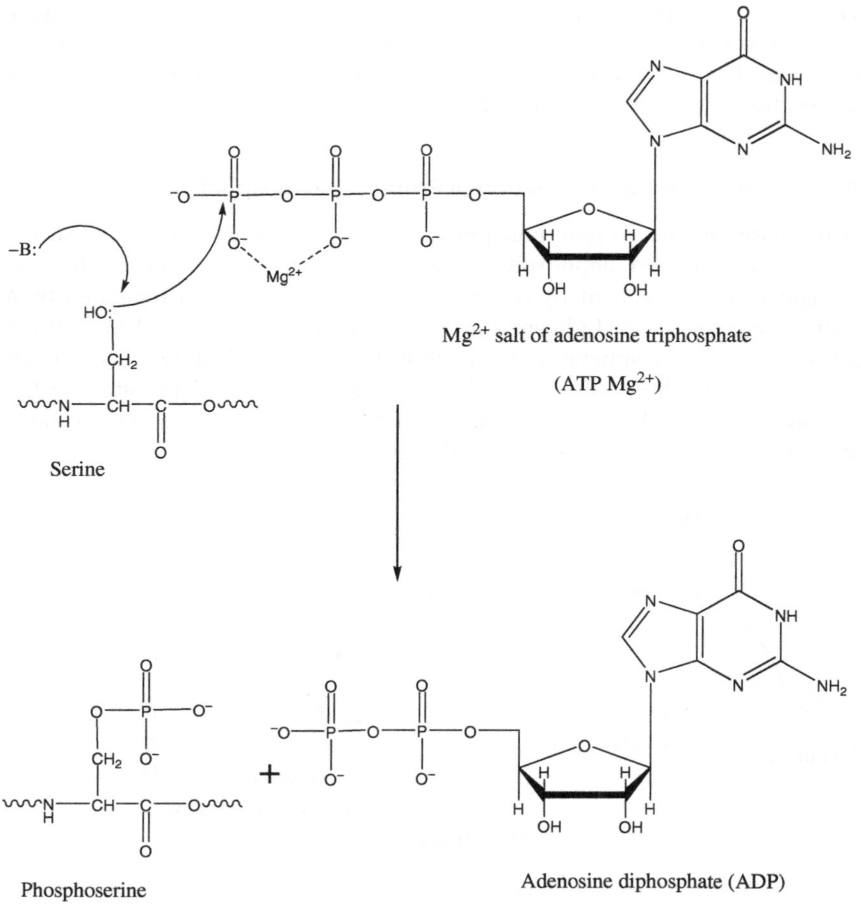

**Figure 13.2** Mechanism of protein phosphorylation in the presence of magnesium ($Mg^{2+}$) ions.

The reaction is nonreversible and a large amount of free energy is released due to the breakage of phosphate–phosphate bond in ATP. Also, during the reaction, ATP converts into adenosine diphosphate (ADP) as shown in Figure 13.2.

### Protein Kinase-Based Mechanism of Protein Phosphorylation

In the living cells, phosphorylation of proteins is also catalyzed by a variety of large number of enzymes. These enzymes carry certain common characteristics and hence fall into a class of proteins called protein kinases. The similarity of kinases is their ability to mediate the transfer of phosphoryl group from ATP to hydroxyl functional side chain of amino acid residues. In this case also, ATP is the chemical energy-carrying universal donor of phosphate group. Kinase-mediated transfer of phosphoryl group from ATP to hydroxyl group of Ser residues is shown in Figure 13.3.

### Mechanism of Dephosphorylation of Proteins by Phosphatases

As discussed earlier, protein phosphorylation is a reversible process in which phosphorylation and dephosphorylation acts as molecular switch for the biological functions of many proteins. Dephosphorylation is mediated by a group of enzymes called phosphatases (Bodansky and Schwartz 1961). There are two kinds of phosphatases, acidic phosphatases and alkaline phosphatases (Bodansky and Schwartz 1961). Both of these are hydrolases that catalyzes the hydrolysis of phosphate groups and converts the phosphate carrying groups back in to hydroxyl groups as shown in Figure 13.3.

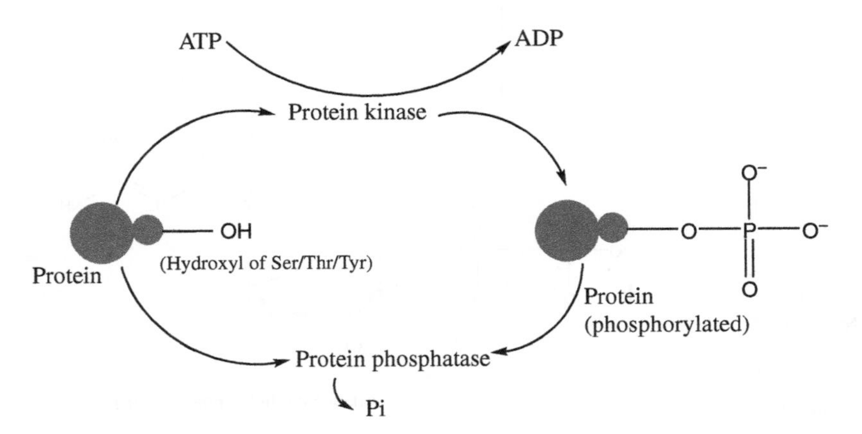


**Figure 13.3** Mechanism of protein phosphorylation mediated by kinases (Ser phosphorylation as an example). (*See insert for color representation of this figure.*)

Extremely large number of protein kinases and phosphatases exists in living cells (Criss 1971). For example, in yeast approximately 3% of all proteins are kinases and phosphatases (Waskiewicz and Cooper 1995). Many of the kinases and phosphatases are highly specific, but some kinases and phosphatases have very broad specificity. Known targets of phosphorylation include enzymes, receptors, ion channel proteins, structural proteins, signaling molecules, etc. Phosphorylation and dephosphorylation may occur continually on a protein, and this cycle allows protein to rapidly switch from active state to inactive state and vice versa (Kim and Asmis 2017). The phosphorylation and dephosphorylation have many unique advantages for the living cells that include (i) the process is rapid; (ii) does not require cells to make new proteins or degrade the old proteins; and (iii) the reactions are easily reversible. The phosphorylation and dephosphorylation occur four times during glycolysis (Weber et al. 1966).

**Protein Kinases**

Protein kinases are a broad group of enzymes involved in the transfer of phosphoryl group from ATP to biomolecules including proteins, carbohydrates, lipids, etc. (Cabib 1963; Sutherland 1972). The enzymes that are involved in transferring phosphoryl group from ATP to hydroxyl groups of amino acids of proteins are all collectively called protein kinases. The other kinases that transfer phosphoryl group to carbohydrates, lipids, etc., are termed as carbohydrate kinases, lipid kinases, etc., respectively. Specific protein kinases are involved in transferring phosphoryl group to hydroxyl groups of specific amino acids such as Ser, and Tyr residues. They are independently called as serine kinases, tyrosine kinases, etc. The protein kinases that dependent on cyclin are called as cyclin-dependent kinases (Freeman and Donoghue 1991). The systematic name of all these kinase enzymes is: ATP: protein phosphotransferases (VerPlank and Goldberg 2017). Serine kinases and tyrosine kinases are described below in detail.

**Serine Kinases**

Ser kinases are a group of enzymes that specifically transfer phosphoryl group from ATP to hydroxyl group of Ser. The Ser kinases also transfer phosphoryl group from ATP to hydroxyl group of Thr and hence, the serine kinases are also called as Ser/Thr kinases (London et al. 1980–1981). The two substrates of these enzymes are ATP and protein, and their two products are ADP and phosphoprotein. The catalytic domain of these enzymes is highly conserved and the sequence variation observed in the kinase genome (kenome) provides recognition to different substrates for phosphorylation. There are several types of Ser/Thr kinases that include casein kinase 2, protein kinase A, protein kinase C, and pelle (Bossemeyer 1995). The casein kinase 2 was discovered in 1954 by Burnett and Kennedy and is also called as CK2 (Burnett and Kennedy 1954).

The protein kinase A consists of two domains: one is a small domain with β-sheet and the other is a large domain with α-helices (Sowadski et al. 1999). The substrate and ATP-binding sites in protein kinase A is located in the cleft between the two domains (Francis et al. 2002). Upon binding to ATP and substrates, the two domains rotate so that the phosphate group of ATP and hydroxyl group containing amino acid of substrate molecule move into a correct position for the phosphorylation reaction to occur. The protein kinase C is a group of 10 isoenzymes that are also called as mitogen-activated protein kinases (MAP kinases or MAPKs) (Khorasanizadeh et al. 2017). The MAPK enzymes respond to extracellular stimulation by mitogen (Ai et al. 2016). MAPKs regulate many cellular functions such as gene expression, differentiation, survival of cells, or apoptosis and mitosis (Zou and Blank 2017). The kinase pelle is involved in phosphorylating itself along with phosphorylating tube and toll proteins (Gish and Clark 2011).

Expression of Ser/Thr kinases is altered in many malignant tumors (Zambrano et al. 2017). Some Ser/Thr kinases are dependent on binding to cyclin, and hence these are called cyclin-dependent kinases (CDKs). The CDKs binds to regulatory cyclin with considerable specificity (Jiménez et al. 2013). At least nine different CDKs are found in animal cells, and these are active only when they bound to cyclin (Malumbres 2014). The CDKs are involved in the activation of cell signaling pathways and transcription factors (Miller and Flaherty 2014).

#### Tyrosine Kinases

Compared to Ser/Thr phosphorylation, Tyr phosphorylation is relatively low in abundance (Jessus and Ozon 1995). However, Tyr phosphorylation is well studied because Tyr phosphorylated proteins are relatively easy to purify using affinity purification methods with the aid of antibodies against Tyr phosphate epitopes (Pendergast and Witte 1987). Tyr kinases are the enzymes involved in transferring phosphoryl group from ATP to Tyr residues. These are an important family of cell surface receptors and hence are also called as receptor tyrosine kinases (Rajabi and Hedayati 2017). A monomeric receptor tyrosine kinase binds to a ligand that stabilizes the interaction between the two monomers and hence forms dimers. The dimeric receptors phosphorylate Tyr residues to trans. The receptor tyrosine kinases are involved in the extracellular signal transduction including hormones, growth factors, and cytokines. The signaling of receptor tyrosine kinases is through the epidermal growth factor receptor (EGFR) (Su et al. 2017). This signaling is critical for the development of organ systems including brain, heart, lung, and skin. In many human cancers, excessive signaling of receptor tyrosine kinase through EGFR pathway is found (Proto et al. 2017).

## Physicochemical and Biological Functions of Protein Phosphorylation

Phosphoryl group is a negatively charged functional group and is hydrophilic in nature (Molnar and Lornad 1962). Hence, phosphorylation introduces a very strong negative charge to protein that influences the changes in protein interactions with nearby amino acids as well as with other proteins (Firzlaff et al. 1991). Phosphorylation of a hydrophobic protein domain converts into highly polar and hydrophilic domain of the protein. This change in hydrophobic into polar and hydrophilic nature may also induce conformational change in higher-order structure of proteins.

Phosphorylation regulates many protein functions including regulation of thermodynamics of biological reactions that involves energy, protein–protein interactions, protein degradation, enzyme inhibition, etc. For example, phosphorylation helps to maintain homeostasis of the human body's water content by phosphorylating $Na^{+}/K^{+}$-ATPase (Lopez-Mejia et al. 2017). During osmoregulation, phosphorylation helps the transport of $Na^{+}$ and $K^{+}$ ions across the cell membrane (Holmes 1965). Protein kinase B phosphorylate the GSK-3 enzyme which is part of the insulin signaling pathway. The reversible phosphorylation is an "on" or "off" switch for a wide range of cellular process. Phosphorylation and dephosphorylation reactions influence the change in protein conformation that results in the activation or deactivation of many enzymes, receptors, etc. For example, p53 is a tumor suppressor protein and contains about eight phosphorylation sites. Phosphorylation activates p53 that leads to the arrest of cell cycle (Meek 2015). Dephosphorylation of p53 reverses this process of cell cycle arrest (Wang et al. 2017).

## Phosphorylation in RTPs

Phosphorylation of protein backbone in recombinant therapeutic proteins (RTPs) and monoclonal antibodies (mAbs) has not been reported widely. However, many RTPs are being used as enzyme replacement therapies to treat many orphan diseases (Zhang et al. 2011). Many of these enzyme replacement therapies are acidic enzymes and are glycosylated in which the glycans may contain phosphate groups. The phosphorylated sugar moieties are derived during the biosynthesis of *N*-glycans (see *N*-glycan chapter i.e. Chapter 8) and are due to incomplete processing of precursor *N*-glyans. For many of these acidic enzymes, phosphorylated glycans are required for their enzymatic activity and also to target them to lysosomes. rhDNase is another RTP that

**Table 13.1** Partial list of RTPs containing phosphorylated glycans.[a]

| Name of RTP | Expressions system | Indications | Literature references |
|---|---|---|---|
| Agalsidase alfa (Replagal) | Mammalian | Anderson–Fabry disease | Shen et al. (2016) |
| Agalsidase beta (Fabrazyme) | Mammalian | Fabry disease | Lee et al. (2003) |
| Alglucosidase alfa (Lumizyme) | Mammalian | Pompe disease | Park et al. (2018) |
| Cerliponase alfa (Brineura) | Mammalian (CHO cells) | Late infantile neuronal ceroid lipofuscinosis type 2 | Brineura prescribing information at www.brineura.com; Markham (2017) |
| Dornase alfa (Pulmozyme) | Mammalian | Cystic fibrosis | Quan et al. (2011) |
| Idursulfase (Elaprase) | Mammalian | Hunter syndrome | Chung et al. (2014) |
| Imiglucerase (Cerezyme) | Mammalian | Gaucher's disease | Togawa et al. (2014) |
| Laronidase (Aldurazyme) | Mammalian | MPS disease | Tsukimura et al. (2008) |
| Modified α-*N*-acetylgalactosaminidase | Mammalian | Fabry disease | Tajima et al. (2009) |

a) In addition to the RTPs listed in this table, many of the RTPs that are being used to treat lysosomal storage diseases and other rare diseases contain phosphorylated sugars.

contains phosphorylated oligosaccharides (Cacia et al. 1998). Table 13.1 shows the list of RTPs containing phosphorylated glycans.

## Methods to Analyze Protein Phosphorylation

The most current methods to analyze protein phosphorylation includes electrophoresis with chemical detection or immunodetection, phosphoprotein enrichment, phosphopeptide detection, assays to detect kinase activity and/or phosphatases activity, mass spectrometry, etc. These methods are discussed briefly below.

### Gel Electrophoresis

Phosphorylated proteins can be separated by one- or two-dimensional gel electrophoresis and protein bands or spots can be detected by using chemical dyes or antibodies against specific phosphorylated amino acids. The gel electrophoresis can be run under either native or denaturing conditions. Using proper molecular weight standard, it is possible to obtain good information on phosphorylation state of the protein of interest.

### Mass Spectrometry

High-resolution mass spectrometry has been used in proteomics to detect phosphoproteins and also to identify the sites of phosphorylation (Duan et al. 2017). Mass spectrometry is one of the very versatile methods to detect PTMs of proteins including phosphorylation. The proteomic method using mass spectrometry (MS) gives the analyst ability to scan through lots of peptide fragments to identify the sites of phosphorylation throughout the protein of interest. In addition, stable isotope labeling by amino acids during cell culture (SILAC) also helps analysts to identify the relative abundance of changes in phosphorylation.

### Enrichment of Phosphoproteins

Prior to analysis of phosphoproteins, they can be enriched to improve the signal-to-noise ratio. The methods to enrich phosphoproteins include metal oxide affinity chromatography (MOAC) and immobilized metal affinity chromatography (IMAC). Since phosphoryl group is negatively charged, they bind strongly to positively charged metal ions and using immobilized metal ions it is possible to enrich phosphorylated proteins.

### Enrichment of Phosphorylated Peptides

Phosphorylated proteins can be digested with proteases to obtain peptides along with phosphorylated peptides (Duan et al. 2017). The peptides with or without phosphorylation can be separated using MOAC and/or IMAC. The enriched or purified phosphorylated peptides are very useful to identify the sites of phosphorylation as well as to generate antibodies against phosphorylated amino acid epitopes. In addition, in vitro labeling of peptides with tandem mass tags also helps in the relative identification of changes in phosphorylation.

In addition to these methods, in vitro kinase assays and specific assays for phosphatases can also be effectively used to identify and measure protein phosphorylation. For example, kinase assays are useful to measure the enzyme activation, whereas phosphatase assays are useful to measure inactivation of enzymes. Further, phosphatases are very useful to identify the sites of phosphorylation using mass spectrometry. In addition to phosphatases, hydrogen fluoride has been used to dephosphorylate glycans containing phosphate groups to identify the location of phosphate group (Aspinall et al. 1993).

## References

Ai, X., Yan, J., Carrillo, E., and Ding, W. (2016). The stress-response MAP kinase signaling in cardiac arrhythmias. *Rev. Physiol. Biochem. Pharmacol.* 172: 77–100.

Aspinall, G.O., McDonald, A.G., Raju, T.S. et al. (1993). Chemical structure of the core region of *Campylobacter jejuni* serotype O:2 lipopolysaccharide. *Eur. J. Biochem.* 213 (3): 1029–1037.

Bodansky, O. and Schwartz, M.K. (1961). Alkaline and acid phosphatases. *Methods Med. Res.* 9: 79–98.

Bossemeyer, D. (1995). Protein kinases – structure and function. *FEBS Lett.* 369 (1): 57–61.

Burnett, G. and Kennedy, E.P. (1954). The enzymatic phosphorylation of proteins. *J. Biol. Chem.* 211 (2): 969–980.

Cabib, E. (1963). Carbohydrate metabolism. *Annu. Rev. Biochem.* 32: 321–354.

Cacia, J., Quan, C.P., Pai, R., and Frenz, J. (1998). Human DNase I contains mannose 6-phosphate and binds the cation-independent mannose 6-phosphate receptor. *Biochemistry* 37 (43): 15154–15161.

Chung, Y.K., Sohn, Y.B., Sohn, J.M. et al. (2014). A biochemical and physicochemical comparison of two recombinant enzymes used for enzyme replacement therapies of hunter syndrome. *Glycoconjugate J.* 31 (4): 309–315. https://doi.org/10.1007/s10719-014-9523-0.

Cowgill, R.W. and Cori, C.F. (1955). The conversion of inactive phosphorylase to phosphorylase b and phosphorylase a in lobster muscle extract. *J. Biol. Chem.* 216 (1): 133–140.

Criss, W.E. (1971). A review of isozymes in cancer. *Cancer Res.* 31 (11): 1523–1542.

Duan, Q., Zhao, H., Zhang, Z. et al. (2017). Mechanistic evaluation and translational signature of gemcitabine-induced chemoresistance by quantitative phosphoproteomics analysis with iTRAQ labeling mass spectrometry. *Sci. Rep.* 7 (1): 12891. https://doi.org/10.1038/s41598-017-13330-2.

Firzlaff, J.M., Lüscher, B., and Eisenman, R.N. (1991). Negative charge at the casein kinase II phosphorylation site is important for transformation but not for Rb protein binding by the E7 protein of human papillomavirus type 16. *Proc. Natl. Acad. Sci. U.S.A.* 88 (12): 5187–5191.

Fischer, E.H. and Krebs, E.G. (1955). Conversion of phosphorylase b to phosphorylase a in muscle extracts. *J. Biol. Chem.* 216 (1): 121–132.

Francis, S.H., Poteet-Smith, C., Busch, J.L. et al. (2002). Mechanisms of autoinhibition in cyclic nucleotide-dependent protein kinases. *Front. Biosci.* 7: d580–d592.

Freeman, R.S. and Donoghue, D.J. (1991). Protein kinases and protooncogenes: biochemical regulators of the eukaryotic cell cycle. *Biochemistry* 30 (9): 2293–2302.

Gish, L.A. and Clark, S.E. (2011). The RLK/Pelle family of kinases. *Plant J.* 66 (1): 117–127. https://doi.org/10.1111/j.1365-313X.2011.04518.x.

Helmreich, E., Michaelides, M.C., and Cori, C.F. (1967). Effects of substrates and a substrate analog on the binding of 5′-adenylic acid to muscle phosphorylase a*. *Biochemistry* 6 (12): 3695–3710.

Holmes, W.N. (1965). Some aspects of osmoregulation in reptiles and birds. *Arch. Anat. Microsc. Morphol. Exp.* 54: 491–513.

Jessus, C. and Ozon, R. (1995). Function and regulation of cdc25 protein phosphate through mitosis and meiosis. *Prog. Cell Cycle Res.* 1: 215–228.

Jiménez, J., Ricco, N., Grijota-Martínez, C. et al. (2013). Redundancy or specificity? The role of the CDK Pho85 in cell cycle control. *Int. J. Biochem. Mol. Biol.* 4 (3): 140–149.

Keller, P.J. and Cori, G.T. (1953). Enzymic conversion of phosphorylase a to phosphorylase b. *Biochim. Biophys. Acta* 12 (1–2): 235–238.

Khorasanizadeh, M., Eskian, M., Gelfand, E.W., and Rezaei, N. (2017). Mitogen-activated protein kinases as therapeutic targets for asthma. *Pharmacol. Ther.* 174: 112–126. https://doi.org/10.1016/j.pharmthera.2017.02.024.

Kim, H.S. and Asmis, R. (2017). Mitogen-activated protein kinase phosphatase 1 (MKP-1) in macrophage biology and cardiovascular disease. A redox-regulated master controller of monocyte function and macrophage phenotype. *Free Radical Biol. Med.* 109: 75–83. https://doi.org/10.1016/j.freeradbiomed.2017.03.020.

Krebs, E.G. and Beavo, J.A. (1979). Phosphorylation–dephosphorylation of enzymes. *Annu. Rev. Biochem.* 48: 923–959.

Lee, K., Jin, X., Zhang, K. et al. (2003). A biochemical and pharmacological comparison of enzyme replacement therapies for the glycolipid storage disorder Fabry disease. *Glycobiology* 13 (4): 305–313.

Levene, P.A. and Alsberg, C.L. (1906). The cleavage products of vitellin. *J. Biol. Chem.* 2 (1): 127–133.

Lipmann, F.A. and Levene, P.A. (1932). Sererinephosphoric acid obtained on hydrolysis of vitellinic acid. *J. Biol. Chem.* 98 (1): 109–114.

London, I.M., Ernst, V., Fagard, R. et al. (1980–1981). Role of heme and of protein kinases in the regulation of eukaryotic protein synthesis. *Tex. Rep. Biol. Med.* 40: 101–110.

Lopez-Mejia, I.C., Castillo-Armengol, J., Lagarrigue, S., and Fajas, L. (2017). Role of cell cycle regulators in adipose tissue and whole body energy homeostasis. *Cell. Mol. Life Sci.* https://doi.org/10.1007/s00018-017-2668-9.

Malumbres, M. (2014). Cyclin-dependent kinases. *Genome Biol.* 15 (6): 122.

Markham, A. (2017). Cerliponase alfa: first global approval. *Drugs* 77 (11): 1247–1249. https://doi.org/10.1007/s40265-017-0771-8.

Meek, D.W. (2015). Regulation of the p53 response and its relationship to cancer. *Biochem. J* 469 (3): 325–346. https://doi.org/10.1042/BJ20150517.

Miller, D.M. and Flaherty, K.T. (2014). Cyclin-dependent kinases as therapeutic targets in melanoma. *Pigment Cell Melanoma Res.* 27 (3): 351–365. https://doi.org/10.1111/pcmr.12211.

Molnar, J. and Lornad, L. (1962). A phosphoryl group acceptor attached to the microsomal fraction of muscle. *Arch. Biochem. Biophys.* 98: 356–363.

Morrison, J.F. and Heyde, E. (1972). Enzymic phosphoryl group transfer. *Annu. Rev. Biochem.* 41 (10): 29–54.

Oliva, C. and Hassan, B.A. (2017). Receptor tyrosine kinases and phosphatases in neuronal wiring: insights from *Drosophila*. *Curr. Top. Dev. Biol.* 123: 399–432. https://doi.org/10.1016/bs.ctdb.2016.10.003.

Park, H., Kim, J., Lee, Y.K. et al. (2018). Four unreported types of glycans containing mannose-6-phosphate are heterogeneously attached at three sites (including newly found Asn 233) to recombinant human acid alpha-glucosidase that is the only approved treatment for Pompe disease. *Biochem. Biophys. Res. Commun.* 495 (4): 2418–2424. https://doi.org/10.1016/j.bbrc.2017.12.101.

Pendergast, A.M. and Witte, O.N. (1987). Role of the ABL oncogene tyrosine kinase activity in human leukaemia. *Baillieres Clin. Haematol.* 1 (4): 1001–1020.

Proto, C., Lo Russo, G., Corrao, G. et al. (2017). Treatment in EGFR-mutated non-small cell lung cancer: how to block the receptor and overcome resistance mechanisms. *Tumori* 103 (4): 325–337. https://doi.org/10.5301/tj.5000663.

Quan, C.P., Cacia, J., Frenz, J., and O'Connor, J.V. (2011). rhDNase I is a well characterized protein. *Pharm. Pharmacol. Commun.* 3: 53–57. https://doi.org/10.1111/j.2042-7158.1997.tb00475.x.

Rajabi, S. and Hedayati, M. (2017). Medullary thyroid cancer: clinical characteristics and new insights into therapeutic strategies targeting tyrosine kinases. *Mol. Diagn. Ther.* 21 (6): 607–620. https://doi.org/10.1007/s40291-017-0289-5.

Sasai, Y. (1965). Histochemical demonstration of phosphorylase a, phosphorylase b and phosphorylase kinase in normal human skin. *Tohoku J. Exp. Med.* 85: 160–170.

Shen, J.S., Busch, A., Day, T.S. et al. (2016). Mannose receptor-mediated delivery of moss-made α-galactosidase A efficiently corrects enzyme deficiency in Fabry mice. *J. Inherit. Metab. Dis.* 39 (2): 293–303. https://doi.org/10.1007/s10545-015-9886-9.

Sowadski, J.M., Epstein, L.F., Lankiewicz, L., and Karlsson, R. (1999). Conformational diversity of catalytic cores of protein kinases. *Pharmacol. Ther.* 82 (2–3): 157–164.

Su, C., Zhou, F., Shen, J. et al. (2017). Treatment of elderly patients or patients who are performance status 2 (PS2) with advanced non-small cell lung cancer without epidermal growth factor receptor (EGFR) mutations and anaplastic lymphoma kinase (ALK) translocations – still a daily challenge. *Eur. J. Cancer* 83: 266–278. https://doi.org/10.1016/j.ejca.2017.07.002.

Sutherland, I.W. (1972). Bacterial exopolysaccharides. *Adv. Microb. Physiol.* 8: 143–213.

Sutherland, E.W. Jr. and Wosilait, W.D. (1955). Inactivation and activation of liver phosphorylase. *Nature* 175 (4447): 169–170.

Tajima, Y., Kawashima, I., Tsukimura, T. et al. (2009). Use of a modified alpha-*N*-acetylgalactosaminidase in the development of enzyme replacement therapy for Fabry disease. *Am. J. Hum. Genet.* 85 (5): 569–580. https://doi.org/10.1016/j.ajhg.2009.09.016.

Togawa, T., Takada, M., Aizawa, Y. et al. (2014). Comparative study on mannose 6-phosphate residue contents of recombinant lysosomal enzymes. *Mol. Genet. Metab.* 111 (3): 369–373. https://doi.org/10.1016/j.ymgme.2013.12.296.

Tsukimura, T., Tajima, Y., Kawashima, I. et al. (2008). Uptake of a recombinant human alpha-L-iduronidase (laronidase) by cultured fibroblasts and osteoblasts. *Biol. Pharm. Bull.* 31 (9): 1691–1695.

VerPlank, J.J.S. and Goldberg, A.L. (2017). Regulating protein breakdown through proteasome phosphorylation. *Biochem. J.* 474 (19): 3355–3371. https://doi.org/10.1042/BCJ20160809.

Wang, Z.P., Tian, Y., and Lin, J. (2017). Role of wild-type p53-induced phosphatase 1 in cancer. *Oncol. Lett.* 14 (4): 3893–3898. https://doi.org/10.3892/ol.2017.6685.

Waskiewicz, A.J. and Cooper, J.A. (1995). Mitogen and stress response pathways: MAP kinase cascades and phosphatase regulation in mammals and yeast. *Curr. Opin. Cell Biol.* 7 (6): 798–805.

Weber, G., Singhal, R.L., Stamm, N.B. et al. (1966). Synchronous behavior pattern of key glycolytic enzymes: glucokinase, phosphofructokinase, and pyruvate kinase. *Adv. Enzyme Regul.* 4: 59–81.

Zambrano, J.N., Neely, B.A., and Yeh, E.S. (2017). Hormonally up-regulated neu-associated kinase: a novel target for breast cancer progression. *Pharmacol. Res.* 119: 188–194. https://doi.org/10.1016/j.phrs.2017.02.007.

Zhang, X.S., Brondyk, W., Lydon, J.T. et al. (2011). Biotherapeutic target or sink: analysis of the macrophage mannose receptor tissue distribution in murine models of lysosomal storage diseases. *J. Inherit. Metab. Dis.* 34 (3): 795–809. https://doi.org/10.1007/s10545-011-9285-9.

Zou, X. and Blank, M. (2017). Targeting p38 MAP kinase signaling in cancer through post-translational modifications. *Cancer Lett.* 384: 19–26. https://doi.org/10.1016/j.canlet.2016.10.008.

# 14

# Prenylation of Proteins

## Introduction

2-Methyl-1,3-butadiene is an organic molecule that is also often referred to as isoprene unit and is an unsaturated hydrocarbon with branched chain substitution of methyl groups with an elemental formula of $C_5H_8$ (Silver and Fall 1991). Unsaturated hydrocarbons are organic molecules that contain one or more carbon–carbon double bonds (C=C bonds). Isoprene, as an unsaturated hydrocarbon, contains two carbon–carbon double bonds as shown in Figure 14.1. Isoprenoids contain two or more covalently linked isoprene units (Nogueira et al. 2017). Farnesyl is a 15-carbon isoprenoid, whereas geranylgeranyl is a 20-carbon isoprenoid (O'Donnell et al. 1997). Accordingly, farnesyl group contains three isoprene units, whereas geranylgeranyl contains four isoprene units. In biological systems, both farnesyl and geranylgeranyl molecules are derived from the cholesterol biosynthetic pathway (Maltese 1990). Prenylation refers to the addition of farnesyl or geranylgeranyl molecules to the Cys residue of proteins in the C-termini. The family of proteins with prenylation includes numerous heterotrimeric G-proteins, Ras superfamily of small GTPases, nuclear lamins, protein kinases, phosphatases, etc. (Boudeau et al. 2003; Labbé et al. 2012; Reddy and Comai 2012; Shinde and Maddika 2017). So far more than 120 proteins of human origin have been identified as prenylated with either farnesyl or geranylgeranyl groups (Diaz-Rodriguez and Distefano 2017).

## Mechanism of Protein Prenylation

Farnesyl or geranylgeranyl groups are covalently linked to C-terminal Cys residues of protein via thioether linkage, i.e. C–S–C linkage. Prenylation of proteins occurs at a common consensus sequence of CAAX motif in which C refers to Cys, A is referred for any aliphatic amino acid except Ala, and X is a terminal uncharged amino acid at the C-termini of proteins (Xu et al. 2015).

*Co- and Post-Translational Modifications of Therapeutic Antibodies and Proteins*, First Edition. T. Shantha Raju.

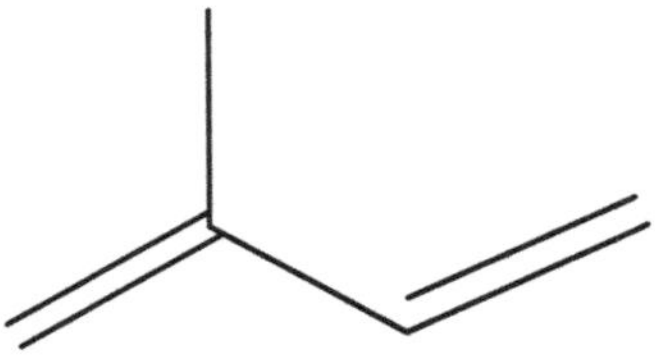

Figure 14.1 Chemical structure of isoprene unit.

The terminal X amino acid plays an important role in directing the protein to undergo either farnesylation or geranylgeranylation. If X is Ser, Met, Ala, Gln, or Cys, then the protein is directed for farnesylation. If X is Leu or Glu or sometimes Met, then the protein is subjected to geranylgeranylation (Xu et al. 2015).

Prenylation is an enzymatic reaction in which prenyltransferases are involved in transferring the prenyl groups (Maurer-Stroh et al. 2003). For example, farnesylation is mediated by farnesyltransferases and geranylgeranylation is mediated by type 1 geranylgeranyltransferases (de Gunzburg 1991). These transferases are heterodimeric enzymes consisting of two subunits, a common α-subunit, and a distinct β-subunit. In the case of farnesyltransferase, the common α-subunit is the product of FNTA gene (farnesyltransferase, CAAX box, α) and the β-subunit is encoded by FNTB gene (farnesyltransferase, CAAX bob, β). The farnesyltransferase enzyme mediates the transfer of farnesyl group from farnesylpyrophosphate donor to Cys residue in the CAAX motif (Tsimberidou et al. 2010; Shen et al. 2015). The α-subunit in geranylgeranyltransferase is encoded by FNTA gene, and the β-subunit is encoded by PGGT1B gene (protein geranylgeranyltransferase type 1 subunit β) (Shen et al. 2015). The transferase enzyme, geranylgeranyltransferase, transfers the geranylgeranyl group to Cys group in the CAAX motif from geranylgeranylpyrophosphate donor (Moorthy et al. 2013; Shen et al. 2015).

Both farnesyltransferase and geranylgeranyltransferase specifically recognize proteins with CAAX motif in the C-terminus (Fu and Casey 1999; Shen et al. 2015). Upon prenylation, the terminal three amino acids, i.e. AAX are removed by proteolysis (Ashby 1998). The proteases involved in removing the terminal AAX residues are termed as CAAX proteases (Ashby 1998; Rusiñol and Sinensky 2006). In humans, the major CAAX protease is encoded by Ras and α-factor converting enzyme 1 (RCE1) gene (Bergo et al. 2008; Hampton et al. 2018). The removal of the three AAX amino acid residues is followed by the methylation of carboxyl group of terminal prenylated cysteine residues (see Methylation in Chapter 11 also). This carboxymethylation is mediated by isoprenylcysteine carboxyl methyltransferases that transfers the methyl group from *S*-adenosylmethionine to carboxyl group of terminal prenylated Cys residue (Volker and Stock 1995). Three enzymes of isoprenylcysteine carboxyl methyltransferases are expressed in humans, and the major enzyme is encoded by ICMT gene (Febo-Ayala et al. 2006).

## Biological Significance of Prenylation of Proteins

Prenylated proteins are involved in signaling pathways and hence control intracellular membrane trafficking (Ray et al. 2017). Some of the prenylated proteins are involved in the modulation of immune responses (Mruk et al. 2005). The modulation of immune response by prenylated proteins includes leukocyte proliferation, activation, motility, etc. (Resh 2006). Prenylated proteins may also impact the immune functions of endothelial cells (Seabra 1998). Prenyl groups are also involved in protein–protein binding through prenyl binding domains. Anti-inflammatory properties of statins are partly due to their activity in reducing the synthesis of prenyl donors such as farnesylpyrophosphate and geranylpyrophosphate (Garcia-Ruiz et al. 2012). In addition, Ras proteins play an important role in cancer development. In humans, Ras proteins are among the few proteins that undergo prenylation and hence, inhibiting the transferases such as farnesyltransferases may help to control the tumor growth. Prenyltransferase inhibitors are also effective in inhibiting prenylation in parasites such as trypanosoma brucii and malaria (Gisselberg et al. 2017).

## Analysis of Prenylation of Proteins

Both enzymatic and gas-chromatographic methods are used to detect and quantitate prenylation of proteins (Sorek and Yalovsky 2010; Shi et al. 2013; Sorek et al. 2013). The gas-chromatographic method was also used in conjunction with mass spectrometry (Sorek and Yalovsky 2010; Sorek et al. 2013). The enzymatic analysis is designed to measure the activity of prenylation enzymes that involves purification of the prenylating proteins and in vitro prenylation of the proteins using radiolabeled prenyl substrates followed by autoradiographic detection of prenylation. The method is both qualitative and semiquantitative (Shi et al. 2013). The gas-chromatography and mass spectrometry (GC–MS) method involves the cleavage of prenyl groups using catalysts followed by the derivation of the carboxyl group of the lipid moiety into ester and analysis by GC–MS. The cleavage followed by esterification of prenyl groups is achieved by Raney-nickel catalyst and then hydrogenation with platinum (IV) oxide (Sorek et al. 2013).

## References

Ashby, M.N. (1998). CaaX converting enzymes. *Curr. Opin. Lipidol.* 9 (2): 99–102.

Bergo, M.O., Wahlstrom, A.M., Fong, L.G., and Young, S.G. (2008). Genetic analyses of the role of RCE1 in RAS membrane association and transformation.

*Methods Enzymol.* 438: 367–389. https://doi.org/10.1016/S0076-6879(07)38026-9.

Boudeau, J., Sapkota, G., and Alessi, D.R. (2003). LKB1, a protein kinase regulating cell proliferation and polarity. *FEBS Lett.* 546 (1): 159–165.

Diaz-Rodriguez, V. and Distefano, M.D. (2017). a-Factor: a chemical biology tool for the study of protein prenylation. *Curr. Top. Pept. Protein Res.* 18: 133–151.

Febo-Ayala, W., Morera-Félix, S.L., Hrycyna, C.A., and Thompson, D.H. (2006). Functional reconstitution of the integral membrane enzyme, isoprenylcysteine carboxyl methyltransferase in synthetic bolalipid membrane vesicles. *Biochemistry* 45 (49): 14683–14694.

Fu, H.W. and Casey, P.J. (1999). Enzymology and biology of CaaX protein prenylation. *Recent Prog. Horm. Res.* 54: 315–342.

Garcia-Ruiz, C., Morales, A., and Fernandez-Checa, J.C. (2012). Statins and protein prenylation in cancer cell biology and therapy. *Anticancer Agents Med. Chem.* 12 (4): 303–315.

Gisselberg, J.E., Zhang, L., Elias, J.E., and Yeh, E. (2017). The prenylated proteome of plasmodium falciparum reveals pathogen-specific prenylation activity and drug mechanism-of-action. *Mol. Cell. Proteomics* 16 (4 Suppl. 1): S54–S64. https://doi.org/10.1074/mcp.M116.064550.

de Gunzburg, J. (1991). Protein farnesyl and geranylgeranyl transferases. *C.R. Seances Soc. Biol. Fil.* 185 (5): 290–305.

Hampton, S.E., Dore, T.M., and Schmidt, W.K. (2018). Rce1: mechanism and inhibition. *Crit. Rev. Biochem. Mol. Biol.* 53 (2): 157–174. https://doi.org/10.1080/10409238.2018.1431606.

Labbé, D.P., Hardy, S., and Tremblay, M.L. (2012). Protein tyrosine phosphatases in cancer: friends and foes! *Prog. Mol. Biol. Transl. Sci.* 106: 253–306. https://doi.org/10.1016/B978-0-12-396456-4.00009-2.

Maltese, W.A. (1990). Posttranslational modification of proteins by isoprenoids in mammalian cells. *FASEB J.* 4 (15): 3319–3328.

Maurer-Stroh, S., Washietl, S., and Eisenhaber, F. (2003). Protein prenyltransferases. *Genome Biol.* 4 (4): 212.

Moorthy, N.S., Sousa, S.F., Ramos, M.J., and Fernandes, P.A. (2013). Farnesyltransferase inhibitors: a comprehensive review based on quantitative structural analysis. *Curr. Med. Chem.* 20 (38): 4888–4923.

Mruk, D.D., Lau, A.S., and Conway, A.M. (2005). Crosstalk between Rab GTPases and cell junctions. *Contraception* 72 (4): 280–290.

Nogueira, M., Enfissi, E.M., Almeida, J., and Fraser, P.D. (2017). Creating plant molecular factories for industrial and nutritional isoprenoid production. *Curr. Opin. Biotechnol.* 49: 80–87. https://doi.org/10.1016/j.copbio.2017.08.002.

O'Donnell, M.P., Massy, Z.A., Guijarro, C. et al. (1997). Isoprenoids, Ras and proliferative glomerular disease. *Contrib. Nephrol.* 120: 219–227.

Ray, A., Jatana, N., and Thukral, L. (2017). Lipidated proteins: spotlight on protein-membrane binding interfaces. *Prog. Biophys. Mol. Biol.* 128: 74–84. https://doi.org/10.1016/j.pbiomolbio.2017.01.002.

Reddy, S. and Comai, L. (2012). Lamin A, farnesylation and aging. *Exp. Cell Res.* 318 (1, 1): 1–7. https://doi.org/10.1016/j.yexcr.2011.08.009.

Resh, M.D. (2006). Trafficking and signaling by fatty-acylated and prenylated proteins. *Nat. Chem. Biol.* 2 (11): 584–590.

Rusiñol, A.E. and Sinensky, M.S. (2006). Farnesylated lamins, progeroid syndromes and farnesyl transferase inhibitors. *J. Cell Sci.* 119 (Pt. 16): 3265–3272.

Seabra, M.C. (1998). Membrane association and targeting of prenylated Ras-like GTPases. *Cell Signal.* 10 (3): 167–172.

Shen, M., Pan, P., Li, Y. et al. (2015). Farnesyltransferase and geranylgeranyltransferase I: structures, mechanism inhibitors and molecular modeling. *Drug Discov. Today* 20 (2): 267–276. https://doi.org/10.1016/j.drudis.2014.10.002.

Shi, W., Zeng, Q., and Running, M.P. (2013). In vitro prenylation assay of *Arabidopsis* proteins. *Methods Mol. Biol.* 1043: 147–160. https://doi.org/10.1007/978-1-62703-532-3_16.

Shinde, S.R. and Maddika, S. (2017). Post-translational modifications of Rab GTPases. *Small GTPases* 9 (1–2): 49–56. https://doi.org/10.1080/21541248.2017.1299270.

Silver, G.M. and Fall, R. (1991). Enzymatic synthesis of isoprene from dimethylallyl diphosphate in aspen leaf extracts. *Plant Physiol.* 97 (4): 1588–1591.

Sorek, N. and Yalovsky, S. (2010). Analysis of protein S-acylation by gas chromatography-coupled mass spectrometry using purified proteins. *Nat. Protoc.* 5 (5): 834–840. https://doi.org/10.1038/nprot.2010.33.

Sorek, N., Akerman, A., and Yalovsky, S. (2013). Analysis of protein prenylation and S-acylation using gas chromatography-coupled mass spectrometry. *Methods Mol. Biol.* 1043: 121–134. https://doi.org/10.1007/978-1-62703-532-3_13.

Tsimberidou, A.M., Chandhasin, C., and Kurzrock, R. (2010). Farnesyltransferase inhibitors: where are we now? *Expert Opin. Invest. Drugs* 19 (12): 1569–1580. https://doi.org/10.1517/13543784.2010.535516.

Volker, C. and Stock, J.B. (1995). Carboxyl methylation of Ras-related proteins. *Methods Enzymol.* 255: 65–82.

Xu, N., Shen, N., Wang, X. et al. (2015). Protein prenylation and human diseases: a balance of protein farnesylation and geranylgeranylation. *Sci. China Life Sci.* 58 (4): 328–335. https://doi.org/10.1007/s11427-015-4836-1.

# 15

# Proteolysis of Proteins

## Introduction

Proteolysis of crystallized globulin was described in 1894 by Chittenden and Mendel. Since then numerous researchers have published about proteolysis of proteins (Kryczka and Boncela 2017). Chemically, proteolysis is a hydrolysis reaction in which the peptide bonds present in proteins are broken down by the action of chemicals or enzymes or higher temperature or a combination thereof (Dean 1979). Hence, proteolysis reactions lead to breakdown of proteins into small peptides and/or into the respective constituent amino acid residues (Heyningen and Trayhurn 1976). Proteolysis or degradation of proteins in the absence of chemicals or enzymes may also occur but such reactions are very slow and may take several years to several decades to degrade proteins (Heyningen and Trayhurn 1976). Proteolysis or protein degradation may also occur at elevated temperatures and rates of such degradation reactions may be dependent on the temperature along with the pH of protein solutions. In the enzymatic proteolysis, proteases mediate the cleavage of peptide bonds, resulting into peptides and/or amino acid residues. Many proteases are zymogens that may require additional activation steps to make them biologically active to cleave the peptide bonds (McClorry 1954). Broadly, there are two categories of proteases, i.e. specific and nonspecific proteases. Both specific and nonspecific proteases are exo- and endo-proteases that are also called as exopeptidases and endopeptidases, respectively. Specific enzymes cleave specific peptide bonds that results in large to small peptides and nonspecific proteases randomly cleaves peptide bonds resulting in smaller peptides and/or into amino acid residues.

In living organisms, proteolysis is a required biological process that also helps in the digestion of food particles as a part of metabolism (Dangin et al. 2002). Proteolysis is also part of the regulatory mechanism to break down the unwanted proteins thus preventing their accumulation in living organisms or cells (Beynon and Bond 1986). Very often many proteins undergo limited proteolysis to attain their active form. Proteolysis is also important

*Co- and Post-Translational Modifications of Therapeutic Antibodies and Proteins*, First Edition. T. Shantha Raju.

for regulation of some cellular and physiological processes. Dis-regulation of proteolysis may lead to diseases that may ultimately lead to death (Kohn 1991). In addition, proteolysis is an important analytical tool to study structure and functions of proteins (Henzel et al. 1989). This chapter describes the chemical and enzymatic proteolysis of proteins in detail.

## Chemical Proteolysis

Chemicals such as mineral acids and inorganic bases are known to catalyze the hydrolysis of proteins into smaller peptides as well as into monomeric amino acid residues (Hill 1965). The rate of acid or base catalyzed degradation of proteins may depend on the reaction conditions such as temperature, concentrations of catalyzing reagents, protein concentration, etc. (Kasper 1970). Hence, chemical proteolysis reactions are either very slow or very fast depending on the conditions of such reactions. The chemical proteolysis reactions might also occur due to low or high pH and/or high temperature. Beside acids and bases, exposure of proteins to certain specific chemicals can also cause hydrolysis of the peptide bonds.

Mineral acid catalyzed hydrolysis or degradation is very commonly used to determine the amino acid composition of proteins (Cohn and Berggren 1924). Specifically, in laboratories acid catalyzed hydrolysis reactions using 6 M HCl at elevated temperature followed by high performance liquid chromatography (HPLC) analysis is commonly used to generate constituent amino acids composition of proteins (Edman and Agren 1947). However, during such harsh acid hydrolysis reactions certain amino acids such as Cys, Met, etc. are lost due to the degradation of their side chains. Also, acid labile amino acid modifications such as acetylation are lost during acid catalyzed hydrolysis. Hence, milder reaction conditions or alternative chemical reactions are used to measure the composition of sensitive amino acid residues or derivatized amino acids (Garnick 1992). Chemical degradation techniques are also used for sequencing of proteins to obtain information on their primary sequence (Smith 1994). Some amino acid residues in the C- and/or N-termini are specifically labile to certain chemicals (Hwang et al. 2014). In addition, specific chemical reagents may also mediate the cleavage of specific peptide bonds in proteins. For example, cyanogen bromide (CNBr) cleaves proteins at unoxidized Met residues in the C-termini (Nika et al. 2014). Hence, CNBr has been used to generate specific peptide fragments of proteins. 3-Bromo-3-methyl-2-(2-nitrophenylthio)-3H-indole (BNPS-skatole) cleaves proteins at Trp residues (Crimmins et al. 2005). Formic acid has been shown to specifically cleave peptide bond between Asp-Pro residues (Zhang et al. 2015). Further, hydroxylamine cleaves the Asn-Gly peptide bonds (Bornstein and Balian 1970). Additionally, 2-nitro-5-thiocyanobenzoic acid

(NTCB) cleaves proteins at cysteine residues (Otieno 1978). In addition to specific chemicals-mediated peptide bond cleavages, proteins may undergo self-cleavage or auto-cleavage (Caravaglios 1958). Such auto-cleavages are sequence-dependent and are likely due to β-elimination. For example, antibodies with a sequence of Ser-Cys-Asp-Lys-Thr-His-Thr-Cys in the hinge region may undergo a very specific self-cleavage at Asp-Lys peptide bond (Cordoba et al. 2005). Such cleavages are nonenzymatic hydrolysis mediated by neighboring groups participation (Harris 2005).

Some proteins are highly resistant to chemical hydrolysis. Even mineral acids at very high concentrations are not capable of hydrolyzing such proteins. For example, ribonuclease A is highly resistant to chemical hydrolysis. Hence, hot sulfuric acid is being normally used to purify ribonuclease A (Stewart and Stevenson 1973). Under such harsh acidic conditions, all other proteins are hydrolyzed leaving only ribonuclease A as an intact protein.

Peptide bonds are highly stable in water at physiological pH and are also stable at ambient temperature. However, the rate of hydrolysis of different peptide bonds may vary with constituent amino acid residues. Hence, the half-life of a peptide bond under physiological conditions may range between several years to several hundreds of years. However, they may undergo slow or rapid cleavage depending on the environment they are exposed and/or stored. Hence, for safety reasons and also to maintain their biological activity, normally, protein solutions are being stored at or below freezing temperature.

## Enzymatic Proteolysis

Enzymes that cleaves proteins or hydrolyze peptide bonds are collectively called as proteases. Most of the proteolytic enzymes are secreted as zymogens and many of the zymogenic proteases undergo autohydrolysis or self-cleavage to attain their active forms (Neurath 1964). For example, fibrinogen is a zymogen and is inactive, which undergoes self-cleavage to become active enzyme called fibrin (Laki and Gladner 1964). However, some enzymes require certain chemical treatment to become active (Skelton 1962). For example, papain is a protease present in papaya, which needs to be activated by treating with cysteine or need the presence of cysteine in solution to cleave peptide bonds of proteins (Polgár 1973). Many proteases can be activated by heating and/or by treating with chemicals. For example, matrix metalloproteases (MMPs) can be activated by heating and/or treating with cysteine (Wetmore and Hardman 1996).

Proteases can be classified into two broad categories: specific and nonspecific enzymes. Most of the proteases are endoproteases (or endopeptidases). But there are some exoproteases (or exopeptidases) also exist in nature. Both endo- and exo-proteases belongs to the category of specific and nonspecific enzymes. Examples of nonspecific proteases are pronase, proteinase-K, pepsin,

**Table 15.1** Category of specific proteases and their specificity.

| Types of proteases | Specificity | Comments |
|---|---|---|
| Aspartic protease | Asp-linkage | Uses of activated water molecule bound to aspartate for the catalysis |
| Asparagine peptide lyase | Asn-linkage | Catalyzes of nucleophilic elimination reaction rather than hydrolysis |
| Cysteine protease | Cys-linkage | Also known as thiol protease |
| Glutamic protease | Glu-linkage | Contain glutamic acid in the enzyme active site |
| Metalloproteases | Vary with the type | Highly specific proteases, also known as metalloproteinase. Most metalloproteases require $Zn^{+}$ for catalysis |
| Serine proteases | Ser-linkage | Found in both eukaryotes and prokaryotes |
| Threonine proteases | Thr-linkage | Catalytic subunits of proteasome |

papain, etc. Examples of specific proteases are trypsin, chymotrypsin, IdeS, glutamase, etc. Some of the nonspecific enzymes may also act as specific enzymes against certain substrates under specific conditions. For example, papain and pepsin are used as specific proteases to generate highly specific cleavages of IgGs to produce antibody fragments (Raju and Scallon 2006). Similarly, specific proteases may also generate minor nonspecific cleavages of proteins. Some specific proteases are further categorized based on their specificity to certain peptide bonds. Some of these specific proteases are presented in Table 15.1. Example of exoprotease (exopeptidase) is carboxypeptidase B.

Enzymatic proteolysis is more common and widely occurs in nature compared to chemical proteolysis. Very common proteolysis is the degradation of proteins present in food particles which undergo hydrolysis in human stomach by pepsin (LeVeen 1946). Acidic environment in the stomach is a perfect reaction condition for pepsin to breakdown proteins present in food particles into small peptides and/or amino acid residues thus making the required amino acid residues available that are essential for the growth and metabolism of living organisms (Porell and Howe 1951). In addition, proteolysis may also occur both inside and outside of the cells. For example, in endoplasmic reticulum (ER), properly folded proteins travels through the cellular compartments within the cells and then secretes out of the cells. However, proteins that are not folded properly often recirculate for proper folding or may eventually undergo degradation within the ER by molecular chaperones (Preissler and Deuerling 2012; see Chapter 8 also). Similarly, many proteases that are secreted out by cells often hydrolyze the secreted proteins (Dorai and Ganguly 2014). In larger organisms, the proteins present in food particles may be digested extracellularly by secreted proteases within specialized digestive organs or guts. However, in microorganisms, the food particles are internalized into cells by phagocytosis that in turn degraded by intracellular proteases (Lovewell et al. 2014).

## Biological Significance of Proteolysis

Chemical proteolysis has limited applicability in which the mineral acid based hydrolysis is a highly useful tool to obtain amino acid composition of proteins. Specific chemical proteolysis is also very useful to obtain sequence information of proteins. However, enzymatic proteolysis has wider applications. Some of the applications of enzymatic proteolysis are described below.

### Post-translational Processing of Proteins by Proteolysis

In living organisms or cells, quite often the proteins are initially translated into inactive forms, which are called preproproteins (Murcha et al. 2012). The preproproteins are first digested by proteases into proproteins that are further digested into active proteins (Bailyes et al. 1991). For example, serum albumin is first translated as preproalbumin (Compere et al. 1981; Elliott et al. 2014). The preproalbumin carries an uncleaved signal peptide at the N-terminus (Peach et al. 1991). Upon cleavage of the signal peptide by signal peptidases, the proalbumin will be formed. Further processing of the proalbumin by proteases removes the propeptide at the N-termini (Campbell 1975). The propeptide is a 6-amino acid component peptide in the N-terminus, and enzymatic proteolysis of this propeptide results in the formation of mature albumin (Campbell 1975).

Proteins that are targeted to a specific location within the cells or targeted for secretion usually contain specific signal peptides (Austen and Ridd 1981). These signal peptides are sequence-specific and are removed by proteolysis upon translocation or secretion (Hortsch and Meyer 1984). The proteases that cleaves signal peptides are called signal peptidases that are highly protein sequence-specific (Dalbey et al. 2017). If the signal peptide sequence does not match with the protein sequence, the signal peptidases do not efficiently cleave the signal peptides.

In eukaryotes, Met residue is the initiator of translation for protein synthesis and in procaryotes, formylmethionine (fMet) is the initiator of translation (Elson et al. 1975). The N-terminal Met residues might be removed by proteolysis during translation and in *Escherichia coli*, the N-terminal fMet will be removed if the penultimate residue is a neutral and small amino acid residue (Petersen et al. 1976). However, if the second residue is bulky and charged, the amino acid residue in the N-termini determine the serum half-life of the protein which is based on N-end rule (Dissmeyer et al. 2017).

Many hormones in eukaryotes are initially synthesized as large precursor proteins known as polyproteins (Chrétien and Seidah 1984). These polyproteins undergo proteolytic cleavages to form polypeptide hormones (Herbert 1980). The cleavage pattern of the polyprotein may vary from tissue to tissue. Hence, proteolysis of a polyprotein may result in generating different hormones depending on its location in the tissues (Rehbein et al. 1986).

Insulin is first synthesized as preproinsulin that contains a signal peptide in the N-termini (Lomedico et al. 1977). Upon cleavage of the signal peptide sequences, peptide proinsulin is formed which undergoes further cleavage at two different sites to form two polypeptide chains that inter-linked by two disulfide bonds. Both intra-chain and inter-chain disulfide bond formation and the protein folding occur in the proinsulin. One of the two polypeptide chains, called B-chain, undergoes further proteolysis to lose two amino acid residues in the C-termini to form the mature insulin (Levine and Mahler 1964).

Most often proteases are expressed in inactive forms as zymogens or as preproteins. These zymogens or preproteins undergo proteolysis to form active proteases (Astrup 1966). For example, the enzymatically active trypsin is expressed as trypsinogen that is enzymatically inactive form of trypsin (Kunitz and Northrop 1934). The inactive trypsinogen undergoes enzymatic proteolysis to form active trypsin (Kunitz 1939). This mechanism is in place for many proteases to store them in their inactive form in cells and released in active form when required of their activity. Such a mechanism is important to make sure that the active proteases are available only when their activity is required. Otherwise, freely available active proteases are detrimental to the survival of organisms. Hence, proteases are also part of regulatory process of organisms.

Blood coagulation cascade involves the sequential activation of many proteases that triggers blood clotting process (Newland 1987). Additionally, the complement cascade-mediated immune responses are also triggered by a sequential activation and interaction of proteases to attack the invading foreign pathogens (Regal et al. 2017).

Normally, viruses produce one single polypeptide chain during translation (Tamm and Eggers 1965). This single polypeptide chain of viruses undergoes proteolysis to form different proteins on the virus particles surface (Yu et al. 2017). For example, Ebola virus produce one single transmembrane protein called GP, which undergoes proteolytic cleavage to form GP1 and GP2 glycoproteins that forms the trimeric spikes on the virus cell surface (Raju 2017). These GP1 and GP2 are furin cleavage products that are both *N*- and *O*-glycosylated and the glycans protect the protein backbone from proteolysis (Martin et al. 2017).

## Intracellular and Extracellular Degradation of Proteins by Proteases

Proteolysis also plays a major role in protein degradation that occurs both extracellularly and intracellularly (Lebraud and Heightman 2017). The extracellular degradation of proteins occurs by secreted proteases. The degradation of proteins within cells occurs constantly, which serves many functions including the removal of damaged or abnormal proteins to prevent their accumulation (Ruggiano et al. 2014). The intracellular degradation of proteins also serves the purpose of removing the enzymes, hormones, and regulatory proteins, etc., that are no longer needed within the cells. The degraded components of the proteins, i.e. amino acids may be reused by cells for the synthesis of new proteins. There are two main pathways that are associated with the intracellular degradation of proteins (Gur et al. 2017). These pathways are proteolysis reactions that occur either in lysosome or proteolysis in proteasome. The lysosome is an acidic compartment within the cells and contains large number of acidic proteases that degrades the proteins that are trafficking through lysosomes. The degradation of proteins in lysosomes is an autophagy–lysosomal pathway and is a highly nonspecific proteolysis (Münz 2017). However, this nonspecific proteolysis might become specific proteolysis under starvation. During starvation proteins containing KFERQ motif or similar motifs undergo selective degradation in lysosome (Mukherjee et al. 2016). Proteasome pathway of proteolysis is ubiquitin-dependent (Barac et al. 2017). Proteasome is an ATP-dependent protease complex. Certain type of ubiquitination (see Chapter 20) targets proteins into proteasome, where they are degraded. One such ubiquitination is polyubiquitination, which is covalently linked to proteins and targets such proteins for degradation in proteasome. While the protein undergoes degradation, the covalently linked polyubiquitin is released and thus reused for similar purposes within the cells (Elliott 2016).

Intracellularly, the rate of degradation of peptide bonds by proteolysis may vary with proteins depending on their intracellular location and their functions. For example, enzymes that are involved in metabolism may degrade proteins much more rapidly than the enzymes that are involved in controlling the constant physiological mechanisms of living organisms. Accordingly, the serum half-life of proteins may vary with their biological functions. For example, ornithine decarboxylase has a serum half-life of ~11 minutes, whereas myosin has a serum half-life of about a month (O'Farrell 1986; Gulve et al. 1991). Actin also has a serum half-life of a month, and hemoglobin has a serum half-life of erythrocyte's life-time (Liem et al. 1975; Tamm et al. 1992). For some proteins, N-end rule may determine their serum-life life (Varshavsky 1996). PEST proteins are the one with the amino acid sequence of Pro, Glu, Ser, and Thr. Such PEST proteins may degrade much more rapidly compared to other proteins (Rechsteiner and Rogers 1996). Post-translational modifications (PTM)s such as deamidation, oxidation, etc., may enhance the protein degradation, whereas glycation, glycosylation, etc., may reduce the rate of protein degradation (Raju and Scallon 2007). The rate of protein degradation may also depend on the physiological state of the organism. The rate of protein degradation may be affected during starvation (Mortimore and Pösö 1987).

## Proteolysis and Food Digestion

Proteolysis plays a major role in digestion of food particles. In humans, digestive enzymes are involved in breaking down of proteins present in food particles (Smith 1967). Initially, proteins are digested into peptides by chymotrypsin, elastase, trypsin, pepsin, etc. These peptides are further digested into amino acids by enzymes such as aminopeptidase, carboxypeptidase, dipeptidase, etc. These digestive enzymes are also expressed as zymogens. For example, pepsin is secreted as pepsinogen in stomach and pepsinogen is converted into active pepsin enzyme in the acidic environment of stomach (Dykes and Kay 1976). Similarly, trypsin is expressed as trypsinogen in pancreas which is activated by enterokinase (Hadorn 1974). The activated trypsin can activate the trypsinogen as well as the chymotrypsin, and it can also undergo self-degradation by autolysis (Neurath and Dixon 1957).

In microorganisms such as bacteria, food particles may be internalized by phagocytosis which is then degraded by proteases inside the cells. Even in bacteria, many proteases are initially expressed as zymogens. For example, in *Bacillus subtilis*, the subtilisin protease is expressed as preprosubtilisin which upon removal of signal peptide undergoes self-activation to form the active enzyme (Wong and Doi 1986). The active substilisin enzyme is involved in the intracellular degradation of proteins within the bacterial cells (Durairaj et al. 2017).

## Role of Proteases in Apoptosis

Proteolysis is also involved in the cell cycle regulation and apoptosis. Cyclins are involved in the activation of kinases that regulates cell division (Minshull et al. 1989). Upon cell division cyclins are degraded by proteases via ubiquitin-mediated pathway that govern the exit of cells from mitosis and to the next cell cycle. Proteases play a major role in apoptosis (Kopeina et al. 2018). Caspases are a group of protease enzymes involved in the apoptosis (Tsapras and Nezis 2017). Caspases are expressed as procaspases that are activated by proteases through their association with apoptosome or death receptors or granzyme B (Ewen et al. 2012). In addition, proteolysis also plays a major role in regulating many other cellular processes (Hickey et al.

2012). Proteolysis is involved in the activation or deactivation of enzymes, transcription factors, receptors, etc. (Salvesen et al. 2016).

## Role of Proteolysis in Human Diseases

Proteolysis plays a major role in defining the status of many human diseases (Labbadia and Morimoto 2015). Overexpression and/or premature activation of proteases are observed in many human diseases. For example, pancreatitis is a human disease, which is due to the abnormal expression and premature activation of proteases in pancreas (Hegyi and Sahin-Tóth 2017). This disease eventually leads to the self-digestion of pancreas. In autoimmune diseases such as rheumatoid arthritis, some lysosomal proteases are released into extracellular environment. Once released into extracellular space, these proteases digest the proteins in the tissues, which results in chronic inflammation (Cuda et al. 2016). Proteolysis also plays a role in Alzheimer, Parkinson, AIDS, muscular dystrophy, etc. diseases. Many of these diseases are controlled by using protease inhibitors as drugs. For example, protease inhibitor cocktail is one of the very effective therapies used to treat AIDS caused by HIV viral infection (Wyen et al. 2008).

Several proteases are associated with inflammation, bacterial infections, and tumor invasion including metastasis. These proteases have the ability to cleave antibodies such as IgGs to make them ineffective to protect host from invasions. Many of these proteases preferentially cleave lower hinge region of IgGs thus making them ineffective from eliciting antibody effector functions to kill the cells (Ryan et al. 2008). For example, MMPs cleave IgGs in the lower hinge region producing either single cleaved IgGs or $F(ab')_2$ and truncated Fc fragments of IgGs (Ryan et al. 2008). Both single-cleaved IgGs and $F(ab')_2$ fragments are not able to kill the pathogenic cells by antibody effector functions (Fan et al. 2012). Similarly, bacterial proteases such as *Staphylococcus aureus* derived from glutaminyl endopeptidase I and IgG degrading enzyme of *Streptococcus pyogenes* (IdeS) cleaves IgGs in the lower hinge region (von Pawel-Rammingen et al. 2002). Such cleavages make IgGs unable to kill the invading pathogens by antibody-mediated cell killing mechanisms. Some examples of the enzymes that cleave IgGs in the lower hinge region are shown in Tables 15.2–15.4.

## Use of Proteases in Laboratories

Many proteases are used in laboratories for numerous applications including protein characterization and structural analyses studies. Amino acids are produced by digesting proteins in mineral acids at elevated temperatures. Amino acids and small peptides are also produced by digesting proteins with nonspecific enzymes such as pronase, proteinase K, pepsin, etc. Many specific proteases are used to determine the primary amino acid sequences of proteins

**Table 15.2** Matrix metalloproteinases that cleave human $IgG_1$ isotype.

| Enzymes[a)] | $IgG_1$ cleavage products |
|---|---|
| MMP-12 (macrophage elastase) | $F(ab')_2$ and truncated Fc fragment |
| MMP-3 (stromelysin) | $F(ab')_2$ and truncated Fc fragment |
| MMP-2 (gelatinase A) | $F(ab')_2$ and truncated Fc fragment |
| MMP-9 (gelatinase B) | $F(ab')_2$ and truncated Fc fragment |
| MMP-7 (matrilysin) | $F(ab')_2$ and truncated Fc fragment |
| MMP-1 (interstitial collagenase) | No cleavage |

a) In Table 15.2, the enzymes are listed in the decreasing order of their cleavage efficiency of human $IgG_1$ isotype and the order is MMP-12 > MMP-3 > MMP-2 = MMP-9 > MMP-7 > MMP-1.

**Table 15.3** Proteases in inflammation that cleave human $IgG_1$ isotype.

| Human proteases | Major proteolytic cleavage products of $IgG_1$ |
|---|---|
| Neutrophil elastase | Fab and Fc fragments |
| Cathepsin G | $F(ab')_2$ and truncated Fc fragments |

**Table 15.4** Proteases in bacterial infections that cleave human $IgG_1$ isotype.

| Bacterial proteases | Major proteolytic cleavage products of $IgG_1$ |
|---|---|
| IdeS | $F(ab')_2$ and truncated Fc fragments |
| Staphylokinase | Fab and Fc fragments |
| Steptokinase | Fab and Fc fragments |

(Dong 1992). In addition, proteases are used to remove contaminating proteins and/or enzymes during the purification of RNA or DNA. Chemically or enzymatically digested protein products are used as supplements during cell culture processes to grow cells and/or to produce recombinant proteins including antibodies using in vitro cell culture systems. Specific proteases are very useful to remove the protein tag from fusion proteins. Proteases are also used to remove unwanted protein domain/s during protein crystallization studies to obtain X-ray crystal structure of proteins.

## Sources of Proteases

Proteases are available from many sources including animals, plants, microorganisms, etc. Venoms are a very rich source of many proteases that exhibit, in addition to their proteolytic activity, cytotoxic, hemotoxic, myotoxic, and hemorrhagic activities. Papaya is a plant and is a rich source of papain. Pepsin is isolated from pig gastric mucus. Many proteases are also produced using recombinant technology for the commercial use.

## Proteolysis in RTPs

Recombinant therapeutic proteins (RTPs) produced using in vitro cell culture processes generally exhibit heterogeneity due to PTMs. The inherent heterogeneity of RTPs has the potential of impacting the product quality, and hence, it may be necessary to understand the mechanisms that are involved in causing the heterogeneity. One such mechanism involved in causing the heterogeneity of RTPs is proteolysis. RTPs may undergo proteolysis during cell culture production, purification, and also during storage (Dorai and Ganguly 2014). Proteolysis of RTPs may be due to both chemical and/or enzymatic proteolysis (Harris 2005). Both the chemical and enzymatic proteolysis will increase heterogeneity in RTPs due to fragmentation in the polypeptide backbone. Chemical proteolysis may be due to stress events such as deamidation, oxidation, etc., or may be due to sequence-specific events (Cordoba et al. 2005). Enzymatic proteolysis may occur due to intracellular and/or extracellularly secreted proteases (Harris 2005). The intracellular proteolysis may occur during cell culture production processes, but the extracellular proteolysis may occur during production, purification, and storage. This section describes the events that trigger proteolysis in RTPs.

### Chemical Proteolysis in RTPs

In RTPs, the chemical-induced proteolysis may occur due to physical or chemical events. The chemical proteolysis of RTPs may also occur due to sequence events (Harris 2005). Physical events such as storage temperature, exposure to light, agitation, etc., may cause aggregation followed by degradation. Chemical events such as oxidation, deamidation, isomerization, etc., may also cause products instability which would lead to the protein degradation. RTPs are stable within a narrow range of pH and any change in temperature may result in change in pH of solution which might cause protein aggregation, dissociation, denaturation, unfolding and/or misfolding, etc. Such events may ultimately induce the proteins to undergo chemical proteolysis of RTPs.

Chemical proteolysis due to sequence events are found to be protein primary sequence-dependent and are likely due to β-elimination. For example, mAbs are highly stable proteins and are resistant to the many degradative pathways. However, the hinge region of mAbs is highly flexible and relatively solvent exposed. Hence, in this region of mAbs, β-elimination followed by the chemical hydrolysis of the peptide bond may occur within the sequence of Ser-Cys-Asp-Lys-Thr-His-Thr-Cys (Cordoba et al. 2005). In this peptide sequence, the peptide bond between Asp and Lys is often cleaved due to β-elimination. Additional sequence events that lead to chemical proteolysis is the pH-induced iso-Asp formation which is mediated by neighboring group participation from His residues (Harris 2005).

### Enzymatic Proteolysis in RTPs

RTPs may undergo enzymatic proteolysis both intracellularly and extracellularly. For example, mAbs may undergo incomplete clipping of C-terminal Lys residues both within the cells and outside the cells during cell culture fermentation process (Luo et al. 2012; Cai et al. 2011). This incomplete clipping is caused by basic carboxypeptidases that include carboxypeptidase-B. In addition, the expected C-terminal Arg residue was found to be absent in some RTPs such as recombinant human erythropoietin (rhEPO) due to proteolysis. Depaolis et al. reported that the C-terminal clipping of Arg residues in rhEPO is also caused by basic carboxypeptidases (Depaolis et al. 1995). Many RTPs are genetically coded with signal peptide sequence in the N-termini for secretion. Upon transcription and also during translocation between the cellular compartments along with the extracellular secretion, the signal peptidases cleave the signal peptide from N-termini of RTPs. If the sequence of signal peptide is not complimentary to RTPs protein sequence, signal peptidases may not efficiently cleave the signal peptides from RTPs thus generating additional heterogeneity to RTPs. The uncleaved signal peptide may cause additional quality attributes issues during the product development in addition to increased heterogeneity. One such quality attribute might be the immunogenicity of the signal peptides (Liu et al. 2014).

During the development of cell culture process for glucagon-like-peptide-1-antibody fusion protein (GLP-1 Fc) Dorai et al. observed that N-termini of GLP-1 region of the fusion protein was proteolytically degraded (Dorai et al. 2009). This proteolytic cleavage removed the biologically active peptide from the GLP1-Fc fusion protein. Upon further investigation, it was found that the serine–threonine class of proteases are involved in the proteolytic cleavage of the biologically active GLP-1 peptide portion from the GLP1-Fc fusion protein (Dorai et al. 2011). Addition of serine–threonine protease inhibitor to the cell culture process prevented the protease clipping of GLP-1 peptide from the fusion protein. Such proteolytic clipping seems to be cell line specific. For example, GLP-1 Fc fusion protein produced in mouse myeloma cells do not appear to contain significant amounts of clipping compared to the GLP-1 Fc fusion protein produced in Chinese hamster ovary (CHO) cells (Dorai et al. 2009). This is because the amount and the nature of proteases expressed may be different in different cell lines. For example, proteases such as elastase, collagenase, plasminogen activator, serine proteases have been reported to be expressed and secreted by CHO cells but not by some mouse myeloma cells (Tsuji and Miyama 1992; Sandberg et al. 2006). Hence, choice of cell lines for the production of RTPs may make a big difference in controlling the product-related heterogeneity of RTPs due to proteolysis.

## Analytical Methods for the Detection of Proteolysis of Proteins

A number of analytical methods are being currently used to identify and to quantitate heterogeneity of proteins due to proteolysis. MALDI-TOF-MS analysis of mAbs is able to show product degradation due to proteolysis. MALDI-TOF-MS has been successfully used to detect degradation of rabbit IgG due to proteolysis (Jordan et al. 2016). However, MALDI-TOF-MS would not be able to detect the heterogeneity of mAbs due to C-terminal clipping. This is because mass accuracy of MALDI-TOF-MS is not good enough to detect such minor changes in the molecular mass of mAb samples. However, other sensitive MS techniques such as ESI-MS would be able to detect minor changes in mass due to proteolysis. Intact mass analysis of mAbs using ESI-MS would give information on the proteolytic clipping of proteins including C-terminal proteolysis due to carboxypeptidases. However, MS techniques are not highly useful to detect heterogeneity due to proteolysis in complex glycoproteins such as rhtPA, rhEPO. However, ion-exchange chromatography (IEC), hydrophobic interaction chromatography (HIC), reversed phase (RP), size-exclusion chromatography (SEC), and HPLC methods can be used to detect and quantitate product heterogeneity due to proteolysis. Both IEC and HIC methods have been successfully used to detect and quantitate C-terminal heterogeneity of mAbs due to proteolysis by carboxypeptidases. In addition to MS and HPLC methods, electrophoretic methods such as SDS-PAGE, capillary electrophoresis with sodium dodecyl sulfate (CE-SDS), capillary gel electrophoresis (CGE), iso-electric focusing (IEF), etc. methods are also being used to detect and quantitate product heterogeneity due to proteolysis.

## References

Astrup, T. (1966). Tissue activators of plasminogen. *Fed Proc.* 25 (1): 42–51.

Austen, B.M. and Ridd, D.H. (1981). The signal peptide and its role in membrane penetration. *Biochem. Soc. Symp.* (46): 235–258.

Bailyes, E.M., Bennett, D.L., and Hutton, J.C. (1991). Proprotein-processing endopeptidases of the insulin secretory granule. *Enzyme* 45 (5–6): 301–313.

Barac, Y.D., Emrich, F., Krutzwakd-Josefson, E. et al. (2017). The ubiquitin-proteasome system: A potential therapeutic target for heart failure. *J. Heart Lung Transplant.* 36 (7): 708–714. https://doi.org/10.1016/j.healun.2017.02.012.

Beynon, R.J. and Bond, J.S. (1986). Catabolism of intracellular protein: molecular aspects. *Am. J. Physiol.* 251 (2 Pt 1): C141–C152.

Bornstein, P. and Balian, G. (1970). The specific nonenzymatic cleavage of bovine ribonuclease with hydroxylamine. *J. Biol. Chem.* 245 (18): 4854–4856.

Cai, B., Pan, H., and Flynn, G.C. (2011). C-terminal lysine processing of human immunoglobulin G2 heavy chain in vivo. *Biotechnol. Bioeng.* 108 (2): 404–412. https://doi.org/10.1002/bit.22933.

Campbell, P.N. (1975). The biosynthesis of serum albumin. *FEBS Lett.* 54 (2): 119–121.

Caravaglios, R. (1958). Effect of ischaemic necrosis and autolysis in vitro on the soluble proteins of rat kidneys. *Biochem. J.* 68 (4): 681–685.

Chittenden, R.H. and Mendel, L.B. (1894). On the proteolysis of crystallized globulin. *J. Physiol.* 17 (1-2): 48–80.

Chrétien, M. and Seidah, N.G. (1984). Precursor polyproteins in endocrine and neuroendocrine systems. *Int. J. Pept. Protein Res.* 23 (4): 335–341.

Cohn, E.J. and Berggren, R.E. (1924). Studies in the physical chemistry of the proteins: III. The relation between the amino acid composition of casein and its capacity to combine with base. *J. Gen. Physiol.* 7 (1): 45–79.

Compere, S.J., Lively, M.O., and MacGillivray, R.T. (1981). Amino-terminal sequence of chicken preproalbumin. *Eur. J. Biochem.* 116 (3): 437–440.

Cordoba, A.J., Shyong, B.J., Breen, D., and Harris, R.J. (2005). Non-enzymatic hinge region fragmentation of antibodies in solution. *J. Chromatogr. B Anal. Technol. Biomed. Life Sci.* 818 (2): 115–121.

Crimmins, D.L., Mische, S.M., and Denslow, N.D. (2005). Chemical cleavage of proteins in solution. *Curr. Protoc. Protein Sci.* Chapter 11:Unit 11.4 https://doi.org/10.1002/0471140864.ps1104s40.

Cuda, C.M., Pope, R.M., and Perlman, H. (2016). The inflammatory role of phagocyte apoptotic pathways in rheumatic diseases. *Nat. Rev. Rheumatol.* 12 (9): 543–558. https://doi.org/10.1038/nrrheum.2016.132.

Dalbey, R.E., Pei, D., and Ekici, Ö.D. (2017). Signal peptidase enzymology and substrate specificity profiling. *Methods Enzymol.* 584: 35–57. https://doi.org/10.1016/bs.mie.2016.09.025.

Dangin, M., Boirie, Y., Guillet, C., and Beaufrère, B. (2002). Influence of the protein digestion rate on protein turnover in young and elderly subjects. *J. Nutr.* 132 (10): 3228S–3233S.

Dean, R.T. (1979). Lysosomes and protein degradation. *Ciba Found. Symp.* (75): 139–149.

DePaolis, A.M., Advani, J.V., and Sharma, B.G. (1995). Characterization of erythropoietin dimerization. *J. Pharm. Sci.* 84 (11): 1280–1284.

Dissmeyer, N., Rivas, S., and Graciet, E. (2017). Life and death of proteins after protease cleavage: protein degradation by the N-end rule pathway. *New Phytol.* https://doi.org/10.1111/nph.14619.

Dong, M.W. (1992). Tryptic mapping by reversed phase liquid chromatography. *Adv. Chromatogr.* 32: 21–51.

Dorai, H. and Ganguly, S. (2014). Mammalian cell-produced therapeutic proteins: heterogeneity derived from protein degradation. *Curr. Opin. Biotechnol.* 30: 198–204. https://doi.org/10.1016/j.copbio.2014.07.007.

Dorai, H., Nemeth, J.F., Cammaart, E. et al. (2009). Development of mammalian production cell lines expressing CNTO736, a glucagon like peptide-1-MIMETIBODY: factors that influence productivity and product quality. *Biotechnol. Bioeng.* 103 (1): 162–176. https://doi.org/10.1002/bit.22217.

Dorai, H., Santiago, A., Campbell, M. et al. (2011). Characterization of the proteases involved in the N-terminal clipping of glucagon-like-peptide-1-antibody fusion proteins. *Biotechnol. Prog.* 27 (1): 220–231. https://doi.org/10.1002/btpr.537.

Durairaj, A., Sabates, A., Nieves, J. et al. (2017). Proprotein convertase subtilisin/kexin type 9 (PCSK9) and its inhibitors: a review of physiology, biology, and clinical data. *Curr. Treat. Options Cardiovasc. Med.* 19 (8): 58. https://doi.org/10.1007/s11936-017-0556-0.

Dykes, C.W. and Kay, J. (1976). Conversion of pepsinogen into pepsin is not a one-step process. *Biochem J.* 153 (1, 1): 141–144.

Edman, P. and Agren, G. (1947). The amino acid composition of secretin. *Arch. Biochem.* 13 (2): 283–286.

Elliott, A.G., Delay, C., Liu, H. et al. (2014). Evolutionary origins of a bioactive peptide buried within Preproalbumin. *Plant Cell* 26 (3): 981–995. https://doi.org/10.1105/tpc.114.123620.

Elliott, P.R. (2016). Molecular basis for specificity of the Met1-linked polyubiquitin signal. *Biochem. Soc. Trans.* 44 (6): 1581–1602.

Elson, N.A., Adams, S.L., Merrick, W.C. et al. (1975). Comparison of fMet-tRNAf and Met-tRNAf from *Escherichia coli* and rabbit liver in initiation of hemoglobin synthesis. *J. Biol. Chem.* 250 (8): 3074–3079.

Ewen, C.L., Kane, K.P., and Bleackley, R.C. (2012). A quarter century of granzymes. *Cell Death Differ.* 19 (1): 28–35. https://doi.org/10.1038/cdd.2011.153.

Fan, X., Brezski, R.J., Fa, M. et al. (2012). A single proteolytic cleavage within the lower hinge of trastuzumab reduces immune effector function and in vivo efficacy. *Breast Cancer Res.* 14 (4): R116. https://doi.org/10.1186/bcr3240.

Garnick, R.L. (1992). Peptide mapping for detecting variants in protein products. *Dev. Biol. Stand.* 76: 117–130.

Gulve, E.A., Mabuchi, K., and Dice, J.F. (1991). Regulation of myosin and overall protein degradation in mouse C2 skeletal myotubes. *J. Cell Physiol.* 147 (1): 37–45.

Gur, E., Korman, M., Hecht, N. et al. (2017). How to control an intracellular proteolytic system: coordinated regulatory switches in the mycobacterial Pup-proteasome system. *Biochim. Biophys. Acta* 1864 (12): 2253–2260. https://doi.org/10.1016/j.bbamcr.2017.08.012.

Hadorn, B. (1974). Pancreatic proteinases; their activation and the disturbances of this mechanism in man. *Med. Clin. North Am.* 58 (6): 1319–1331.

Harris, R.J. (2005). Heterogeneity of recombinant antibodies: linking structure to function. *Dev. Biol. (Basel)* 122: 117–127.

Hegyi, E. and Sahin-Tóth, M. (2017). Genetic risk in chronic pancreatitis: the trypsin-dependent pathway. *Dig. Dis. Sci.* 62 (7): 1692–1701. https://doi.org/10.1007/s10620-017-4601-3.

Henzel, W.J., Stults, J.T., Hsu, C.A., and Aswad, D.W. (1989). The primary structure of a protein carboxyl methyltransferase from bovine brain that selectively methylates L-isoaspartyl sites. *J. Biol. Chem.* 264 (27): 15905–15911.

Herbert, E. (1980). Biosynthesis and processing of cellular and viral polyproteins. *Nature* 288 (5787): 115–116.

Heyningen, R. and Trayhurn, P. (1976). Proteolysis of lens proteins (autolysis). *Exp. Eye Res.* 22 (6): 625–637.

Hickey, C.M., Wilson, N.R., and Hochstrasser, M. (2012). Function and regulation of SUMO proteases. *Nat. Rev. Mol. Cell Biol.* 13 (12): 755–766. https://doi.org/10.1038/nrm3478.

Hill, R.L. (1965). Hydrolysis of proteins. *Adv. Protein Chem.* 20: 37–107.

Hortsch, M. and Meyer, D.I. (1984). Pushing the signal hypothesis: what are the limits? *Biol. Cell* 52 (1 Pt A): 1–8.

Hwang, P.M., Pan, J.S., and Sykes, B.D. (2014). Targeted expression, purification, and cleavage of fusion proteins from inclusion bodies in *Escherichia coli*. *FEBS Lett.* 588 (2): 247–252. https://doi.org/10.1016/j.febslet.2013.09.028.

Jordan, R.E., Fernandez, J., Brezski, R.J. et al. (2016). A peptide immunization approach to counteract a *Staphylococcus aureus* protease defense against host immunity. *Immunol. Lett.* 172: 29–39. https://doi.org/10.1016/j.imlet.2016.02.009.

Kasper, C.B. (1970). Fragmentation of proteins for sequence studies and separation of peptide mixtures. *Mol. Biol. Biochem. Biophys.* 8: 137–184.

Kohn, E.C. (1991). Invasion and metastasis: biology and clinical potential. *Pharmacol. Ther.* 52 (2): 235–244.

Kopeina, G.S., Prokhorova, E.A., Lavrik, I.N., and Zhivotovsky, B. (2018). Alterations in the nucleocytoplasmic transport in apoptosis: caspases lead the way. *Cell Prolif.* e12467. https://doi.org/10.1111/cpr.12467.

Kryczka, J. and Boncela, J. (2017). Proteases revisited: roles and therapeutic implications in fibrosis. *Mediators Inflamm.* 2017: 2570154. https://doi.org/10.1155/2017/2570154.

Kunitz, M. and Northrop, J.H. (1934). The isolation of crystalline trypsinogen and its conversion into crystalline trypsin. *Science* 80 (2083): 505–506.

Kunitz, M. (1939). Formation of trypsin from crystalline trypsinogen by means of enterokinase. *J Gen Physiol.* 22 (4): 429–446.

Laki, K. and Gladner, J.A. (1964). Chemistry and physiology of the fibrinogen-fibrin transition. *Physiol. Rev.* 44: 127–160.

Labbadia, J. and Morimoto, R.I. (2015). The biology of proteostasis in aging and disease. *Annu. Rev. Biochem.* 84: 435–464. https://doi.org/10.1146/annurev-biochem-060614-033955.

Lebraud, H. and Heightman, T.D. (2017). Protein degradation: a validated therapeutic strategy with exciting prospects. *Essays Biochem.* https://doi.org/10.1042/EBC20170030.

LeVeen, H.H. (1946). Active pepsin fraction, resting peptic activity, and percentage of pepsin inhibition in gastric juice. *Proc. Soc. Exp. Biol. Med.* 63 (2): 259–263.

Levine, R. and Mahler, R. (1964). Production, secretion, and availability of insulin. *Annu. Rev. Med.* 15: 413–432.

Liem, H.H., Spector, J.I., Conway, T.P. et al. (1975). Effect of hemoglobin and hematin on plasma clearance of hemopexin, photo-inactivated hemopexin and albumin (38575). *Proc. Soc. Exp. Biol. Med.* 148 (2): 519–522.

Liu, H., Ponniah, G., Zhang, H.M. et al. (2014). In vitro and in vivo modifications of recombinant and human IgG antibodies. *MAbs* 6 (5): 1145–1154. https://doi.org/10.4161/mabs.29883.

Lomedico, P.T., Chan, S.J., Steiner, D.F., and Saunders, G.F. (1977). Immunological and chemical characterization of bovine preproinsulin. *J. Biol. Chem.* 252 (22): 7971–7978.

Luo, J., Zhang, J., Ren, D. et al. (2012). Probing of C-terminal lysine variation in a recombinant monoclonal antibody production using Chinese hamster ovary cells with chemically defined media. *Biotechnol. Bioeng.* 109 (9): 2306–2315. https://doi.org/10.1002/bit.24510.

Lovewell, R.R., Patankar, Y.R., and Berwin, B. (2014). Mechanisms of phagocytosis and host clearance of Pseudomonas aeruginosa. *Am. J. Physiol. Lung. Cell. Mol. Physiol.* 306 (7): L591–L603. https://doi.org/10.1152/ajplung.00335.2013.

Martin, B., Canard, B., and Decroly, E. (2017). Filovirus proteins for antiviral drug discovery: structure/function bases of the replication cycle. *Antiviral Res.* 141: 48–61. https://doi.org/10.1016/j.antiviral.2017.02.004.

McClorry, R.G. (1954). Prothrombin estimation; a review and discussion. *Can. J. Med. Technol.* 16 (2): 37–47.

Minshull, J., Pines, J., Golsteyn, R. et al. (1989). The role of cyclin synthesis, modification and destruction in the control of cell division. *J. Cell. Sci. Suppl.* 12: 77–97.

Mortimore, G.E. and Pösö, A.R. (1987). Intracellular protein catabolism and its control during nutrient deprivation and supply. *Annu. Rev. Nutr.* 7: 539–564.

Mukherjee, A., Patel, B., Koga, H. et al. (2016). Selective endosomal microautophagy is starvation-inducible in Drosophila. *Autophagy* 12 (11): 1984–1999.

Münz, C. (2017). Autophagy proteins in phagocyte endocytosis and exocytosis. *Front. Immunol.* 8: 1183. https://doi.org/10.3389/fimmu.2017.01183.

Murcha, M.W., Wang, Y., and Whelan, J. (2012). A molecular link between mitochondrial preprotein transporters and respiratory chain complexes. *Plant Signal. Behav.* 7 (12): 1594–1597. https://doi.org/10.4161/psb.22250.

Neurath, H. and Dixon, G.H. (1957). Structure and activation of trypsinogen and chymotrypsinogen. *Fed. Proc.* 16 (3): 791–801.

Neurath, H. (1964). Mechanism of zymogen activation. *Fed. Proc.* 23: 1–7.

Newland, J.R. (1987). Blood coagulation: a review. *Am. J. Obstet. Gynecol.* 156 (6): 1420–1422.

Nika, H., Hawke, D.H., and Angeletti, R.H. (2014). N-terminal protein characterization by mass spectrometry after cyanogen bromide cleavage using combined microscale liquid- and solid-phase derivatization. *J. Biomol. Tech.* 25 (1): 19–30. https://doi.org/10.7171/jbt.14-2501-003.

O'Farrell, M.K. (1986). Metabolic turnover of proliferation-related nuclear proteins in serum-stimulated Swiss mouse 3T3 cells. *FEBS Lett.* 204 (2): 233–238.

Otieno, S. (1978). Generation of a free alpha-amino group by Raney nickel after 2-nitro-5-thiocyanobenzoic acid cleavage at cysteine residues: application to automated sequencing. *Biochemistry* 17 (25): 5468–5474.

Peach, R.J., Boswell, D.R., and Brennan, S.O. (1991). Prediction of the site of signal peptidase cleavage in normal and variant human preproalbumin. *Protein. Seq. Data. Anal.* 4 (2): 123–126.

Petersen, H.U., Danchin, A., and Grunberg-Manago, M. (1976). Toward an understanding of the formylation of initiator tRNA methionine in prokaryotic protein synthesis. II. A two-state model for the 70S ribosome. *Biochemistry* 15 (7): 1362–1369.

Polgár, L. (1973). On the mode of activation of the catalytically essential sulfhydryl group of papain. *Eur. J. Biochem.* 33 (1): 104–109.

Porell, W.J. and Howe, C.W. (1951). Assessment of gastric acidity and pepsin in relation to peptic ulcer. I. Acidity of the gastric contents. *BMQ* 2 (3): 73–77.

Preissler, S. and Deuerling, E. (2012). Ribosome-associated chaperones as key players in proteostasis. *Trends Biochem. Sci.* 37 (7): 274–283. https://doi.org/10.1016/j.tibs.2012.03.002.

Raju, T.S. and Scallon, B. (2007). Fc glycans terminated with N-acetylglucosamine residues increase antibody resistance to papain. *Biotechnol. Prog.* 23 (4): 964–971.

Raju, T.S. and Scallon, B.J. (2006). Glycosylation in the Fc domain of IgG increases resistance to proteolytic cleavage by papain. *Biochem. Biophys. Res Commun.* 341 (3): 797–803.

Raju, T.S. (2017). Antibody mixtures to treat Ebola. *Curr. Biotechnol.* 6 (1): 5–8.

Rechsteiner, M. and Rogers, S.W. (1996). PEST sequences and regulation by proteolysis. *Trends Biochem. Sci.* 21 (7): 267–271.

Regal, J.F., Burwick, R.M., and Fleming, S.D. (2017). The complement system and preeclampsia. *Curr. Hypertens Rep.* 19 (11): 87. https://doi.org/10.1007/s11906-017-0784-4.

Rehbein, M., Hillers, M., Mohr, E. et al. (1986). The neurohypophyseal hormones vasopressin and oxytocin. Precursor structure, synthesis and regulation. *Biol. Chem. Hoppe Seyler* 367 (8): 695–704.

Ruggiano, A., Foresti, O., and Carvalho, P. (2014). Quality control: ER-associated degradation: protein quality control and beyond. *J. Cell Biol.* 204 (6): 869–879. https://doi.org/10.1083/jcb.201312042.

Ryan, M.H., Petrone, D., Nemeth, J.F. et al. (2008). Proteolysis of purified IgGs by human and bacterial enzymes in vitro and the detection of specific proteolytic fragments of endogenous IgG in rheumatoid synovial fluid. *Mol. Immunol.* 45 (7): 1837–1846.

Salvesen, G.S., Hempel, A., and Coll, N.S. (2016). Protease signaling in animal and plant-regulated cell death. *FEBS J.* 283 (14): 2577–2598. https://doi.org/10.1111/febs.13616.

Sandberg, H., Lütkemeyer, D., Kuprin, S. et al. (2006). Mapping and partial characterization of proteases expressed by a CHO production cell line. *Biotechnol. Bioeng.* 95 (5, 5): 961–971.

Skelton, G.S. (1962). The effect of various chemicals on papain activation. *Enzymologia* 24: 338–340.

Smith, B.J. (1994). Chemical cleavage of proteins. *Methods Mol. Biol.* 32: 297–309.

Smith, G.H. (1967). A review of gastrointestinal enzymes. *J. Am. Pharm. Assoc.* 7 (10): 523–526.

Stewart, G.R. and Stevenson, K.J. (1973). The isolation and partial characterization of ribonuclease A from Bison bison. *Biochem. J.* 135 (3): 427–441.

Tamm, I. and Eggers, H.J. (1965). Biochemistry of virus reproduction. *Am. J. Med.* 38: 678–698.

Tamm, I., Kikuchi, T., Krueger, J., and Murphy, J.S. (1992). Dissociation between early loss of actin fibres and subsequent cell death in serum-deprived quiescent Balb/c-3T3 cells. *Cell Signal.* 4 (6): 675–686.

Tsapras, P. and Nezis, I.P. (2017). Caspase involvement in autophagy. *Cell Death Differ.* 24 (8): 1369–1379. https://doi.org/10.1038/cdd.2017.43.

Tsuji, T. and Miyama, A. (1992). Chinese hamster ovary cells produce an enzyme that nicks heat-labile enterotoxin from enterotoxigenic *Escherichia coli*. *Eur. J. Epidemiol.* 8 (1): 74–80.

Varshavsky, A. (1996). The N-end rule: functions, mysteries uses. *Proc. Natl. Acad. Sci. U.S.A.* 93 (22): 12142–12149.

von Pawel-Rammingen, U., Johansson, B.P., and Björck, L. (2002). IdeS, a novel streptococcal cysteine proteinase with unique specificity for immunoglobulin G. *EMBO J.* 21 (7): 1607–1615.

Wetmore, D.R. and Hardman, K.D. (1996). Roles of the propeptide and metal ions in the folding and stability of the catalytic domain of stromelysin (matrix metalloproteinase 3). *Biochemistry* 35 (21): 6549–6558.

Wong, S.L. and Doi, R.H. (1986). Determination of the signal peptidase cleavage site in the preprosubtilisin of *Bacillus subtilis*. *J. Biol. Chem.* 261 (22): 10176–10181.

Wyen, C., Fuhr, U., Frank, D. et al. (2008). Effect of an antiretroviral regimen containing ritonavir boosted lopinavir on intestinal and hepatic CYP3A, CYP2D6 and P-glycoprotein in HIV-infected patients. *Clin. Pharmacol. Ther.* 84 (1): 75–82. https://doi.org/10.1038/sj.clpt.6100452.

Yu, D.S., Weng, T.H., Wu, X.X. et al. (2017). The lifecycle of the Ebola virus in host cells. *Oncotarget* 8 (33): 55750–55759. https://doi.org/10.18632/oncotarget.18498.

Zhang, H., Li, M., Shi, S. et al. (2015). Design and identification of a high efficient formic acid cleavage site for separation of fusion protein. *Protein J.* 34 (1): 9–17. https://doi.org/10.1007/s10930-014-9592-8.

# 16

# Selenylation

## Introduction

Selenium is a nonmetal trace element and is produced as a by-product during the refining of metal sulfide ores (Saunders et al. 2016). Selenium (with the elemental symbol Se) is used in the manufacturing of glass and pigments (Barceloux 1999). Elemental Se is a semiconductor and hence used in photocells as well as in certain types of DC power surge protectors. In biological systems, Se is found to be covalently linked to Cys and Met analogues (Shtutman and Chagovets 1971). The Cys and Met analogues containing Se are called as selenocysteine and selenomethionine in which sulfur atom is replaced by Se. In living organisms, Se containing amino acid residues are found in some prokaryotic and eukaryotic enzymes (Burk and Hill 2015). Selenium containing enzymes are mostly involved in redox reactions. For example, glutathione peroxidase (GSH) and thioredoxin reductase are found to contain selenocysteine residues (Brigelius-Flohé and Flohé 2017). Selenomethionine is found in Brazil nuts, cereal grains, soybeans, legumes, etc. GSH is an enzyme that enzymatically reduces peroxide into water to limit the formation of reactive oxygen species (ROS) (Brigelius-Flohé 2006). Glutathione reductase enzyme catalyzes the reduction of disulfide bond in oxidized thioredoxin with nicotinamide adenine dinucleotide phosphate (NADPH). In addition, thioredoxin reductase is involved in many physiological processes (Saccoccia et al. 2014). For example, thioredoxin provides defense mechanism against oxidative stress (Lu and Holmgren 2009).

## Biological Significance of Selenylation of Proteins

In biological processes, GSH family of enzymes act as antioxidants and are involved in destroying ROS producing reagents such as hydrogen peroxides and related organic peroxides as shown in Figure 16.1 (Brigelius-Flohé 2006).

*Co- and Post-Translational Modifications of Therapeutic Antibodies and Proteins*, First Edition. T. Shantha Raju.

$$H_2O_2 \xrightarrow[-GSSG]{2GSH} 2H_2O$$

**Figure 16.1** Conversion of hydrogen peroxide into water mediated by glutathione peroxidase (GSH = Glutathione; GSSG = Glutathione disulfide).

In humans, GSH is involved in protecting red blood cells from oxidation by ROS (Mohanty et al. 2014). There are eight different GSH encoding genes present in humans and are named as GPX1–GPX8 (Cardoso et al. 2017). In humans, nearly all types of cells contain the enzyme, GPx1, encoded by GPX1 gene (Brigelius-Flohé and Maiorino 2013). The GPx1 enzyme almost exclusively acts as reducing agent converting hydrogen peroxide into water (Sunde et al. 2018). The GPX3 gene encodes the GPx3 enzyme and is an extracellular plasma protein (Brown and Arthur 2001). The GPx4 enzyme, encoded by GPX4 gene, is also an extracellular protein and is localized in intestines.

The enzyme thioredoxin reductase is involved in reducing the oxidized disulfide bonds in thioredoxin (Monteiro et al. 2017). Thioredoxin is also involved in the reduction of disulfide bonds in other proteins. Since the reduction involves the oxidation, the reduction of disulfide bonds results in the oxidation of thioredoxin. The oxidized thioredoxin is reduced by thioredoxin reductase (Couto et al. 2016). In humans, three thioredoxin reductase genes encode three different enzymes named as TrxR1, TrxR2, and TrxR3. The TrxR1 enzyme is present in the cytosol, and its main function is the maintenance of ribonucleotide reductase cycle (Prigge et al. 2017). The TrxR2 enzyme mainly functions in the detoxification of ROS in the organelle and is present in mitochondria (Patenaude et al. 2004). The TrxR3 enzyme is a testes-specific enzyme and is an enzyme isoform (Urig and Becker 2006).

In addition, deiodinase family of enzymes also contains selenocysteine (Lu and Holmgren 2009). The thyroid deiodinases are involved in the maturation and catabolism of thyroid hormones (Prummel and Wiersinga 2004). Three different thyroid deiodinases are found in humans. The iodothyronine deiodinase type I enzyme converts the thyroxine to bioactive tri-iodothyronine in the peripheral tissues. In addition, iodothyronine deiodinase type I enzyme is also involved in the catabolism of thyroid hormones. The iodothyronine deiodinase type II enzyme is also involved in the conversion of inactive thyroxine into biologically active tri-iodothyronine in thyroid gland. This enzyme is known to be associated with Graves' disease (Yoshihara et al. 2015). The iodothyronine deiodinase type III enzyme is mainly involved in the catabolism of thyroxine and tri-iodothyronine. The type III enzyme regulates the thyroid hormone levels and its functions during pregnancy and also during early fetal development. The enzyme expression levels are high in the female uterus during pregnancy.

## References

Barceloux, D.G. (1999). Selenium. *J. Toxicol. Clin. Toxicol.* 37 (2): 145–172.

Brigelius-Flohé, R. (2006). Glutathione peroxidases and redox-regulated transcription factors. *Biol. Chem.* 387 (10–11): 1329–1335.

Brigelius-Flohé, R. and Flohé, L. (2017). Selenium and redox signaling. *Arch. Biochem. Biophys.* 617: 48–59. https://doi.org/10.1016/j.abb.2016.08.003.

Brigelius-Flohé, R. and Maiorino, M. (2013). Glutathione peroxidases. *Biochim. Biophys. Acta* 1830 (5): 3289–3303. https://doi.org/10.1016/j.bbagen.2012.11.020.

Brown, K.M. and Arthur, J.R. (2001). Selenium, selenoproteins and human health: a review. *Public Health Nutr.* 4 (2B): 593–599.

Burk, R.F. and Hill, K.E. (2015). Regulation of selenium metabolism and transport. *Annu. Rev. Nutr.* 35: 109–134. https://doi.org/10.1146/annurev-nutr-071714-034250.

Cardoso, B.R., Hare, D.J., Bush, A.I., and Roberts, B.R. (2017). Glutathione peroxidase 4: a new player in neurodegeneration? *Mol. Psychiatry* 22 (3): 328–335. https://doi.org/10.1038/mp.2016.196.

Couto, N., Wood, J., and Barber, J. (2016). The role of glutathione reductase and related enzymes on cellular redox homoeostasis network. *Free Radical Biol. Med.* 95: 27–42. https://doi.org/10.1016/j.freeradbiomed.2016.02.028.

Lu, J. and Holmgren, A. (2009). Selenoproteins. *J. Biol. Chem.* 284 (2): 723–727. https://doi.org/10.1074/jbc.R800045200.

Mohanty, J.G., Nagababu, E., and Rifkind, J.M. (2014). Red blood cell oxidative stress impairs oxygen delivery and induces red blood cell aging. *Front. Physiol.* 5: 84. https://doi.org/10.3389/fphys.2014.00084.

Monteiro, H.P., Ogata, F.T., and Stern, A. (2017). Thioredoxin promotes survival signaling events under nitrosative/oxidative stress associated with cancer development. *Biomed. J.* 40 (4): 189–199. https://doi.org/10.1016/j.bj.2017.06.002.

Patenaude, A., Ven Murthy, M.R., and Mirault, M.E. (2004). Mitochondrial thioredoxin system: effects of TrxR2 overexpression on redox balance, cell growth, and apoptosis. *J. Biol. Chem.* 279 (26): 27302–27314.

Prigge, J.R., Coppo, L., Martin, S.S. et al. (2017). Hepatocyte hyperproliferation upon liver-specific co-disruption of thioredoxin-1, thioredoxin reductase-1, and glutathione reductase. *Cell Rep.* 19 (13): 2771–2781. https://doi.org/10.1016/j.celrep.2017.06.019.

Prummel, M.F. and Wiersinga, W.M. (2004). Thyroid autoimmunity and miscarriage. *Eur. J. Endocrinol.* 150 (6): 751–755.

Saccoccia, F., Angelucci, F., Boumis, G. et al. (2014). Thioredoxin reductase and its inhibitors. *Curr. Protein Pept. Sci.* 15 (6): 621–646.

Saunders, J.A., Pivetz, B.E., Voorhies, N., and Wilkin, R.T. (2016). Potential aquifer vulnerability in regions down-gradient from uranium in situ recovery (ISR) sites. *J. Environ. Manage.* 183: 67–83. https://doi.org/10.1016/j.jenvman.2016.08.049.

Shtutman, T.S.M. and Chagovets, R.V. (1971). Metabolic relations between vitamin E, selenium and sulfur-containing amino acids (literature review). *Vopr. Pitan.* 30 (5): 13–20.

Sunde, R.A., Zemaitis, E.T. II, Blink, A.B., and Lawinger, J.A. (2018). Impact of glutathione peroxidase-1 (Gpx1) genotype on selenoenzyme and transcript expression when repleting selenium-deficient mice. *Biol. Trace Elem. Res.* https://doi.org/10.1007/s12011-018-1281-6.

Urig, S. and Becker, K. (2006). On the potential of thioredoxin reductase inhibitors for cancer therapy. *Semin. Cancer Biol.* 16 (6): 452–465.

Yoshihara, A., Noh, J.Y., Mukasa, K. et al. (2015). Serum human chorionic gonadotropin levels and thyroid hormone levels in gestational transient thyrotoxicosis: is the serum hCG level useful for differentiating between active Graves' disease and GTT? *Endocr. J.* 62 (6): 557–560. https://doi.org/10.1507/endocrj.EJ14-0596.

# 17

# Signal Peptides

## Introduction

The membrane-bound proteins and secreted proteins are synthesized by ribosomes. These proteins are mostly associated with the endoplasmic reticulum (ER) membranes (Larson 1979). The ER membranes associated with ribosomes are also called as rough ER (RER) (Sabatini et al. 1975). The N-terminus of membrane-bound proteins and secreted proteins contain peptide regions that are termed as signal peptides or signal sequences (Austen and Ridd 1981). These signal peptides or signal sequences are usually consisting 13–36 amino acid residues and are predominantly hydrophobic peptides. Signal peptides are usually cleaved by signal peptidases upon translocation (Paetzel 2014). The primary amino acid sequence of signal peptides needs to be highly complementary to the protein for signal peptidases to cleave them completely (Auclair et al. 2012; Dalbey et al. 2017). The signal peptides are often used for the efficient secretion of many recombinant therapeutic proteins (RTPs) (Brady et al. 2015). Designing the signal peptide sequences that are complementary to the protein is very important for the efficient removal of signal peptides by the signal peptidases. Otherwise, the uncleaved residual signal peptides in RTPs might cause safety risks such as immunogenicity (Fonseca et al. 2018).

## Biological Significance of Signal Peptides

In the nascent polypeptides, the signal peptides are recognized by ribonucleoprotein complex (Zwieb 1989). The ribonucleoprotein complex is also called as signal recognition particle (SRP) (Zwieb 1989). The SRP recognizes the signal peptide on the nascent protein as it emerges from the ribosome exit side (Nagai et al. 2003). In eukaryotes, SRP consists of six proteins, i.e. SRP9, SRP14, SRP19, SRP54, SRP68, and SRP72 along with an RNA which is called 7SL RNA (Walter and Johnson 1994). In the ER, the complex of SRP and signal peptide induces translational elongation arrest so that the translational

*Co- and Post-Translational Modifications of Therapeutic Antibodies and Proteins*, First Edition. T. Shantha Raju.

complex and SRP binds to the SRP receptor (SR) (Koch et al. 2003). The SR is a heterodimeric complex that consists of two subunits, an α-subunit (SR-α) and a β-subunit (SR-β) (Jiang et al. 2008). The SR-α is encoded by SRPRA gene, whereas the SR-β is encoded by SRPRB gene. The translocation channel called translocon is associated with SR (Johnson and van Waes 1999; Halic and Beckmann 2005). Through the association of SR and translocon, the nascent polypeptide extrudes into the ER lumen (Jomaa et al. 2017). Following the passage through the ER, the signal peptide of the elongating protein is removed (Paetzel 2014). The signal peptide removal is mediated by proteases called signal peptidases (York et al. 1999). The signal peptidases are Ser proteases and are highly specific proteolytic enzymes (Dalbey et al. 2017). The signal peptidase present in the ER is a multiprotein complex often referred to as signal peptidase complex (SPC). The human SPC consists of five subunits. These five subunits are encoded by SPCS1, SPCS2, SPCS3, SEC11A, and SEC11C genes. The catalytic portion of SPC protein is encoded by SEC11A gene (Auclair et al. 2012). In Drosophila the SPC member, Spase12, is required for the development and cell differentiation (Haase Gilbert et al. 2013).

## Signal Peptides in RTPs

Most of the RTPs including mAbs are engineered to secret out of the host cells (Kerekes et al. 2017). Hence, most of recombinant proteins and mAbs gene are designed to contain signal peptide sequences. The signal peptide sequences must be complementary to the protein sequence so that the hydrophobicity of the signal peptide is maintained for it to interact with SRP, in turn with SR as well as with the signal peptidases. Failure to design the proper signal peptide sequence will result in incomplete removal of the signal peptides by signal peptidases. Hence, very often some RTPs and some mAbs may contain residual amounts of signal peptides containing proteins as product-related impurities. The uncleaved and/or the left-over residual signal peptide on RTPs and mAbs might be immunogenic (Kou et al. 2017; Fonseca et al. 2018). Hence, proper specifications may need to be set for signal peptide sequences for RTPs and mAbs during the quality control (QC) release specification setting (Brady et al. 2015).

## References

Auclair, S.M., Bhanu, M.K., and Kendall, D.A. (2012). Signal peptidase I: cleaving the way to mature proteins. *Protein Sci.* 21 (1): 13–25. https://doi.org/10.1002/pro.757.

Austen, B.M. and Ridd, D.H. (1981). The signal peptide and its role in membrane penetration. *Biochem. Soc. Symp.* (46): 235–258.

Brady, L.J., Scott, R.A., and Balland, A. (2015). An optimized approach to the rapid assessment and detection of sequence variants in recombinant protein products. *Anal. Bioanal. Chem.* 407 (13): 3851–3860. https://doi.org/10.1007/s00216-015-8618-1.

Dalbey, R.E., Pei, D., and Ekici, Ö.D. (2017). Signal peptidase enzymology and substrate specificity profiling. *Methods Enzymol.* 584: 35–57. https://doi.org/10.1016/bs.mie.2016.09.025.

Fonseca, J.A., McCaffery, J.N., Caceres, J. et al. (2018). Inclusion of the murine IgGκ signal peptide increases the cellular immunogenicity of a simian adenoviral vectored *Plasmodium vivax* multistage vaccine. *Vaccine* 36 (20): 2799–2808. https://doi.org/10.1016/j.vaccine.2018.03.091.

Haase Gilbert, E., Kwak, S.J., Chen, R., and Mardon, G. (2013). *Drosophila* signal peptidase complex member Spase12 is required for development and cell differentiation. *PLoS ONE* 8 (4): e60908. https://doi.org/10.1371/journal.pone.0060908.

Halic, M. and Beckmann, R. (2005). The signal recognition particle and its interactions during protein targeting. *Curr. Opin. Struct. Biol.* 15 (1): 116–125.

Jiang, Y., Cheng, Z., Mandon, E.C., and Gilmore, R. (2008). An interaction between the SRP receptor and the translocon is critical during cotranslational protein translocation. *J. Cell Biol.* 180 (6): 1149–1161. https://doi.org/10.1083/jcb.200707196.

Johnson, A.E. and van Waes, M.A. (1999). The translocon: a dynamic gateway at the ER membrane. *Annu. Rev. Cell Dev. Biol.* 15: 799–842.

Jomaa, A., Fu, Y.H., Boehringer, D. et al. (2017). Structure of the quaternary complex between SRP, SR, and translocon bound to the translating ribosome. *Nat. Commun.* 8: 15470. https://doi.org/10.1038/ncomms15470.

Kerekes, A., Hoffmann, O.I., Iski, G. et al. (2017). Secretion of a recombinant protein without a signal peptide by the exocrine glands of transgenic rabbits. *PLoS ONE* 12 (10): e0187214. https://doi.org/10.1371/journal.pone.0187214.

Koch, H.G., Moser, M., and Müller, M. (2003). Signal recognition particle-dependent protein targeting, universal to all kingdoms of life. *Rev. Physiol. Biochem. Pharmacol.* 146: 55–94.

Kou, Y., Xu, Y., Zhao, Z. et al. (2017). Tissue plasminogen activator (tPA) signal sequence enhances immunogenicity of MVA-based vaccine against tuberculosis. *Immunol. Lett.* 190: 51–57. https://doi.org/10.1016/j.imlet.2017.07.007.

Larson, B.L. (1979). Biosynthesis and secretion of milk proteins: a review. *J. Dairy Res.* 46 (2): 161–174.

Nagai, K., Oubridge, C., Kuglstatter, A. et al. (2003). Structure, function and evolution of the signal recognition particle. *EMBO J.* 22 (14): 3479–3485.

Paetzel, M. (2014). Structure and mechanism of *Escherichia coli* type I signal peptidase. *Biochim. Biophys. Acta* 1843 (8): 1497–1508. https://doi.org/10.1016/j.bbamcr.2013.12.003.

Sabatini, D.D., Ojakian, G., Lande, M.A. et al. (1975). Structural and functional aspects of the protein synthesizing apparatus in the rough endoplasmic reticulum. *Adv. Exp. Med. Biol.* 62: 151–180.

Walter, P. and Johnson, A.E. (1994). Signal sequence recognition and protein targeting to the endoplasmic reticulum membrane. *Annu. Rev. Cell Biol.* 10: 87–119.

York, I.A., Goldberg, A.L., Mo, X.Y., and Rock, K.L. (1999). Proteolysis and class I major histocompatibility complex antigen presentation. *Immunol. Rev.* 172: 49–66.

Zwieb, C. (1989). Structure and function of signal recognition particle RNA. *Prog. Nucleic Acid Res. Mol. Biol.* 37: 207–234.

# 18

# Sulfation of Proteins and Glycoproteins

## Introduction

Sulfur as an element is similar to oxygen and is available in nature in the native elemental form as well as in the form of inorganic salts such as sulfates, sulfides, etc. (Mozingo et al. 1945). In the periodic table of elements, sulfur is situated below oxygen. Atomic sulfur is abundant in nature and is a multivalent nonmetal inorganic element. It is an essential element for all life, and the organosulfur or inorganic sulfides are useful compounds in many biological systems (Davies and Sexton 1946). Cystine, Cys, and Met are the three-sulfur containing amino acids (White 1949). Biotin and thymine are the two sulfur-containing vitamins (Sigel et al. 1969). In addition, many cofactors such as glutathione, thioredoxin also contain sulfur as a constituent element (de la Torre et al. 1982).

Sulfur-containing amino acids, specifically Cys and Met, are present in almost all proteins as constituent monomeric units. In addition, many proteins carry sulfate group as substitution to side chain functional groups of some amino acids. For example, hydroxyl functional group present in Tyr, Thr, and Ser may carry a sulfate group. In addition, amine group of the amino acid residues present at the N-termini of some proteins may also carry a sulfate group (Renskers et al. 1980). Besides proteins, sulfate group is also present in heparin, heparin sulfate, chondroitin sulfate, and dermatan sulfate. In these biomolecules, sulfate group exists as a substitution on the hydroxyl groups of sugar molecules (Calatroni et al. 1969).

The presence of sulfate group in biomolecules is the result of sulfation reactions (Huttner 1988). Involvement of sulfate conjugation reactions in biotransformations were described more than 100 years ago (Jordan 1899). In the beginning, sulfation was thought to occur mainly with drug inactivation and metabolism (Bousquet 1962). In these processes, oxidative reactions initiate the drug inactivation and metabolism during phase I. In phase II, conjugation of a drug with sulfate or glutathione occurs that increases the water solubility of drugs. The increased water solubility facilitates the excretion of the drug's end products. In addition to drug inactivation and metabolism, sulfation is

*Co- and Post-Translational Modifications of Therapeutic Antibodies and Proteins*, First Edition. T. Shantha Raju.

also involved in the inactivation of hormones, catecholamines, activation, and inactivation of xenobiotics, removal of endogenous products such as bile salts, along with the modulation of structure and function of proteins (Mulder 1986).

Sulfation reactions of biomolecules occur both in the cytosol and in the Golgi compartments of living cells (Walter and Haass 2000). Both the cytosolic and the Golgi sulfation reactions are mediated by sulfotransferases (SulTs). Accordingly, the substrates for SulTs are available both in the cytosol as well as in the Golgi compartments (Falany 1997). However, the nature of substrates available for sulfation in the cytosol is different than the nature of substrates available for sulfation in the Golgi compartments. Also, the SulTs involved in the sulfation in the cytosol and the Golgi compartments are different (Brix et al. 1998). But the donor compound for sulfation is the same for both the cytosolic and the Golgi compartments sulfation reactions. This donor compound is an organosulfur molecule and is biosynthesized with in the cells (see section "Biosynthesis of PAPS").

The sulfur-containing inorganic molecules become available to living cells and organisms mainly through food sources and water (Garland et al. 1999). In living cells and organisms sulfation of biomolecules originates mainly from the organic sulfate molecules (Kline and Schoenhard 1970). These organic sulfate molecules are sulfated by inorganic sulfate molecules with in the living cells and organisms. The conversion of inorganic sulfate molecules into organic sulfate molecules and the subsequent sulfation of biomolecules in living organisms involved many different enzymes (Stanley et al. 1975). These biological processes of sulfation and the impact of sulfation on biomolecules are discussed in this chapter.

## Biosynthesis of PAPS

PAPS is an organic sulfate molecule that is also a high-energy cosubstrate (Vargas et al. 1994). PAPS is a sulfate donor for most of the biological sulfation reactions (Negishi et al. 2001). Biosynthesis of PAPS involves inorganic and organic molecules along with the enzymes as shown in Figure 18.1.

Biosynthesis of PAPS is a two-step process as depicted in Figure 18.1. In the first step of PAPS biosynthesis, adenosine triphosphate sulfurylase (ATPS) catalyzes the biosynthesis of adenosine 5'-phosphosulfate (APS) from adenosine triphosphate (ATP) and inorganic sulfate. In the second step, phosphorylation of APS is catalyzed by APS kinase (APK) that leads to the formation of PAPS. In mammalian cells, PAPS is synthesized in the cytosol and is translocated to the lumen of the Golgi by a PAPS-specific translocase (Ozeran et al. 1996). The limiting factor in the biosynthesis of PAPS is the synthesis of APS. The APS concentrations are tissue-specific and exhibits relatively low steady-state concentrations. PAPS concentrations are high in the liver, but even at the high

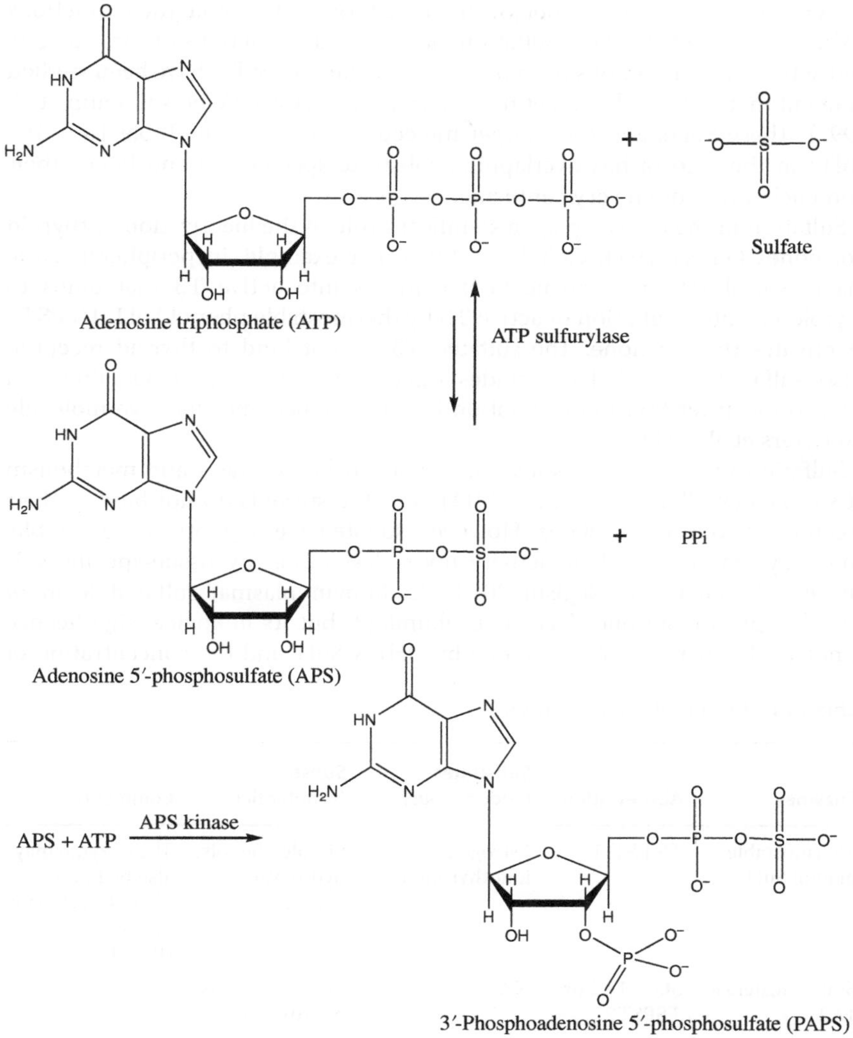

Adenosine triphosphate (ATP)

Sulfate

ATP sulfurylase

PPi

Adenosine 5′-phosphosulfate (APS)

APS + ATP → APS kinase

3′-Phosphoadenosine 5′-phosphosulfate (PAPS)

**Figure 18.1** Biosynthesis of PAPS (sulfate = inorganic sulfate from food sources).

concentrations, the amounts can be depleted in a matter of minutes during higher rate of sulfation of biomolecules.

## Sulfation Reactions in the Cytosol

Sulfation reactions or the sulfation conjugation reactions in the cytosol have been studied by endocrinologists as well as by pharmacologists. Sulfation in the cytosol involves a number of different targets that alter their functions (Weinshilboum et al. 1997). Sulfation inactivates the functions of some targets and activates the functions of some targets. So far, five SulTs have been studied in detail that sulfates the target molecules in the cytosol (Weinshilboum et al. 1997). These SulTs and their target molecules are shown in Table 18.1. The SulTs in the cytosol has overlapping substrate specificities and hence their nomenclature is often very confusing.

Sulfation in the cytosol plays a significant role in the inactivation of thyroid hormones (Visser 1994; Coughtrie 2016). For example, in peripheral tissue inactive prohormone tyroxine (T4) converts into active T3 that binds to thyroid receptor. Sulfation of active T3 by thermostable phenol SulT TS PST1 inactivates the hormone. The sulfated T3 do not bind to thyroid receptor. Also sulfated T3 and T4 degrades rapidly (Visser 1994). Dopamine is a neurotransmitter that upon sulfation by TL PST becomes inactive molecule (Renskers et al. 1980).

Sulfation in the cytosol is also important in homeostasis and metabolism of steroids (Wollam and Antebi 2011). Sulfated steroids do not bind to their receptors (Cook et al. 2009). However, sulfated steroids are more soluble, and they may regain their activity upon desulfation by tissue-specific sulfatases (Kodama and Negishi 2013). In human plasma, sulfated form of dehydroepiandrosterone (DHEA) is abundant, but its biological significance is not yet known. DHEA is sulfated by DHEA SulT and the concentration of

**Table 18.1** Human SulTs in the cytosol.

| Enzymes | Abbreviations | Substrates (endogenous) | Substrates (xenobiotic) | Comments |
|---|---|---|---|---|
| Thermostable phenol SulT-1 | TS PST-1 | Estrogen, iodothyronine | Simple phenols, hydroxyary-lamines | The enzyme may also be known as phenol-preferring phenol ST (P-PST) |
| Sulfotransferase 1A2 | SULT1A2 or TSPST2 | NA | Simple phenols, hydroxyary-lamines | — |
| Thermolabile phenol SulT | TL PST | Catecholamines, iodothyronine | Minoxidil, 1-napthol, salbutamol | Targets activated upon sulfation |
| Estrogen SulT | EST | Estrogen | Equilenin, ethylastradiol | — |
| Dehydroepian-drosterone SulT | DHEAST | Androgens, bile salts, cholesterol, estrogen | Aliphatic alcohols, benzylic alcohols | The enzyme is also known as hdroxysteroid SulT (HST) |

sulfated DHEA decreases with age (Prough et al. 2016). The DHEA SulT is also involved in the sulfation of cholesterol and the secondary bile salts (Alnouti 2009). The secondary bile salts, secreted by the liver and by the bacteria present in the intestine, are hepato-toxic (Rodrigues et al. 2014). The sulfated bile salts excrete through urine (van de Laarschot et al. 2016).

In addition to inactivation and subsequent excretion of sulfated molecules, sulfation in the cytosol is also involved in the activation of many different molecules. For example, minoxidil (6-(piperidin-1-yl) pyrimidine-2,4-diamine 3-oxide) is an antihypertensive agent (Baker et al. 1994). Native form of minoxidil is a prodrug and is not active. TL PST is involved in the sulfation of minoxidil to its sulfated form. The minoxidil sulfate is the actual active molecule that is used as antihypertensive agent (Baker et al. 1994).

## Sulfation in the Golgi Compartments

In the Golgi compartments, the biomolecules that undergo sulfation include proteins, glycoproteins, glycosylaminoglycans (or glycosaminoglycans (GAGs), and proteoglycans (Walter and Haass 2000). Unlike the sulfation in the cytosol, the sulfation in the Golgi is the actual post-translational modification (PTM). Most of the protein sulfation in the Golgi occurs on the Tyr residues. Sulfation of GAGs and proteoglycans also occurs in the Golgi compartments. The SulTs involved in the sulfation of Tyr residues are different than the SulTs involved in the sulfation of glycan moieties of glycoproteins, glycosylaminoglycans, and proteoglycans. Hence, the SulTs located in the Golgi compartments do not have overlapping specificity for acceptor molecules. In addition, the mechanism of sulfation of Tyr residues and glycan residues is also different. Hence, these topics are discussed separately in the following sections.

## Sulfation of Tyrosine Residues

More than 50 years ago Bettelheim (1954) first reported the *O*-sulfation of Tyr residues in bovine fibrinopeptide. Later on, Huttner et al. discovered that proteins isolated from cells and tissues of different species contain sulfated Tyr residues (Stadler et al. 1983). These pioneering research work suggested that the Tyr sulfation is a very common PTM of proteins. Later on, it was discovered that the Tyr sulfation is found to be present in animals and plants but not present in yeast and prokaryotes (Moore 2003). The protein Tyr sulfation is a stable modification and in animals it excretes through the urine (Christensen et al. 2008). So far, no desulfation of Tyr sulfation by sulfatases has been observed.

## Mechanism of Tyrosine Sulfation

Conversion of inorganic sulfate molecules into organic sulfate molecules in living organisms is an enzymatic process in which sulfurylases are involved (Seubert et al. 1979). As shown in Figure 18.1, ATP sulfurylase is involved in the sulfation of ATP into APS and the APK converts the APS into PAPS. Specific SulTs are involved in the transfer of activated sulfate group from PAPS to different biomolecules.

Tyr sulfation occurs in the trans-Golgi network of the Golgi compartments. The Tyr sulfation reactions are mediated by an enzyme called tyrosylprotein sulfotransferase (TPST, EC 2.8.2.20). TPST is a membrane-bound protein and the active site of the enzyme is located on the luminal side (Yang et al. 2015). TPST catalyzes the transfer of sulfate from PAPS to the hydroxyl functional group on tyrosine residues (see Figure 18.1). The end products of this enzyme reactions are tyrosine *O*-sulfate as an ester and 3′,5′-adenosine diphosphate (3′,5′-ADP).

In 1990, Niehrs and Huttner purified TPST from bovine adrenal medulla and characterized its enzymatic activity (Niehrs and Huttner 1990; Niehrs et al. 1990). Later on, it was determined that there are two TPST enzymes termed as TPST1 and TPST2 (Westmuckett et al. 2008). These two enzymes appear to contain a highly conserved domain that is responsible for SulT activity. Sequence analysis showed that both TPST1 and TPST2 are similar enzymes with different substrate specificity (Tanaka et al. 2017). For example, sulfation site peptides of human C4 and heparin cofactor II were sulfated more efficiently by TPST1 than TPST2 (Hoffhines et al. 2006). Both TPST1 and TPST2 enzymes are glycoproteins. Protein sequence of TPST1 enzyme contains 370 amino acid residues, whereas TPST2 contains 377 amino acid residues. Both enzymes contain a short, 7-amino acid residue, cytosolic tail. This short cytosolic tail may be involved in the intracellular transport and localization of the enzymes (Tanaka et al. 2017).

Tyr sulfation occurs on the secretory proteins, membrane proteins, and also on the lysosomal proteins (Stone et al. 2009). More specifically, human proteins such as adhesion molecules, G-protein-coupled receptors, coagulation factors, serine protease inhibitors, extracellular matrix (ECM) proteins, hormones, etc. are known to contain Tyr sulfation (Sasaki 2012). In any given organism, Tyr sulfation occurs up to 1% of Tyr residues of proteins. Hence, protein Tyr sulfation is one of the most common PTM and is ubiquitous in all tissues of living organisms (Yang et al. 2015).

## Biological Functions of Tyr Sulfation

Knock-out experiments of *TPST* genes in mice has shown that the Tyr sulfation affects the growth of the mice (Ouyang et al. 2002). Specifically, absence of Tyr sulfation affects the body weight, fecundity, and also postnatal viability. Mammalian sulfatases cannot easily degrade Tyr sulfation, and hence, it is an irreversible modification in vivo. Tyr sulfation is involved in protein–protein interactions (Niehrs et al. 1994). For example, Tyr sulfation is required for cholecystokinin, a peptide hormone, to bind to its receptor (Vishnuvardhan and Beinfeld 2000). Tyr sulfation increases binding affinity of hirudin to thrombin and factor VIII to von Willerbrand factor (Leyte et al. 1991; Hsieh et al. 2012). Tyr sulfation increases the intracellular processing of gastrin, a peptide hormone. Secretion kinetics of Drosophila yolk protein 2 is altered upon Tyr sulfation. Tyr sulfation is required for P-selectin to bind to P-selectin glycoprotein ligand 1 (PSGL1) (Westmuckett et al. 2011).

## Sulfation of Glycosaminoglycans (GAGs) and Proteoglycans

In addition to Tyr sulfation, another class of biomolecules that undergo sulfation in the Golgi compartments are GAGs and proteoglycans. The GAGs are long unbranched polysaccharides containing disaccharide repeating units of either GalNAc or GlcNAc and uronic acid such as glucuronic acid or iduronic acid. The GAGs are also called as mucopolysaccharides. These are highly negatively charged polysaccharides with high viscosity in solution. Primarily, the GAGs are located on the cell surface and/or in the ECM. In some cell types, the GAGs are also found in secretory vesicles. There are seven categories of GAGs: hyaluronic acid, dermatan sulfate, chondroitin sulfate (chondroitin-4-sulfate and chondroitin-6-sulfate), heparin, heparin sulfate, and keratin sulfate (Valcarcel et al. 2017). Other than hyaluronic acid, the remaining six categories of GAGs are sulfated. Unlike other GAGs, hyaluronic acid is also not found to be covalently linked to proteins. Many of the other GAGs are linked to a protein core and hence are also called as proteoglycans. The GAGs are linked by xylose to Ser residues of core proteins such as perlecan, syndecan, and glypican (Rhiner and Hengartner 2006).

Heparin, heparin sulfate, chondroitin sulfate, and dermatan sulfate are biosynthesized in the Golgi apparatus (Muir 1969). The core proteins are biosynthesized in the rough endoplasmic reticulum that undergo O-glycosylation to form the GAGs. The *O*-xylosyltransferase is the initiator of the biosynthesis of GAGs on the core proteins. However, keratin sulfates are different as they contain poly-*N*-acetyllactosamine chains on either *N*-linked glycans or *O*-linked glycans. Additionally, hyaluronic acids are not synthesized in the Golgi. They are synthesized by the integral membrane synthases and are immediately secreted as dynamically elongated disaccharide repeating units containing polysaccharide chains.

SulTs specific to different GAGs add sulfate groups in the Golgi by transferring sulfate from PAPS to sugar residues. With the exception of heparan sulfate

**Table 18.2** GAGs composition and sulfation.

| GAGs | Disaccharide composition | Sulfation | Comments |
|---|---|---|---|
| Chondroitin 4-sulfate | GlcA and GalNAc | GalNAc-4-sulfate | — |
| Chondroitin 6-sulfate | GlcA and GalNAc | GalNAC-6-sulfate | — |
| Dermatan sulfates | IdoA or GlcA and GalNAc | GalNAc-4-sulfate | — |
| Heparin | IdoA or GlcA and GlcNAc | IdoA-2-sulfate GlcNAc-6-sulfate | — |
| Heparan sulfates | IdoA or GlcA an GlcNAC | IdoA-2-sulfate GlcNAC-6-sulfate | Heparan sulfates contain less sulfate than heparins |
| Keratan sulfates | Gal and GlcNAc | GlcNAc-6-sulfate | — |

specific SulT, 3-*O*-sulfotransferase-1, the GAGs-specific SulTs are type II transmembrane proteins (Baldwin et al. 2013). The sugar composition and the sulfate linkage sites for GAGs are summarized in Table 18.2. The addition of sulfate groups by SulTs to sugar residues starts with the initiation of biosynthesis of GAGs. However, *N*-sulfation in heparan sulfate and sulfation in keratin sulfate follows different pathways.

The *N*-sulfation of selected GlcNAc residues in heparan sulfate is carried out by a bifunctional *N*-deacetylase/*N*-sulfotransferase (NDST) enzymes (Dagälv et al. 2015). So far four mammalian NDST enzymes, i.e. NDST1, NDST2, NDST3, and NDST4, have been identified (Aikawa et al. 2001). These enzymes first deacetylate the GlcNAc residues and then sulfate them by transferring sulfate group from PAPS. *N*-Sulfation is very unique to heparan sulfate and heparan that do not occur on other GAGs. The NDST1 and NDST2 enzymes have a very wide tissue distribution. However, NDST3 and NDST4 are mostly expressed during the embryonic period (Jao et al. 2016). The four different NDSTs may have different specificities for substrates (Aikawa et al. 2001).

Following the *N*-sulfation, epimerization of GlcA at C5 into IdoA takes place. After this epimerization, 2-*O*-sulfation of mostly IdoA residues and to some extent GlcA residues occurs that is mediated by 2-*O*-SulT. The 2-*O*-SulT forms complexes with C5-epimerase and this might be the reason epimerization follows with the 2-*O*-sulfation (Pinhal et al. 2001). In addition to 2-*O*-sulfation, heparan sulfate also undergoes 6-*O*-sulfation of GlcNAc residues that is mediated by GlcNAc 6-*O*-SulT (Grunwell and Bertozzi 2002). There are three heparan sulfate-specific GlcNAc 6-*O*-SulTs exists, and they have very similar substrate specificities (Bowman and Bertozzi 1999). Further, 3-*O*-sulfation of GlcNAc residues have been observed which is mediated by 3-*O*-SulTs (Kusche et al. 1990). At least six 3-*O*-SulTs specific to heparan sulfate have been identified. Some of the 3-*O*-SulTs have very distinct substrate specificities.

Chondroitin sulfates contain repeating disaccharide units that are composed of GalNAc and GlcA residues (Campo 1970). Based on the sulfation pattern, they are distinguished as chondroitin 4-sulfate and chondroitin 6-sulfate (Nandini and Sugahara 2006). GalNAc 4-*O*-SulT mediates the transfer of sulfate from PAPS to GalNAc residues at O-4 position, whereas GalNAc 6-*O*-SulT mediates the transfer of sulfate from PAPS to GalNAc residues at O-6 position of chondroitin 4-sulfate and chondroitin 6-sulfate, respectively. Three distinct GalNAc 4-*O*-STs have been identified in humans, whereas two different GalNAc 6-*O*-SulTs have been identified in human (Honke and Taniguchi 2002).

Dermatan sulfate (DS) is composed of repeating disaccharide units containing GalNAc and IdoA or GlcA (Prydz 2015). The proportion of IdoA in DS varies with the GAGs chain and some DS chains may contain up to 100% of IdoA. DS contains sulfate groups at O-4 position of GalNAc residues and the SulT's involved in the transfer of sulfate from PAPS are the same as the one involved in CS. Both CS and DS may also contain sulfate at O-2 position of GlcA and/or IdoA residues (Honke and Taniguchi 2002). Only one 2-*O*-SulT is involved in the biosynthesis of 2-*O*-sulfate containing CS and DS.

Keratan sulfate (KS) contains polylactosamine chains that consists of repeating disaccharide units composed of Gal and GlcNAc residues (Chakrabarti and Park 1980). These repeating disaccharide units are found on both *N*- and *O*-linked oligosaccharides of KS. The KS chains are predominantly found in cornea and cartilage (Pomin 2015). The KS found in cornea are linked to core protein through *N*-linked oligosaccharide chains and are highly sulfated (Funderburgh 2002). However, the KS found in cartilage is linked to core proteins through *O*-linked oligosaccharides to Ser or Thr and are shorter chains than the KS found in cornea (Funderburgh 2002). Both *N*- and *O*-linked KS chains are sulfated by the same SulTs (Pomin 2015).

## Biological Functions of GAGs

In the beginning, GAGs were considered as the structural components of connective tissues and ECM (Schiller 1966; Hay 1981). But later on, GAGs have been shown to play very important biological functions (Quittot et al. 2017). GAGs are large molecules and contain very high negative charges. The high negative charges are central to their biological functions and for their interactions with other biomolecules including proteins. For example, binding of heparin to antithrombin III is dependent on sulfation at specific position of heparin molecules (Rosenberg 1978). GAGs are involved in cell adhesion, cell migration, cell proliferation, neural development, wound healing, etc. Heparin is used as anticoagulant in the clinic and also used to treat thromboembolic diseases in humans (Rosenberg 1977). Interactions of dermatan sulfate with fibroblast growth factor (FGF), specifically FGF-2 and FGF-7 are implicated in cell proliferation and wound repair (Asada et al. 2009). CS plays critical roles in inhibiting axonal growth, pathogen infection, etc. (Dyck and Karimi-Abdolrezaee 2015). KS is involved in the maintenance of tissue hydration (Watson and Young 2004). KS is also involved in macrophage adhesion, regulation of embryo implantation, etc. In addition, KS also affects the motility of endothelial cells of cornea.

## Sulfation in RTPs

Sulfation in many recombinant therapeutic proteins (RTPs) has been reported (White et al. 1998). Yolk protein 2 (YP2) of *Drosophila melanogaster* was expressed in mouse fibroblasts and the recombinant YP2 (rYP2) was shown to contain Tyr sulfate residues similar to endogenous YP2 purified from the flies suggesting that it is possible to produce sulfated RTPs (Friederich et al. 1988). However, recombinant hirudin produced in yeast was found to be not sulfated and nonsulfated hirudin showed at least 10 times lower affinity to bind to thrombin compared to the endogenous sulfated hirudin (Niehrs et al. 1990). The endogenous hirudin contain sulfated Tyr residues at position 63. The in vitro sulfation of recombinant hirudin using TPST restored its binding affinity to thrombin (Niehrs et al. 1990). Plasma-derived human FVIII has been shown to contain three Tyr sulfation sites at residues Tyr718, Tyr719, and Tyr723 (Hortin 1990). The rFVIII produced in Chinese hamster ovary (CHO) cells is heterogeneous and the heterogeneity is shown to be partly due to incomplete sulfation (Mikkelsen et al. 1991). The incomplete sulfation of rFVIII affects its ability to bind to thrombin and hence affects its biological properties (Mikkelsen et al. 1991). Recombinant human fibrinogen expressed in baby hamster kidney (BHK) cells contain Tyr sulfate similar to the endogenous protein and is biologically active (Farrell et al. 1991). PSGL-1 is a mucin-like glycoprotein found on the cell surface of leukocytes (Vandendries et al. 2004). PSGL-1 is a dimeric molecule that binds to P- and L-selectins. Accordingly, binding of neutrophils to selectins requires PSGL-1 which promotes cell adhesion during the inflammatory response. The amino terminus of PSGL-1 contains a sulfated Tyr residue that is required for P-selectin binding along with the presence of sialyl Lewis x epitope containing *O*-glycans (Fieger et al. 2003). Hence, the recombinant PSGL-1 must be produced in appropriate cell lines that contain proper glycosylation and sulfation machinery so that it will bind to P-selectin. For example, PSGL-1 produced in the presence of inhibitors that inhibits sulfation binds weakly to P-selectin compared to the sulfated PSGL-1 (Pouyani and Seed 1995). Reduced sulfation of Tyr residues of recombinant factor IX has been observed that might contribute to its reduced biological activity (White et al. 1997).

In addition to Tyr sulfation, xylose-based core structures of GAGs have been observed to be present in some RTPs (Spahr et al. 2014). Immunoglobulin G (IgG) Fc fusion proteins containing G-S-G-G-G-G (GSG4) peptide linker expressed in CHO cells are reported to contain xylose type glycans (Spahr et al. 2013). About 50% of such molecules appears to contain tetrasaccharide core structures of GAGs (Spahr et al. 2013). A very low levels of sulfation, and phosphorylation has been observed in these molecules. Another IgG fusion protein with GS linker has been shown to contain xylose-based glycans that contain sulfated glycans (Spencer et al. 2013).

GAGs have many therapeutic applications and are being used to treat many human diseases. Modified GAGs such as low molecular weight heparins are also used as therapeutics to treat many human diseases. The GAGs are purified from endogenous sources and often modified, chemically or enzymatically. The recombinant GAGs are rare because of the complexity of production of these molecules. However, efforts are underway to produce recombinant GAGs using a combination of recombinant technology, cell engineering, and chemoselective and/or chemoenzymatic synthesis. For example, using *Escherichia coli* K5 and some biosynthetic enzymes, a bioengineered heparin was produced to use as a replacement for endogenous heparin isolated from animal sources (Restaino et al. 2013). In this case heparosan, a version of heparin without sulfation, was produced using *E. coli*, which was *N*-sulfated using chemoselective methods and then the sulfate at O-6 position of GlcNAc was added using 6-OST-1 enzyme (Restaino et al. 2013).

## Analysis of Sulfation in Biomolecules

Sulfation introduces a strong negative charge to biomolecules such as proteins, peptides, and glycans. Such a strong negative charge can alter the physicochemical and biological properties of the biomolecules. The change in physicochemical and biological properties can be used to purify the sulfated molecules from the nonsulfated molecules. The negative charge also changes the electrochemical mobility of the biomolecules. Hence, sulfated molecules can be separated by sodium dodecyl sulfate – polyacrylamide gel electrophoresis (SDS-PAGE), capillary electrophoresis (CE), ion-exchange chromatography (IEC), etc., methods. In earlier days, sulfation was mostly detected by radiolabeling. Sulfation in proteins was also detected using antibodies that bind to sulfated protein epitopes. However, the limitation of such antibodies was that they do not detect a wide array of sulfated protein epitopes. Recent advances in the analysis of biomolecules aided the detection of sulfation much more efficiently. These recent methods are described below, separately for Tyr sulfation and GAGs sulfation.

## Analysis of Tyr Sulfation

Tyr sulfation was used to be characterized by radiolabeling in which radiolabeled sulfate was incorporated into target proteins within the cells. The radiolabeled target proteins were subsequently purified, degraded by proteases or by alkaline hydrolysis followed by thin-layer electrophoresis to detect the radiolabeled Tyr. However, this method has many disadvantages and hence, recent advances have been made to characterize Tyr sulfation more conveniently. Specific immunoassays have been developed to detect and quantitate Tyr sulfation (Hoffhines et al. 2006). However, difficulty of this immunoassay method is the discrepancy of antibodies to distinguish the structurally similar phosphorylated Tyr residues (Ross et al. 1981). Since sulfation introduces a highly negatively charged sulfate group to proteins, many chromatographic techniques were used to separate the sulfated proteins from nonsulfated proteins. Among these chromatographic techniques, ion-exchange high performance liquid chromatography (IE-HPLC) and reversed-phase high performance liquid chromatography (RP-HPLC) methods have been extensively used. Additionally, digestion with aryl sulfatases before and after chromatographic separation is also very useful to characterize Tyr sulfation. In addition to chromatographic methods, mass spectral methods are also being used to detect Tyr sulfation. Although sulfate and phosphate groups have similar mass, sulfate groups are more unstable for mass spectral analysis compared to phosphate groups. However, during the MS/MS analysis, phosphate group forms a stable immonium ion, whereas the immonium ion of sulfate group has not been detected thus amenable for MS analyses.

## Analysis of Sulfated Glycoconjugates

Different analytical methods are being used to analyze sulfated glycoconjugates. Many of these methods were developed to measure GAGs as they are widely distributed and play both physiological and pathological roles in the living organisms. The analytical methods used to analyze GAGs can be broadly categorized as, colorimetric methods, electrophoretic methods, chromatographic methods, and mass spectral methods.

### Colorimetric Methods to Analyze Sulfated Glycoconjugates

The colorimetric methods used to analyze GAGs include toluidine blue staining, alcian blue staining, and dimethylmethylene blue staining. All these three colorimetric methods are mostly used in conjunction with some separation methods such as electrophoretic separation methods to separate GAGs from other biomolecules as well as by themselves. Toluidine blue is a thiazine and a cationic dye. The positive charge of the dye neutralizes the negative charge of sulfate groups of GAGs and produces colors of different wavelength depending on the concentration of GAGs. The method has been used to detect low levels of GAGs in tissue extracts. The toluidine blue staining method has also been widely used in pathohistology to detect GAGs in tissue sections. Since toluidine blue nonspecifically recognizes negatively charged molecules, the method is not very useful for quantitation. Alcian blue is a copper-based tetravalent cationic dye and contains a hydrophobic core. The reagent interacts with GAGs through its cation and is also used in conjunction with silver staining of PAGE gels. Similar to toluidine blue, alcian blue also nonspecifically binds other negatively charged molecules. Dimethylmethylene blue is also a cationic dye that binds to GAGs through ionic interactions and exhibits a characteristic absorbance shift upon binding to GAGs. The advantage of this dye is that it can be used both in solution as well as in solid-phase.

## Electrophoretic Methods to Analyze Sulfated Glycoconjugates

Among the electrophoretic methods, both gel-electrophoresis and CE methods are used to separate GAGs. PAGE-electrophoresis separation followed by detection using the colorimetric dyes are used to analyze GAGs. More recently, CE methods are being used to separate GAGs based on size, purity, and charge. Following the CE separation, the GAGs are detected by direct or indirect UV, generation of metal complexes such as copper complexes and also by mass spectrometry.

## Chromatographic Methods to Analyze Sulfated Glycoconjugates

Chromatographic methods used to analyze GAGs are paper chromatography, thin-layer chromatography, gas chromatography or gas–liquid chromatography, and high performance liquid chromatography (HPLC). Both paper and thin-layer chromatographic methods are used to separate GAGs, which were detected using colorimetric spraying reagents. For gas chromatography or gas-liquid chromatography, the GAGs are hydrolyzed and the resulting monosaccharides were derivatized using derivatization chemistry such as acetylation or methylation or converting them into trimethylsilyl ethers followed by separation and detection by gas-chromatographic analysis. Recently, HPLC methods have been adopted to analyze and also to quantitate GAGs. For HPLC analysis, GAGs are depolymerized by treating with enzymes such as heparinase, keratanase, chondroitinase, followed by HPLC separation of the resulting disaccharides. Often, the disaccharides are labeled with fluorescent tags for detection. Liquid chromatography–mass spectrometry (LC–MS) methods are also used to separate and to identify the GAGs disaccharides.

### Analysis of Sulfated Glycoconjugates by Mass Spectrometry

More recently, high-resolution mass spectrometric methods are also used to analyze GAGs, qualitatively and quantitatively. For MS analysis, the GAGs polysaccharides are digested with enzymes such as heparinase, keratanase, chondroitinase, etc. followed by separation using liquid-chromatography and detection by mass spectrometry. The mass spectral methods are used to detect and quantitate all 23 different disaccharides (8 from CS/DS, 1 from HA, 12 from HS, and 2 from KS) derived from GAGs (Gill et al. 2013; Osago et al. 2014).

## References

Aikawa, J., Grobe, K., Tsujimoto, M., and Esko, J.D. (2001). Multiple isozymes of heparan sulfate/heparin GlcNAc N-deacetylase/GlcN N-sulfotransferase. Structure and activity of the fourth member, NDST4. *J. Biol. Chem.* 276 (8): 5876–5882.

Alnouti, Y. (2009). Bile acid sulfation: a pathway of bile acid elimination and detoxification. *Toxicol. Sci.* 108 (2): 225–246. https://doi.org/10.1093/toxsci/kfn268.

Asada, M., Shinomiya, M., Suzuki, M. et al. (2009). Glycosaminoglycan affinity of the complete fibroblast growth factor family. *Biochim. Biophys. Acta* 1790 (1, 1): 40–48. https://doi.org/10.1016/j.bbagen.2008.09.001.

Baker, C.A., Uno, H., and Johnson, G.A. (1994). Minoxidil sulfation in the hair follicle. *Skin Pharmacol.* 7 (6): 335–339.

Baldwin, J., Shukla, D., and Tiwari, V. (2013). Members of 3-*O*-sulfotransferases (3-*O*ST) family: a valuable tool from zebrafish to humans for understanding herpes simplex virus entry. *Open Virol. J.* 7: 5–11. https://doi.org/10.2174/1874357901307010005.

Bettelheim, F.R. (1954). Tyrosine-*O*-sulfate in a peptide from fibrinogen. *J. Am. Chem. Soc.* 76 (10): 2838–2839. https://doi.org/10.1021/ja01639a073.

Bousquet, W.F. (1962). Pharmacology and biochemistry of drug metabolism. *J. Pharm. Sci.* 51: 297–309.

Bowman, K.G. and Bertozzi, C.R. (1999). Carbohydrate sulfotransferases: mediators of extracellular communication. *Chem. Biol.* 6 (1): R9–R22.

Brix, L.A., Nicoll, R., Zhu, X., and McManus, M.E. (1998). Structural and functional characterisation of human sulfotransferases. *Chem. Biol. Interact.* 109 (1–3): 123–127.

Calatroni, A., Donnelly, P.V., and Di Ferrante, N. (1969). The glycosaminoglycans of human plasma. *J. Clin. Invest.* 48 (2): 332–343.

Campo, R.D. (1970). Protein—polysaccharides of cartilage and bone in health and disease. *Clin. Orthop. Relat. Res.* 68: 182–209.

Chakrabarti, B. and Park, J.W. (1980). Glycosaminoglycans: structure and interaction. *CRC Crit. Rev. Biochem.* 8 (3): 225–313.

Christensen, B., Petersen, T.E., and Sørensen, E.S. (2008). Post-translational modification and proteolytic processing of urinary osteopontin. *Biochem. J.* 411 (1, 1): 53–61.

Cook, I.T., Duniec-Dmuchowski, Z., Kocarek, T.A. et al. (2009). 24-Hydroxycholesterol sulfation by human cytosolic sulfotransferases: formation of monosulfates and disulfates, molecular modeling, sulfatase sensitivity, and inhibition of liver x receptor activation. *Drug Metab. Dispos.* 37 (10): 2069–2078. https://doi.org/10.1124/dmd.108.025759.

Coughtrie, M.W. (2016). Function and organization of the human cytosolic sulfotransferase (SULT) family. *Chem. Biol. Interact.* 259 (Pt A): 2–7. https://doi.org/10.1016/j.cbi.2016.05.005.

Dagälv, A., Lundequist, A., Filipek-Górniok, B. et al. (2015). Heparan sulfate structure: methods to study N-sulfation and NDST action. *Methods Mol. Biol.* 1229: 189–200. https://doi.org/10.1007/978-1-4939-1714-3_17.

Davies, W.H. and Sexton, W.A. (1946). Chemical constitution and fungistatic action of organic sulphur compounds. *Biochem. J.* 40 (3): 331–334.

Dyck, S.M. and Karimi-Abdolrezaee, S. (2015). Chondroitin sulfate proteoglycans: key modulators in the developing and pathologic central nervous system. *Exp. Neurol.* 269: 169–187. https://doi.org/10.1016/j.expneurol.2015.04.006.

Falany, C.N. (1997). Sulfation and sulfotransferases. Introduction: changing view of sulfation and the cytosolic sulfotransferases. *FASEB J.* 11 (1): 1–2.

Farrell, D.H., Mulvihill, E.R., Huang, S.M. et al. (1991). Recombinant human fibrinogen and sulfation of the gamma' chain. *Biochemistry* 30 (39): 9414–9420.

Fieger, C.B., Sassetti, C.M., and Rosen, S.D. (2003). Endoglycan, a member of the CD34 family, functions as an L-selectin ligand through modification with tyrosine sulfation and sialyl Lewis x. *J. Biol. Chem.* 278 (30): 27390–27398.

Friederich, E., Baeuerle, P.A., Garoff, H. et al. (1988). Expression, tyrosine sulfation, and secretion of yolk protein 2 of *Drosophila melanogaster* in mouse fibroblasts. *J. Biol. Chem.* 263 (29): 14930–14938.

Funderburgh, J.L. (2002). Keratan sulfate biosynthesis. *IUBMB Life* 54 (4): 187–194.

Garland, S.A., Hoff, K., Vickery, L.E., and Culotta, V.C. (1999). *Saccharomyces cerevisiae* ISU1 and ISU2: members of a well-conserved gene family for iron-sulfur cluster assembly. *J. Mol. Biol.* 294 (4): 897–907.

Gill, V.L., Aich, U., Rao, S. et al. (2013). Disaccharide analysis of glycosaminoglycans using hydrophilic interaction chromatography and mass spectrometry. *Anal. Chem.* 85 (2): 1138–1145. https://doi.org/10.1021/ac3030448.

Grunwell, J.R. and Bertozzi, C.R. (2002). Carbohydrate sulfotransferases of the GalNAc/Gal/GlcNAc6ST family. *Biochemistry* 41 (44): 13117–13126.

Hay, E.D. (1981). Extracellular matrix. *J. Cell Biol.* 91 (3 Pt. 2): 205s–223s.

Hoffhines, A.J., Damoc, E., Bridges, K.G. et al. (2006). Detection and purification of tyrosine-sulfated proteins using a novel anti-sulfotyrosine monoclonal antibody. *J. Biol. Chem.* 281 (49): 37877–37887.

Honke, K. and Taniguchi, N. (2002). Sulfotransferases and sulfated oligosaccharides. *Med. Res. Rev.* 22 (6): 637–654.

Hortin, G.L. (1990). Sulfation of tyrosine residues in coagulation factor V. *Blood* 76 (5): 946–952.

Hsieh, Y.S., Taleski, D., Wilkinson, B.L. et al. (2012). Effect of O-glycosylation and tyrosine sulfation of leech-derived peptides on binding and inhibitory activity against thrombin. *Chem. Commun. (Camb.)* 48 (10): 1547–1549. https://doi.org/10.1039/c1cc14773k.

Huttner, W.B. (1988). Tyrosine sulfation and the secretory pathway. *Annu. Rev. Physiol.* 50: 363–376.

Jao, T.M., Li, Y.L., Lin, S.W. et al. (2016). Alteration of colonic epithelial cell differentiation in mice deficient for glucosaminyl *N*-deacetylase/*N*-sulfotransferase 4. *Oncotarget* 7 (51): 84938–84950. https://doi.org/10.18632/oncotarget.12915.

Jordan, E.O. (1899). bacillus pyocyaneus and its pigments. *J. Exp. Med.* 4 (5–6): 627–647.

Kline, B.C. and Schoenhard, D.E. (1970). Biochemical characterization of sulfur assimilation by *Salmonella pullorum*. *J. Bacteriol.* 102 (1): 142–148.

Kodama, S. and Negishi, M. (2013). Sulfotransferase genes: regulation by nuclear receptors in response to xeno/endo-biotics. *Drug Metab. Rev.* 45 (4): 441–449. https://doi.org/10.3109/03602532.2013.835630.

Kusche, M., Torri, G., Casu, B., and Lindahl, U. (1990). Biosynthesis of heparin. Availability of glucosaminyl 3-*O*-sulfation sites. *J. Biol. Chem.* 265 (13): 7292–7300.

van de Laarschot, L.F., Jansen, P.L., Schaap, F.G., and Olde Damink, S.W. (2016). The role of bile salts in liver regeneration. *Hepatol. Int.* 10 (5): 733–740. https://doi.org/10.1007/s12072-016-9723-8.

Leyte, A., van Schijndel, H.B., Niehrs, C. et al. (1991). Sulfation of Tyr1680 of human blood coagulation factor VIII is essential for the interaction of factor VIII with von Willebrand factor. *J. Biol. Chem.* 266 (2): 740–746.

Mikkelsen, J., Thomsen, J., and Ezban, M. (1991). Heterogeneity in the tyrosine sulfation of Chinese hamster ovary cell produced recombinant FVIII. *Biochemistry* 30 (6): 1533–1537.

Moore, K.L. (2003). The biology and enzymology of protein tyrosine *O*-sulfation. *J. Biol. Chem.* 278 (27): 24243–24246.

Mozingo, R., Harris, S.A. et al. (1945). Hydrogenation of compounds containing divalent sulfur. *J. Am. Chem. Soc.* 67: 2092–2095.

Muir, H. (1969). The structure and metabolism of mucopolysaccharides (glycosaminoglycans) and the problem of the mucopolysaccharidoses. *Am. J. Med.* 47 (5): 673–690.

Mulder, G.J. (1986). Sex differences in drug conjugation and their consequences for drug toxicity. Sulfation, glucuronidation and glutathione conjugation. *Chem. Biol. Interact.* 57 (1): 1–15.

Nandini, C.D. and Sugahara, K. (2006). Role of the sulfation pattern of chondroitin sulfate in its biological activities and in the binding of growth factors. *Adv. Pharmacol.* 53: 253–279.

Negishi, M., Pedersen, L.G., Petrotchenko, E. et al. (2001). Structure and function of sulfotransferases. *Arch. Biochem. Biophys.* 390 (2): 149–157.

Niehrs, C. and Huttner, W.B. (1990). Purification and characterization of tyrosylprotein sulfotransferase. *EMBO J.* 9 (1): 35–42.

Niehrs, C., Huttner, W.B., Carvallo, D., and Degryse, E. (1990). Conversion of recombinant hirudin to the natural form by in vitro tyrosine sulfation. Differential substrate specificities of leech and bovine tyrosylprotein sulfotransferases. *J. Biol. Chem.* 265 (16): 9314–9318.

Niehrs, C., Beisswanger, R., and Huttner, W.B. (1994). Protein tyrosine sulfation, 1993 — an update. *Chem. Biol. Interact.* 92 (1–3): 257–271.

Osago, H., Shibata, T., Hara, N. et al. (2014). Quantitative analysis of glycosaminoglycans, chondroitin/dermatan sulfate, hyaluronic acid, heparan sulfate, and keratan sulfate by liquid chromatography-electrospray ionization-tandem mass spectrometry. *Anal. Biochem.* 467: 62–74. https://doi.org/10.1016/j.ab.2014.08.005.

Ouyang, Y.B., Crawley, J.T., Aston, C.E., and Moore, K.L. (2002). Reduced body weight and increased postimplantation fetal death in tyrosylprotein sulfotransferase-1-deficient mice. *J. Biol. Chem.* 277 (26): 23781–23787.

Ozeran, J.D., Westley, J., and Schwartz, N.B. (1996). Identification and partial purification of PAPS translocase. *Biochemistry* 35 (12): 3695–3703.

Pinhal, M.A., Smith, B., Olson, S. et al. (2001). Enzyme interactions in heparan sulfate biosynthesis: uronosyl 5-epimerase and 2-*O*-sulfotransferase interact in vivo. *Proc. Natl. Acad. Sci. U.S.A.* 98 (23): 12984–12989.

Pomin, V.H. (2015). Keratan sulfate: an up-to-date review. *Int. J. Biol. Macromol.* 72: 282–289. https://doi.org/10.1016/j.ijbiomac.2014.08.029.

Pouyani, T. and Seed, B. (1995). PSGL-1 recognition of P-selectin is controlled by a tyrosine sulfation consensus at the PSGL-1 amino terminus. *Cell* 83 (2): 333–343.

Prough, R.A., Clark, B.J., and Klinge, C.M. (2016). Novel mechanisms for DHEA action. *J. Mol. Endocrinol.* 56 (3): R139–R155. https://doi.org/10.1530/JME-16-0013.

Prydz, K. (2015). Determinants of glycosaminoglycan (GAG) structure. *Biomolecules* 5 (3): 2003–2022. https://doi.org/10.3390/biom5032003.

Quittot, N., Sebastiao, M., and Bourgault, S. (2017). Modulation of amyloid assembly by glycosaminoglycans: from mechanism to biological significance. *Biochem. Cell Biol.* 95 (3): 329–337. https://doi.org/10.1139/bcb-2016-0236.

Renskers, K.J., Feor, K.D., and Roth, J.A. (1980). Sulfation of dopamine and other biogenic amines by human brain phenol sulfotransferase. *J. Neurochem.* 34 (6): 1362–1368.

Restaino, O.F., Bhaskar, U., Paul, P. et al. (2013). High cell density cultivation of a recombinant *E. coli* strain expressing a key enzyme in bioengineered heparin production. *Appl. Microbiol. Biotechnol.* 97 (9): 3893–3900. https://doi.org/10.1007/s00253-012-4682-z.

Rhiner, C. and Hengartner, M.O. (2006). Sugar antennae for guidance signals: syndecans and glypicans integrate directional cues for navigating neurons. *Sci. World J.* 6: 1024–1036.

Rodrigues, A.D., Lai, Y., Cvijic, M.E. et al. (2014). Drug-induced perturbations of the bile acid pool, cholestasis, and hepatotoxicity: mechanistic considerations beyond the direct inhibition of the bile salt export pump. *Drug Metab. Dispos.* 42 (4): 566–574. https://doi.org/10.1124/dmd.113.054205. Review. Erratum in: (2014). *Drug Metab. Dispos.* 42 (11): 1978.

Rosenberg, R.D. (1977). Chemistry of the hemostatic mechanism and its relationship to the action of heparin. *Fed. Proc.* 36 (1): 10–18.

Rosenberg, R.D. (1978). Heparin, antithrombin, and abnormal clotting. *Annu. Rev. Med.* 29: 367–378.

Ross, A.H., Baltimore, D., and Eisen, H.N. (1981). Phosphotyrosine-containing proteins isolated by affinity chromatography with antibodies to a synthetic hapten. *Nature* 294 (5842): 654–656.

Sasaki, N. (2012). Current status and future prospects for research on tyrosine sulfation. *Curr. Pharm. Biotechnol.* 13 (14): 2632–2641.

Schiller, S. (1966). Connective and supporting tissues: mucopolysaccharides of connective tissues. *Annu. Rev. Physiol.* 28: 137–158.

Seubert, P.A., Grant, P.A., Christie, E.A. et al. (1979). Kinetic and chemical properties of ATP sulphurylase from *Penicillin chrysogenum*. *Ciba Found. Symp.* 72: 19–47.

Sigel, H., McCormick, D.B., Griesser, R. et al. (1969). Metal ion complexes with biotin and biotin derivatives. Participation of sulfur in the orientation of divalent cations. *Biochemistry* 8 (7): 2687–2695.

Spahr, C., Kim, J.J., Deng, S. et al. (2013). Recombinant human lecithin-cholesterol acyltransferase Fc fusion: analysis of N- and O-linked glycans and identification and elimination of a xylose-based O-linked tetrasaccharide core in the linker region. *Protein Sci.* 22 (12): 1739–1753. https://doi.org/10.1002/pro.2373.

Spahr, C., Shi, S.D., and Lu, H.S. (2014). O-glycosylation of glycine-serine linkers in recombinant Fc-fusion proteins: attachment of glycosaminoglycans and other intermediates with phosphorylation at the xylose sugar subunit. *mAbs* 6 (4): 904–914. https://doi.org/10.4161/mabs.28763.

Spencer, D., Novarra, S., Zhu, L. et al. (2013). O-xylosylation in a recombinant protein is directed at a common motif on glycine-serine linkers. *J. Pharm. Sci.* 102 (11): 3920–3924. https://doi.org/10.1002/jps.23733.

Stadler, J., Gerisch, G., Bauer, G. et al. (1983). In vivo sulfation of the contact site A glycoprotein of *Dictyostelium discoideum*. *EMBO J.* 2 (7): 1137–1143.

Stanley, P.E., Kelley, B.C., Tuovinen, O.H., and Nicholas, D.J. (1975). A bioluminescence method for determining adenosine 3′-phosphate 5′-phosphate (PAP) and adenosine 3′-phosphate 5′-sulfatophosphate (PAPS) in biological materials. *Anal. Biochem.* 67 (2): 540–551.

Stone, M.J., Chuang, S., Hou, X. et al. (2009). Tyrosine sulfation: an increasingly recognised post-translational modification of secreted proteins. *New Biotechnol.* 25 (5): 299–317.

Tanaka, S., Nishiyori, T., Kojo, H. et al. (2017). Structural basis for the broad substrate specificity of the human tyrosylprotein sulfotransferase-1. *Sci. Rep.* 7 (1): 8776. https://doi.org/10.1038/s41598-017-07141-8.

de la Torre, A., Lara, C., Yee, B.C. et al. (1982). Physiochemical properties of ferralterin. A regulatory iron-sulfur protein functional in oxygenic photosynthesis. *Arch. Biochem. Biophys.* 213 (2): 545–550.

Valcarcel, J., Novoa-Carballal, R., Pérez-Martín, R.I. et al. (2017). Glycosaminoglycans from marine sources as therapeutic agents. *Biotechnol. Adv.* 35 (6): 711–725. https://doi.org/10.1016/j.biotechadv.2017.07.008.

Vandendries, E.R., Furie, B.C., and Furie, B. (2004). Role of P-selectin and PSGL-1 in coagulation and thrombosis. *Thromb. Haemost.* 92 (3): 459–466.

Vargas, F., Frerot, O., Brion, F. et al. (1994). 3′-Phosphoadenosine 5′-phosphosulfate biosynthesis and the sulfation of cholecystokinin by the tyrosylprotein-sulfotransferase in rat brain tissue. *Chem. Biol. Interact.* 92 (1-3): 281–291.

Vishnuvardhan, D. and Beinfeld, M.C. (2000). Role of tyrosine sulfation and serine phosphorylation in the processing of procholecystokinin to amidated cholecystokinin and its secretion in transfected AtT-20 cells. *Biochemistry* 39 (45): 13825–13830.

Visser, T.J. (1994). Role of sulfation in thyroid hormone metabolism. *Chem. Biol. Interact.* 92 (1-3): 293–303.

Walter, J. and Haass, C. (2000). Posttranslational modifications of amyloid precursor protein: ectodomain phosphorylation and sulfation. *Methods Mol. Med.* 32: 149–168. https://doi.org/10.1385/1-59259-195-7:149.

Watson, P.G. and Young, R.D. (2004). Scleral structure, organisation and disease. A review. *Exp. Eye Res.* 78 (3): 609–623.

Weinshilboum, R.M., Otterness, D.M., Aksoy, I.A. et al. (1997). Sulfation and sulfotransferases 1: sulfotransferase molecular biology: cDNAs and genes. *FASEB J.* 11 (1): 3–14.

Westmuckett, A.D., Hoffhines, A.J., Borghei, A., and Moore, K.L. (2008). Early postnatal pulmonary failure and primary hypothyroidism in mice with combined TPST-1 and TPST-2 deficiency. *Gen. Comp. Endocrinol.* 156 (1): 145–153. https://doi.org/10.1016/j.ygcen.2007.

Westmuckett, A.D., Thacker, K.M., and Moore, K.L. (2011). Tyrosine sulfation of native mouse Psgl-1 is required for optimal leukocyte rolling on P-selectin in vivo. *PLoS ONE* 6 (5): e20406. https://doi.org/10.1371/journal.pone.0020406.

White, B.V. (1949). Lipotropic agents and sulfur-containing amino acids. *Conn. State Med. J.* 13 (4): 349–352.

White, G.C. II, Beebe, A., and Nielsen, B. (1997). Recombinant factor IX. *Thromb. Haemost.* 78 (1): 261–265.

White, G.C. II, Pickens, E.M., Liles, D.K., and Roberts, H.R. (1998). Mammalian recombinant coagulation proteins: structure and function. *Transfus. Sci.* 19 (2): 177–189.

Wollam, J. and Antebi, A. (2011). Sterol regulation of metabolism, homeostasis and development. *Annu. Rev. Biochem.* 80: 885–916. https://doi.org/10.1146/annurev-biochem-081308-165917.

Yang, Y.S., Wang, C.C., Chen, B.H. et al. (2015). Tyrosine sulfation as a protein post-translational modification. *Molecules* 20 (2): 2138–2164. https://doi.org/10.3390/molecules20022138.

# 19

# SUMOylation

## Introduction

SUMOylation is a post-translational modification (PTM) in which small ubiquitin-like modifier (SUMO) proteins are covalently attached or detached to proteins within the living cells or organisms (Wilson and Rangasamy 2001). SUMO proteins are similar to ubiquitin, but SUMOylated proteins are not marked for degradation by proteolysis in the proteasome unlike some type of ubiquitinated proteins (Müller et al. 2001; Gill 2004). Covalent attachment or detachment of SUMO proteins changes protein functions (Jentsch and Psakhye 2013). SUMOylation is an enzymatic modification of proteins (Lowrey et al. 2017). SUMOylation affects many cellular processes such as apoptosis, nuclear–cytosolic transport, protein stability, progression through the cell cycle, response to stress, transcriptional regulation, etc. (Hay 2005).

SUMO proteins are small and contain about 100 amino acids with an apparent molecular weight of ~12 kDa. The SUMO proteins encoding SUMO genes are highly conserved in eukaryotes. Yeast contains only one SUMO encoding gene (SMT3), whereas plants contain at least eight genes and mammals contains four genes (Kurepa et al. 2003; Johnson 2004; Enserink 2015). The SUMO encoding four genes of mammals translates SUMO-1, SUMO-2, SUMO-3, and SUMO-4 proteins (Kurepa et al. 2003; Johnson 2004; Enserink 2015). Accordingly, the exact molecular weight of SUMO protein varies with the species and family members. For example, in mammals, SUMO-1 is a 101-amino acid protein with an apparent molecular weight of ~11.6 kDa, whereas SUMO-2 and SUMO-3 differs by only three amino acid residues in the N-termini. The SUMO-2 and SUMO-3 shares about 95% sequence homology. SUMO-4 is the least characterized protein compared to SUMO-1, SUMO-2, and SUMO-3.

*Co- and Post-Translational Modifications of Therapeutic Antibodies and Proteins*, First Edition. T. Shantha Raju.

## Mechanism of SUMOylation

SUMOylation is an enzymatic conjugation reaction similar to ubiquitination (Cui et al. 2014; also see Chapter 20). SUMO proteins are synthesized as propeptides and require enzymatic cleavage by sentrin-specific protease 1 (SENP) to reveal C-terminal diglycine motif (Klein and Nigg 2009). The enzymatically cleaved SUMOs are activated by SUMO-activating enzyme subunits 1 and 2 called SAE1 and SAE2 (Wang 2009). The activated SUMO protein is then conjugated to proteins that contain a Ψ-K-x-D/E tetra peptide motif, where Ψ is a hydrophobic residue, K is the lysine residue, x is any amino acid residue, and D/E is an acidic amino acid residue (Wang 2009). The C-termini of SUMO protein is conjugated to ε-amino group of Lys residue through an enzyme cascade similar to ubiquitination (Yang et al. 2017; Zhang et al. 2017).

## Biological Significance of SUMOylation

SUMOylation affects many functions of proteins (Truong et al. 2012). SUMOylation impacts the protein stability, intracellular localization, and binding partners (Droescher et al. 2013; Zhao 2018). SUMOylation also affects the nuclear-cytosolic transport and transcriptional regulation (Gill 2004; Yang et al. 2017). SUMOylation and deSUMOylation are specifically involved in cell stress response (Guo and Henley 2014). SUMOylation can alter biochemical properties of proteins and also plays a major role in the DNA repair (Garvin and Morris 2017).

## References

Cui, Z., Scruggs, S.B., Gilda, J.E. et al. (2014). Regulation of cardiac proteasomes by ubiquitination, SUMOylation and beyond. *J. Mol. Cell. Cardiol.* 71: 32–42. https://doi.org/10.1016/j.yjmcc.2013.10.008.

Droescher, M., Chaugule, V.K., and Pichler, A. (2013). SUMO rules: regulatory concepts and their implication in neurologic functions. *Neuromol. Med.* 15 (4): 639–660. https://doi.org/10.1007/s12017-013-8258-6.

Enserink, J.M. (2015). SUMO and the cellular stress response. *Cell Div.* 10 (4): https://doi.org/10.1186/s13008-015-0010-1.

Garvin, A.J. and Morris, J.R. (2017). SUMO, a small, but powerful, regulator of double-strand break repair. *Philos. Trans. R. Soc. London, Ser. B* 372 (1731): https://doi.org/10.1098/rstb.2016.0281.

Gill, G. (2004). SUMO and ubiquitin in the nucleus: different functions, similar mechanisms? *Genes Dev.* 18 (17): 2046–2059.

Guo, C. and Henley, J.M. (2014). Wrestling with stress: roles of protein SUMOylation and deSUMOylation in cell stress response. *IUBMB Life* 66 (2): 71–77. https://doi.org/10.1002/iub.1244.

Hay, R.T. (2005). SUMO a history of modification. *Mol. Cell.* 18 (1): 1–12.

Jentsch, S. and Psakhye, I. (2013). Control of nuclear activities by substrate-selective and protein-group SUMOylation. *Annu. Rev. Genet.* 47: 167–186. https://doi.org/10.1146/annurev-genet-111212-133453.

Johnson, E.S. (2004). Protein modification by SUMO. *Annu. Rev. Biochem.* 73: 355–382.

Klein, U.R. and Nigg, E.A. (2009). SUMO-dependent regulation of centrin-2. *J. Cell Sci.* 122 (Pt. 18): 3312–3321. https://doi.org/10.1242/jcs.050245.

Kurepa, J., Walker, J.M., Smalle, J. et al. (2003). The small ubiquitin-like modifier (SUMO) protein modification system in *Arabidopsis*. Accumulation of SUMO1 and -2 conjugates is increased by stress. *J. Biol. Chem.* 278 (9): 6862–6872.

Lowrey, A.J., Cramblet, W., and Bentz, G.L. (2017). Viral manipulation of the cellular SUMOylation machinery. *Cell Commun. Signal.* 15 (1): 27. https://doi .org/10.1186/s12964-017-0183-0.

Müller, S., Hoege, C., Pyrowolakis, G., and Jentsch, S. (2001). SUMO, ubiquitin's mysterious cousin. *Nat. Rev. Mol. Cell Biol.* 2 (3): 202–210.

Truong, K., Lee, T.D., Li, B., and Chen, Y. (2012). SUMOylation of SAE2 C terminus regulates SAE nuclear localization. *J. Biol. Chem.* 287 (51): 42611–42619. https://doi.org/10.1074/jbc.M112.420877.

Wang, J. (2009). SUMO conjugation and cardiovascular development. *Front. Biosci. (Landmark Ed.)* 14: 1219–1229.

Wilson, V.G. and Rangasamy, D. (2001). Intracellular targeting of proteins by SUMOylation. *Exp. Cell. Res.* 271 (1): 57–65.

Yang, Y., He, Y., Wang, X. et al. (2017). Protein SUMOylation modification and its associations with disease. *Open Biol.* 7 (10): https://doi.org/10.1098/rsob .170167.

Zhang, Y., Li, Y., Tang, B., and Zhang, C.Y. (2017). The strategies for identification and quantification of SUMOylation. *Chem. Commun. (Camb.)* 53 (52): 6989–6998. https://doi.org/10.1039/c7cc00901a.

Zhao, X. (2018). SUMO-mediated regulation of nuclear functions and signaling processes. *Mol. Cell* 71 (3): 409–418. https://doi.org/10.1016/j.molcel.2018.07 .027.

# 20

# Ubiquitination

## Introduction

Ubiquitin is a small protein or a large peptide consisting of 76 amino acid residues (Watson et al. 1978). The apparent molecular weight of ubiquitin is about 8.5 kDa. In humans, ubiquitin is present in almost all types of cellular tissues. Hence, ubiquitin is ubiquitous in nature. Ubiquitin is also present in other eukaryotic organisms but not in prokaryotic organisms. Ubiquitin was first discovered by Goldstein et al. (1975).

Ubiquitin is highly conserved in eukaryotes, and its sequence similarity between human and yeast is about 96%. The human genome carries four ubiquitin genes, i.e. UBB, UBC, UBA52, and RPS27A. All these four genes encode ubiquitin (Schlesinger and Bond 1987). The amino acid composition of ubiquitin in humans is $M_1Q_{6}I_7F_2V_3K_7T_7G_6L_8E_6P_3S_3D_5R_4Y_1$ and its primary amino acid sequence is shown in Figure 20.1.

## Mechanism of Ubiquitination

Ubiquitination is the addition of ubiquitin polypeptide to proteins and is a post-translational modification (PTM) (Wilkinson 1987). During initial ubiquitination, the carboxyl group of C-terminal Gly residue of ubiquitin is covalently linked to the side chain primary amine functional group of one of the seven Lys residue or Cys, Ser, and Thr residues in the N-termini of target proteins (Hochstrasser 1995). Following the monoubiquitination, additional ubiquitin peptides may be added to form polyubiquitinated proteins. The polyubiquitination may occur via one of the seven Lys residues or through the N-terminal Met residue present in ubiquitin.

Ubiquitination occurs in three steps and each step involves a specific group of enzymes (Swatek and Komander 2016). Initial step is the activation with ubiquitin-activating enzymes (E1s), which is followed by the second step of conjugation with ubiquitin-conjugating enzymes (E2s) and the final third step

*Co- and Post-Translational Modifications of Therapeutic Antibodies and Proteins*, First Edition. T. Shantha Raju.

MQIFVKTLTGKTITLEVEPSDTIENVKAKIQDKEGIPPDQQRLIFAGKQLEDGRTLSDYNIQKEST
LHLVLRLRGG

**Figure 20.1** Primary amino acid sequence of ubiquitin.

is ligation with ubiquitin ligases (E3s) (Chen and Pickart 1990). All these three steps lead to the formation of isopeptide bond between ubiquitin and target protein via Lys residues or the amino acid residue at the N-terminus of target proteins (Chau et al. 1989). Ubiquitination may also occur at the Cys residues of the target protein via thioester bond (Schwartzkopff et al. 2015). In addition, ubiquitin can form ester bonds with the hydroxyl groups of Ser or Thr residues of the target protein (Bhogaraju et al. 2016). During ubiquitination reaction, the carboxyl group of C-terminal Gly residue of ubiquitin forms an isopeptide bond with the primary amine group of Lys residues or with the primary amine group of N-terminal amino acid residues of the target protein. Similarly, the carboxyl group of C-terminal Gly residue of ubiquitin forms a thioester bond with the thiol group of Cys residues of the target protein. In the similar fashion, but a different way, the carboxyl group of C-terminal Gly residue of ubiquitin forms an ester bond with the hydroxyl group of Ser or Thr residues of target proteins (Swatek and Komander 2016).

## Biological Significance of Ubiquitination

Ubiquitination may affect the protein functions in many different ways (Cai et al. 2017). Ubiquitination may cause protein degradation via proteasome, may affect their cellular location, may alter their functions, or may impact their interactions with other proteins (Rechsteiner 1987). The impact of ubiquitin on the proteins depends on the site of attachment (Popovic et al. 2014). For example, in some proteins, ubiquitin linked at Lys 29 or Lys 48 is marked for degradation by proteolysis in the proteasome (Nakasone et al. 2014). Upon degradation of protein by proteasome, the ubiquitin that is not degraded is recycled. Ubiquitin linked through other amino acid residues such as Lys 6, Lys 11, Lys 63, or Met 1 may be involved in regulating cellular process such as DNA repair, endocytic trafficking, inflammation, translation, etc. (Gao et al. 2014). In addition, ubiquitin is involved in cell division and multiplication, apoptosis, viral infection, ribosome biogenesis, receptor modulation, degeneration of neurons and muscular cells, etc. (Mansour 2018).

## References

Bhogaraju, S., Kalayil, S., Liu, Y. et al. (2016). Phosphoribosylation of ubiquitin promotes serine ubiquitination and impairs conventional ubiquitination. *Cell* 167 (6): 1636–1649.e13. https://doi.org/10.1016/j.cell.2016.11.019.

Cai, J., Culley, M.K., Zhao, Y., and Zhao, J. (2017). The role of ubiquitination and deubiquitination in the regulation of cell junctions. *Protein Cell* https://doi.org/10.1007/s13238-017-0486-3.

Chau, V., Tobias, J.W., Bachmair, A. et al. (1989). A multiubiquitin chain is confined to specific lysine in a targeted short-lived protein. *Science* 243 (4898): 1576–1583.

Chen, Z. and Pickart, C.M. (1990). A 25-kilodalton ubiquitin carrier protein (E2) catalyzes multi-ubiquitin chain synthesis via lysine 48 of ubiquitin. *J. Biol. Chem.* 265 (35): 21835–21842.

Gao, C., Huang, W., Kanasaki, K., and Xu, Y. (2014). The role of ubiquitination and SUMOylation in diabetic nephropathy. *Biomed. Res. Int.* 2014: 160692. https://doi.org/10.1155/2014/160692.

Goldstein, G., Scheid, M., Hammerling, U. et al. (1975). Isolation of a polypeptide that has lymphocyte-differentiating properties and is probably represented universally in living cells. *Proc. Natl. Acad. Sci. U.S.A.* 72 (1): 11–15.

Hochstrasser, M. (1995). Ubiquitin, proteasomes, and the regulation of intracellular protein degradation. *Curr. Opin. Cell Biol.* 7 (2): 215–223.

Mansour, M.A. (2018). Ubiquitination: friend and foe in cancer. *Int. J. Biochem. Cell Biol.* 101: 80–93. https://doi.org/10.1016/j.biocel.2018.06.001.

Nakasone, N., Nakamura, Y.S., Higaki, K. et al. (2014). Endoplasmic reticulum-associated degradation of Niemann-Pick C1: evidence for the role of heat shock proteins and identification of lysine residues that accept ubiquitin. *J. Biol. Chem.* 289 (28): 19714–19725. https://doi.org/10.1074/jbc.M114.549915.

Popovic, D., Vucic, D., and Dikic, I. (2014). Ubiquitination in disease pathogenesis and treatment. *Nat. Med.* 20 (11): 1242–1253. https://doi.org/10.1038/nm.3739.

Rechsteiner, M. (1987). Ubiquitin-mediated pathways for intracellular proteolysis. *Annu. Rev. Cell Biol.* 3: 1–30.

Schlesinger, M.J. and Bond, U. (1987). Ubiquitin genes. *Oxf. Surv. Eukaryot. Genes* 4: 77–91.

Schwartzkopff, B., Platta, H.W., Hasan, S. et al. (2015). Cysteine-specific ubiquitination protects the peroxisomal import receptor Pex5p against proteasomal degradation. *Biosci. Rep.* 35 (3): https://doi.org/10.1042/BSR20150103.

Swatek, K.N. and Komander, D. (2016). Ubiquitin modifications. *Cell Res.* 26 (4): 399–422. https://doi.org/10.1038/cr.2016.39.

Watson, D.C., Levy, W.B., and Dixon, G.H. (1978). Free ubiquitin is a non-histone protein of trout testis chromatin. *Nature* 276 (5684): 196–198.

Wilkinson, K.D. (1987). Protein ubiquitination: a regulatory post-translational modification. *Anti-Cancer Drug Des.* 2 (2): 211–229.

# 21

# Other CTMs and PTMs of Proteins

## Adenylylation or AMPylation

Adenylylation is the addition of adenylyl group to proteins. Adenylyl group is an adenosine monophosphate molecule and hence, adenylylation is now called as AMPylation. Adenylylation or AMPylation involves the covalent addition of adenosine monophosphate to proteins (Hedberg and Itzen 2015). In many proteins, AMP is added to tyrosine residues via *O*-linkage and also to histidine and lysine residues via *N*-linkage. AMPylation regulates the activity of glutamine synthetase (Woolery et al. 2010).

## ADP-Ribosylation

Adenosine diphosphate (ADP)-ribosylation is a post-translational modification (PTM) in which one or more ADP-ribose moieties are covalently linked to Arg residues of proteins via an amide linkage (Liu and Yu 2015). ADP-ribosylation is a reversible modification that plays many functional roles including a role in cellular processing, cell signaling, DNA repair, gene regulation, apoptosis, etc. (Bütepage et al. 2015). Improper ADP-ribosylation may lead to cancer (Cohen and Chang 2018). ADP-ribosylation is also implicated to the toxicity of many bacterial toxins such as cholera toxin, diphtheria toxin, etc. (Aktories et al. 2017).

## Amidation

Amidation is commonly a C-terminal modification of proteins. Amidation of proteins is normally formed by oxidative dissociation of Gly residues in the C-terminus (Prigge et al. 2000). An additional Gly residue is present in the biosynthetic precursor of proteins which undergoes hydroxylation to form amide. The hydroxyglycine dissociates to form the amide at the C-termini

*Co- and Post-Translational Modifications of Therapeutic Antibodies and Proteins,* First Edition. T. Shantha Raju.

of protein. Amidation at C-termini of hormones, neurotransmitters, etc., is essential for their biological activity (Bradbury and Smyth 1991).

## Arginylation

Arginylation is a tRNA-dependent addition of Arg residues to proteins and is mediated by arginyltransferase (Kashina 2015). Arginylation is involved in multiple physiological pathways and is also involved in embryogenesis along with the adulthood (Saha and Kashina 2011). Arginylation is a novel biological regulator of contractility (Kurosaka et al. 2012; Rai et al. 2008).

## Butyrylation

Butyrylation is observed in histone proteins and is a PTM. Butyrylation is a protein acylation that involves the covalent linkage of butyryl group to Lys residues via an amide bond formation in histones (Sabari et al. 2017). Butyryl-coenzyme A (CoA) is structurally similar to acetyl-CoA and differs by two $CH_2$ units (Lee 2013). Butyrylation is involved in the regulation of many biological processes (Sabari et al. 2017; Lee 2013).

## Carbamylation

Carbamylation is the addition of isocyanic acid to the N-terminus of proteins or to the Lys side chain of proteins via an amide bond (Jaisson et al. 2018). Isocyanic acid is an organic compound with a molecular formula of HNCO (H—N=C=O). It is a volatile colorless and poisonous chemical compound with a boiling point of 23.5 °C. It is commonly present in smog and cigarette smoke. Carbamylation is a nonenzymatic modification of proteins that impact the structural and functional properties of proteins. Carbamylation is involved in the molecular aging of proteins (Jaisson et al. 2018). In kidney diseases, accumulation of urea may lead to carbamylation of proteins (Badar et al. 2018).

## Carbonylation

Carbonylation is the addition of carbon monoxide to proteins and other compounds. Carbonylation commonly occur on His, Cys, and Lys residues of proteins. Oxidative stress may lead to metal-catalyzed carbonylation. Carbonylation may play a role in the regulation of Na/K-ATPase signaling and sensitivity to salts (Shah et al. 2016). Carbonylation is one of the biomarkers for multiple myeloma and other autoimmune diseases (Ibitoye et al. 2016).

## γ-Carboxylation

γ-Carboxylation is a vitamin K-dependent modification of proteins and is a PTM (Wallin and Hutson 2004). γ-Carboxylation is a carboxylation of glutamic acid residues. γ-Carboxylation is found in many clotting factors and other proteins that are involved in the coagulation cascade (Zhu et al. 2007). γ-Carboxylation is involved in calcium binding and the subsequent exposure of the hydrophobic regions of proteins to the cells lipid bilayer (Kalafatis et al. 1996).

## Citrullination

Citrullination is also known as deamination in which arginine residues converts into citrulline residues. During citrullination, Arg residues undergo hydrolysis to convert into citrulline residues and in the process eliminates ammonia molecules. Arginine deiminases catalyzes the conversion of Arg into citrulline (Thompson and Fast 2006). Citrullination may impact the protein folding and hence impacting the structure and function of proteins. Immune system recognizes citrullinated proteins as foreign molecules which might lead to autoimmune and age-related diseases (Olsen et al. 2018).

## Diphthamide

Diphthamide is a histidine derivative found in archaeal and eukaryotic elongation factor 2 (Schaffrath et al. 2014). It is a post-translationally modified form of histidine and its IUPAC name is 2-amino-3-[2-(3-carbamoyl-3-trimethylammonio-propyl)-3H-imidazol-4yl]-propanoate (Schaffrath and Stark 2014). Both histidine and *S*-adenosyl methionine are involved in the biosynthesis of diphthamide (Narrowe et al. 2018). Diphthamide affects the death receptor pathways (Su et al. 2013). Loss of diphthamide renders MCF7 cells hypersensitive to tumor necrosis factor (Stahl et al. 2015).

## Formylation

Addition of formyl group (carbonyl group covalently bonded to hydrogen atom) is called formylation. In 1964, Marcker and Sanger discovered formylated methionine (*N*-formylmethionine) in *Escherichia coli* (Marcker and Sanger 1964). Recently, formylated lysine residues are found in histone proteins (Jiang et al. 2007). Formylation is catalyzed by formyltransferases

(Meinnel et al. 1993). Formylmethionine is involved in the initiation of protein synthesis in bacteria and organelles but not in eukaryotes and Archaea (Mangroo et al. 1995). Formylation in histones modulate chromatin and is also involved in gene activation (Wisniewski et al. 2008).

## Glypiation

Glypiation is the formation of glycosylphosphatidylinositol (GPI) anchor and is discussed in Chapter 7. GPI anchor is linked to proteins via an amide bond to C-terminal tail and widely occur on the cell surface glycoproteins of eukaryotes as well as in some Archaea (Canut et al. 2016).

## Hypusine Formation

Hypusine is a derivative of Lys and is also called as N6-(4-amino-2-hydroxybutyl)lysine. The name hypusine is derived from *hy*droxy*pu*trescine and ly*sine*. The IUPAC name of hypusine is (2S)-2-amino-6-{[(2R)-4-amino-2hydroxybutyl]amino}hexanoic acid. Hypusine was first purified from the brain of bovine (Shiba et al. 1971). Hypusine is an unusual amino acid. Hypusine is found in all eukaryotes and also in some archaea (Michael 2016). However, hypusine is not found in bacteria. In eukaryotes, hypusine is found only in one protein that is eukaryotic translation initiation factor 5A (elF5A). A similar protein found in archaebacteria contains hypusine in a 1 : 1 molar ratio of protein to hypusine (Bartig et al. 1992). The hypusine formation in elF5A is a PTM and is due to the modification of one specific Lys residues of the elF5A protein. The protein domain of elF5A in which the specific Lys residue is modified into hypusine is conserved in eukaryotes. The modification of the specific Lys residue of elF5A involves two steps and two enzymes are involved in this modification. In the first step of modification, deoxyhypusine synthase cleaves spermidine and transfers the 4-aminobutyl residue to the ε-amino functional group of the specific Lys residue of elF5A. Spermidine is a polyamine compound originally isolated from semen (Jänne et al. 1973). Following the transfer of 4-aminobutyl moiety, the deoxyhypusine hydroxylase mediates the addition of hydroxyl group to the deoxyhypusine to form hypusine. Hypusine and elF5A are important for the cell viability and proliferation in eukaryotes. Hypusine and elF5A are also important for translation elongation (Saini et al. 2009). Excretion of hypusine was observed in children and patients with familial hyperlysinemia (Woody and Pupene 1973).

## Iodination

Iodination is the addition of iodine atom to tyrosine residues. Iodination is found in thyroglobulin that is a precursor to thyroid hormones (Ekholm 1981).

The thyroid hormones are produced upon iodination of tyrosine residues of thyroglobulin and its subsequent cleavage. Iodination is mediated by thyroperoxidase in the follicular colloid. Iodination plays a role in apoptosis and necrosis of thyrocytes (Carayanniotis 2011).

## Lipoylation

Lipoylation is an acylation of proteins in which a lipoate group is attached to Lys residues via an amide bond. Lipoate is derived from lipoic acid which is a C8 organosulfur compound and is derived from caprylic acid. Lipoic acid is also known as α-lipoic acid and is synthesized in animals. α-Lipoic acid is essential for aerobic metabolism and is used as dietary supplement.

## Malonylation

Malonic acid is a dicarboxylic acid with chemical structure $CH_2(COOH)_2$ and is a propanedioic acid. Its esters, salts, and ionized form are called malonates. Malonylation refers to the acylation of amine groups of proteins with the ionized form of malonic acid. Malonylation is similar to succinylation and exhibits similar impact on protein properties (Hirschey and Zhao 2015). Malonylation plays a role in the regulation of mitochondrial function (Bowman et al. 2017).

## Myristoylation

Myristoylation is a type of acylation that involves the covalent attachment of myristate group to proteins via an amide bond. Myristate or myristoyl group is derived from myristic acid and is a 14-carbon saturated fatty acid moiety. The systematic name of myristic acid is *n*-tetradecanoic acid. Myristoylation can be a CTM or PTM and is mediated by *N*-myristoyltransferase (Schlott et al. 2018). Myristoylation is commonly found on the α-amino group of an N-terminal Gly residues. Myristoylation is a very common modification found among many organisms such as plants, animals, protozoans, fungi, and viruses. Myristoylation is involved in protein–protein interactions and also involved in protein–lipid interactions. Myristoylation may play a role in membrane targeting and signal transduction (Rioux et al. 2011).

## Neddylation

Neddylation is the covalent bonding of neural precursor cell expressed developmentally downregulated protein (NEDD) to proteins (Enchev et al.

2015). Neddylation is analogous to ubiquitination. Nedd (for example NEDD8) is a ubiquitin like protein and neddylation may relies on E3 ligases. NEDD8 links to proteins through its carboxyl-terminal Gly to the Lys residues of target proteins. Neddylation can cause conformational change in proteins and hence may impact protein–protein interactions. Neddylation may be involved in Alzheimer's disease and in human oral carcinoma (Delgado et al. 2018).

## Palmitoylation

Palmitoylation is an acylation modification in which palmitate groups are covalently attached to proteins (Ross 1995). Palmitate or palmitoyl group is derived from palmitic acid which is a C16 saturated fatty acid (Casey 1994). The systematic name of palmitic acid is *n*-hexadecanoic acid, and its molecular formula is $CH_3(CH_2)_{14}COOH$. Palmitic acid is a major component of palm oil and is derived from the fruit of oil palms. Palmitoylation increases hydrophobicity of proteins and may help proteins with their membrane association. Palmitoylation is also involved in subcellular trafficking of proteins between membrane compartments (Chamberlain et al. 2013). Aberrant palmitoylation is observed in Huntington disease (Sanders and Hayden 2015).

## Polyglutamylation

Polyglutamylation is a covalent linkage of multiple glutamic acid residues to the N-terminus of proteins and is a PTM. Polyglutamylation is a very common occurrence in tubulins and in some other tubulin related proteins (Song and Brady 2015). Polyglutamylation is a reversible PTM and is evolutionarily conserved from protists to mammals. Polyglutamylation may be involved in the regulation of protein functions (Janke et al. 2008).

## Polyglycylation

Polyglycylation is a covalent modification of C-terminal tail of proteins like tubulins with one or more glycine residues and is a PTM. Up to 40 Gly residues may be covalently linked to tubulins at the C-terminus (Vinh et al. 1997). Polyglycylation occurs through the addition of polyglycine peptide to the γ-carboxyl group of glutamic acids of tubulin proteins (Thazhath et al. 2002). Polyglycylation of tubulin may be essential and is involved in cell motility along with the cell division (Xia et al. 2000).

## Propionylation

Propionylation of proteins is similar to butyrylation in which propionyl group is attached to proteins via an amide linkage to Lys residues (Lee 2013). Impact of propionylation on proteins is similar to butyrylation and is involved in regulating many biological processes in living organisms (Choudhary et al. 2014).

## Pupylation

Pupylation is the covalent modification of proteins with Prokaryotic ubiquitin-like protein (Barandun et al. 2012). The prokaryotic ubiquitin like protein (Pup) is found in *Mycobacterium tuberculosis* and is an ubiquitin functional analogue (Akhter and Thakur 2017). Enzymology of ubiquitination and pupylation is different, but their functions are the same (Delley et al. 2017). Hence, like eukaryotes, prokaryotes may use small protein modifiers like Pup to control the stability of proteins (Delley et al. 2017; Samanovic and Darwin 2016).

## Pyroglutamate Formation

Pyroglutamate is the basic form of pyroglutamic acid (PCA). The PCA is also called as pidolic acid and is a 5-oxoproline. Free glutamine residues and glutamic acid residues in the N-termini of proteins can cyclize, spontaneously or by enzymatically, to form pyroglutamate ring structures. This enzymatic cyclization is mediated by glutaminyl cyclases (Schilling et al. 2008). Pyroglutamate is found in many proteins including bacteriorhodopsin and recombinant therapeutic proteins (RTPs) (Yu et al. 2006). Pyroglutamate acts on the cholinergic system in the brain and may be involved in the progression of Alzheimer's disease (Blin et al. 2009; Wu et al. 2014). Magnesium salt of PCA is used as a mineral supplement (De Franceschi et al. 1997).

## S-Glutathionylation

Glutathione is a tripeptide consisting of glutamate, cysteine, and glycine (Harley 1965). In glutathione, the carboxyl group of the glutamate side chain is linked to the amine group of cysteine and the carboxyl group of cysteine in turn linked to the amine group of Gly (Cotgreave 2003). The free thiol group of glutathione acts as a reducing agent and is involved in the reduction of disulfide bonds and in the process the S-glutathione molecule itself is oxidized

**Table 21.1** Amino acids and their commonly observed PTMs.

| Amino acids | | | |
|---|---|---|---|
| **Name** | **Three-letter abbreviation** | **One-letter abbreviation** | **Commonly observed PTMs** |
| Alanine | Ala | A | N-terminal acetylation |
| Arginine | Arg | R | Multiple PTMs including citrullination, deamination, methylation, etc. |
| Asparagine | Asn | N | Deamidation, N-glycosylation, etc. |
| Aspartic acid | Asp | D | Isomerization |
| Cysteine | Cys | C | Disulfide formation, oxidation, palmitoylation, N-terminal acetylation, S-nitrosylation, etc. |
| Glutamine | Gln | Q | Cyclization, deamidation, isomerization, etc. |
| Glutamic acid | Glu | E | Cyclization, γ-carboxylation, etc. |
| Glycine | Gly | G | N-acetylation, N-myristoylation, etc. |
| Histidine | His | H | Phosphorylation |
| Isoleucine | Ile | I | Maybe N-terminal acetylation |
| Leucine | Leu | L | Maybe N-terminal acetylation |
| Lysine | Lys | K | Acetylation, hydroxylation, methylation, SUMOylation, ubiquitination, etc. |
| Methionine | Met | M | N-terminal acetylation, oxidation, etc. |
| Phenylalanine | Phe | F | Hydroxylation, acetylation, etc. |
| Proline | Pro | P | Hydroxylation |
| Serine | Ser | S | N-acetylation, glycosylation, phosphorylation, etc. |
| Threonine | Thr | T | Acetylation, glycosylation, phosphorylation, etc. |
| Tryptophan | Trp | W | Hydroxylation, oxidation, etc. |
| Tyrosine | Tyr | Y | Oxidation, phosphorylation, sulfation, etc. |
| Valine | Val | V | N-terminal acetylation |

to form glutathione disulfide. S-Glutathionylation refers to the addition of glutathione residues to the Cys residues of proteins through disulfide bond formation (Mailloux and Willmore 2014). S-Glutathionylation is involved in preventing oxidation of thiol groups in proteins and is also involved in controlling the cell-signaling pathways (Zhang et al. 2018).

## S-Nitrosylation

S-Nitrosylation is a PTM in which nitric oxide (NO) is covalently linked to the thiol group of Cys residues of proteins to form S-nitrosothiol (SNO). S-Nitrosylation is now described as an enzymatic reaction involving S-nitrosylases (Seth and Stamler 2011). The S-nitrosylation is a reversible modification, and the reverse nitrosylation is called denitrosylation. The denitrosylation is mediated by S-nitrosoglutathione reductase and also by thioredoxin-related proteins (Sengupta and Holmgren 2013). S-Nitrosylation is required for a wide array of cellular responses including red cell mediated autoregulation of blood flow in vertebrates.

Some amino acid residues of proteins may undergo multiple CTMs and PTMs and some amino acids may not have any modifications. Some of the very common CTMs and PTMs of protein linked amino acid residues are presented in Table 21.1.

In addition to all the CTMs and PTMs described in this book, there are many more biomolecular modifications existing in nature. For the CTMs and PTMs that are not covered in this book, readers are encouraged to look for other available resources.

## References

Akhter, Y. and Thakur, S. (2017). Targets of ubiquitin like system in mycobacteria and related actinobacterial species. *Microbiol. Res.* 204: 9–29. https://doi.org/10.1016/j.micres.2017.07.002.

Aktories, K., Schwan, C., and Lang, A.E. (2017). ADP-ribosylation and cross-linking of actin by bacterial protein toxins. *Handb. Exp. Pharmacol.* 235: 179–206. https://doi.org/10.1007/164_2016_26.

Badar, A., Arif, Z., and Alam, K. (2018). Role of carbamylated biomolecules in human diseases. *IUBMB Life* 70 (4): 267–275. https://doi.org/10.1002/iub.1732.

Barandun, J., Delley, C.L., and Weber-Ban, E. (2012). The pupylation pathway and its role in mycobacteria. *BMC Biol.* 10: 95. https://doi.org/10.1186/1741-7007-10-95.

Bartig, D., Lemkemeier, K., Frank, J. et al. (1992). The archaebacterial hypusine-containing protein. Structural features suggest common ancestry with eukaryotic translation initiation factor 5A. *Eur. J. Biochem.* 204 (2): 751–758.

Blin, O., Audebert, C., Pitel, S. et al. (2009). Effects of dimethylaminoethanol pyroglutamate (DMAE p-Glu) against memory deficits induced by scopolamine: evidence from preclinical and clinical studies. *Psychopharmacology (Berl.)* 207 (2): 201–212. https://doi.org/10.1007/s00213-009-1648-7.

Bowman, C.E., Rodriguez, S., Selen Alpergin, E.S. et al. (2017). The mammalian malonyl-CoA synthetase ACSF3 is required for mitochondrial protein malonylation and metabolic efficiency. *Cell Chem. Biol.* 24 (6): 673–684.e4. https://doi.org/10.1016/j.chembiol.2017.04.009.

Bradbury, A.F. and Smyth, D.G. (1991). Peptide amidation. *Trends Biochem. Sci.* 16 (3): 112–115.

Bütepage, M., Eckei, L., Verheugd, P., and Lüscher, B. (2015). Intracellular mono-ADP-ribosylation in signaling and disease. *Cells* 4 (4): 569–595. https://doi.org/10.3390/cells4040569.

Canut, H., Albenne, C., and Jamet, E. (2016). Post-translational modifications of plant cell wall proteins and peptides: A survey from a proteomics point of view. *Biochim. Biophys. Acta* 1864 (8): 983–990. https://doi.org/10.1016/j.bbapap.2016.02.022.

Carayanniotis, G. (2011). Molecular parameters linking thyroglobulin iodination with autoimmune thyroiditis. *Hormones (Athens)* 10 (1): 27–35.

Casey, P.J. (1994). Lipid modifications of G proteins. *Curr. Opin. Cell. Biol.* 6 (2): 219–225.

Chamberlain, L.H., Lemonidis, K., Sanchez-Perez, M. et al. (2013). Palmitoylation and the trafficking of peripheral membrane proteins. *Biochem. Soc. Trans.* 41 (1, 1): 62–66. https://doi.org/10.1042/BST20120243.

Choudhary, C., Weinert, B.T., Nishida, Y. et al. (2014). The growing landscape of lysine acetylation links metabolism and cell signalling. *Nat. Rev. Mol. Cell. Biol.* 15 (8): 536–550. https://doi.org/10.1038/nrm3841.

Cohen, M.S. and Chang, P. (2018). Insights into the biogenesis, function, and regulation of ADP-ribosylation. *Nat. Chem. Biol.* 14 (3): 236–243. https://doi.org/10.1038/nchembio.2568.

Cotgreave, I.A. (2003). Analytical developments in the assay of intra- and extracellular GSH homeostasis: specific protein S-glutathionylation, cellular GSH and mixed disulphide compartmentalisation and interstitial GSH redox balance. *Biofactors* 17 (1–4): 269–277.

De Franceschi, L., Bachir, D., Galacteros, F. et al. (1997). Oral magnesium supplements reduce erythrocyte dehydration in patients with sickle cell disease. *J. Clin. Invest.* 100 (7): 1847–1852.

Delgado, T.C., Barbier-Torres, L., Zubiete-Franco, I. et al. (2018). Neddylation, a novel paradigm in liver cancer. *Transl. Gastroenterol. Hepatol.* 3 (37): https://doi.org/10.21037/tgh.2018.06.05.

Delley, C.L., Müller, A.U., Ziemski, M., and Weber-Ban, E. (2017). Prokaryotic ubiquitin-like protein and its ligase/deligase enyzmes. *J. Mol. Biol.* 429 (22): 3486–3499. https://doi.org/10.1016/j.jmb.2017.04.020.

Ekholm, R. (1981). Iodination of thyroglobulin. An intracellular or extracellular process? *Mol. Cell. Endocrinol.* 24 (2): 141–163.

Enchev, R.I., Schulman, B.A., and Peter, M. (2015). Protein neddylation: beyond cullin-RING ligases. *Nat. Rev. Mol. Cell Biol.* 16 (1): 30–44. https://doi.org/10.1038/nrm3919.

Harley, J.D. (1965). Role of reduced glutathione in human erythrocytes. *Nature* 206 (988): 1054–1055.

Hedberg, C. and Itzen, A. (2015). Molecular perspectives on protein adenylylation. *ACS Chem. Biol.* 10 (1): 12–21. https://doi.org/10.1021/cb500854e.

Hirschey, M.D. and Zhao, Y. (2015). Metabolic regulation by lysine malonylation, succinylation, and glutarylation. *Mol. Cell. Proteomics* 14 (9): 2308–2315. https://doi.org/10.1074/mcp.R114.046664.

Ibitoye, R., Kemp, K., Rice, C. et al. (2016). Oxidative stress-related biomarkers in multiple sclerosis: a review. *Biomark Med.* 10 (8): 889–902. https://doi.org/10.2217/bmm-2016-0097.

Jaisson, S., Pietrement, C., and Gillery, P. (2018). Protein carbamylation: chemistry, pathophysiological involvement, and biomarkers. *Adv. Clin. Chem.* 84: 1–38. https://doi.org/10.1016/bs.acc.2017.12.001.

Janke, C., Rogowski, K., and van Dijk, J. (2008). Polyglutamylation: a fine-regulator of protein function? 'Protein modifications: beyond the usual suspects' review series. *EMBO Rep.* 9 (7): 636–641. https://doi.org/10.1038/embor.2008.114.

Jänne, J., Hölttä, E., Haaranen, P., and Elfving, K. (1973). Polyamines and polyamine-metabolizing enzyme activities in human semen. *Clin. Chim. Acta* 48 (4): 393–401.

Jiang, T., Zhou, X., Taghizadeh, K. et al. (2007). N-formylation of lysine in histone proteins as a secondary modification arising from oxidative DNA damage. *Proc. Natl. Acad. Sci. U.S.A.* 104 (1): 60–65.

Kalafatis, M., Egan, J.O., van't Veer, C., and Mann, K.G. (1996). Regulation and regulatory role of gamma-carboxyglutamic acid containing clotting factors. *Crit. Rev. Eukaryot. Gene Expr.* 6 (1): 87–101.

Kashina, A.S. (2015). Protein arginylation: over 50 years of discovery. *Methods Mol. Biol.* 1337: 1–11. https://doi.org/10.1007/978-1-4939-2935-1_1.

Kurosaka, S., Leu, N.A., Pavlov, I. et al. (2012). Arginylation regulates myofibrils to maintain heart function and prevent dilated cardiomyopathy. *J. Mol. Cell. Cardiol.* 53 (3): 333–341. https://doi.org/10.1016/j.yjmcc.2012.05.007.

Lee, S. (2013). Post-translational modification of proteins in toxicological research: focus on lysine acylation. *Toxicol. Res.* 29 (2): 81–86. https://doi.org/10.5487/TR.2013.29.2.081.

Liu, C. and Yu, X. (2015). ADP-ribosyltransferases and poly ADP-ribosylation. *Curr. Protein Pept. Sci.* 16 (6): 491–501.

Mailloux, R.J. and Willmore, W.G. (2014). S-glutathionylation reactions in mitochondrial function and disease. *Front. Cell Dev. Biol.* 2 (68): https://doi.org/10.3389/fcell.2014.00068.

Mangroo, D., Wu, X.Q., and RajBhandary, U.L. (1995). *Escherichia coli* initiator tRNA: structure-function relationships and interactions with the translational machinery. *Biochem. Cell Biol.* 73 (11–12): 1023–1031.

Marcker, K. and Sanger, F. (1964). *N*-formyl-methionyl-*S*-RNA. *J. Mol. Biol.* 8: 835–840.

Meinnel, T., Mechulam, Y., and Blanquet, S. (1993). Methionine as translation start signal: a review of the enzymes of the pathway in *Escherichia coli. Biochimie* 75 (12): 1061–1075.

Michael, A.J. (2016). Polyamines in eukaryotes bacteria, and archaea. *J. Biol. Chem.* 291 (29): 14896–14903. https://doi.org/10.1074/jbc.R116.734780.

Narrowe, A.B., Spang, A., Stairs, C.W. et al. (2018 Sep 1). Complex evolutionary history of translation elongation factor 2 and diphthamide biosynthesis in archaea and parabasalids. *Genome Biol. Evol.* 10 (9): 2380–2393. https://doi.org/10.1093/gbe/evy154.

Olsen, I., Singhrao, S.K., and Potempa, J. (2018). Citrullination as a plausible link to periodontitis, rheumatoid arthritis, atherosclerosis and Alzheimer's disease. *J. Oral. Microbiol.* 10 (1): 1487742. https://doi.org/10.1080/20002297.2018.1487742.

Prigge, S.T., Mains, R.E., Eipper, B.A., and Amzel, L.M. (2000). New insights into copper monooxygenases and peptide amidation: structure, mechanism and function. *Cell. Mol. Life Sci.* 57 (8–9): 1236–1259.

Rai, R., Wong, C.C., Xu, T. et al. (2008). Arginyltransferase regulates alpha cardiac actin function, myofibril formation and contractility during heart development. *Development* 135 (23): 3881–3889. https://doi.org/10.1242/dev.022723.

Rioux, V., Pédrono, F., and Legrand, P. (2011). Regulation of mammalian desaturases by myristic acid: N-terminal myristoylation and other modulations. *Biochim. Biophys. Acta* 1811 (1): 1–8. https://doi.org/10.1016/j.bbalip.2010.09.005.

Ross, E.M. (1995). Protein modification. Palmitoylation in G-protein signaling pathways. *Curr. Biol.* 5 (2): 107–109.

Sabari, B.R., Zhang, D., Allis, C.D., and Zhao, Y. (2017). Metabolic regulation of gene expression through histone acylations. *Nat. Rev. Mol. Cell Biol.* 18 (2): 90–101. https://doi.org/10.1038/nrm.2016.140.

Saini, P., Eyler, D.E., Green, R., and Dever, T.E. (2009). Hypusine-containing protein eIF5A promotes translation elongation. *Nature* 459 (7243): 118–121. https://doi.org/10.1038/nature08034.

Samanovic, M.I. and Darwin, K.H. (2016). Game of 'Somes: protein destruction for *Mycobacterium tuberculosis* pathogenesis. *Trends Microbiol.* 24 (1): 26–34. https://doi.org/10.1016/j.tim.2015.10.001.

Saha, S. and Kashina, A. (2011). Posttranslational arginylation as a global biological regulator. *Dev. Biol.* 358 (1): 1–8. https://doi.org/10.1016/j.ydbio.2011.06.043.

Sanders, S.S. and Hayden, M.R. (2015). Aberrant palmitoylation in Huntington disease. *Biochem. Soc. Trans.* 43 (2): 205–210. https://doi.org/10.1042/BST20140242.

Schaffrath, R., Abdel-Fattah, W., Klassen, R., and Stark, M.J. (2014). The diphthamide modification pathway from *Saccharomyces cerevisiae* – revisited. *Mol. Microbiol.* 94 (6): 1213–1226. https://doi.org/10.1111/mmi.12845.
Schaffrath, R. and Stark, M.J. (2014). Decoding the biosynthesis and function of diphthamide, an enigmatic modification of translation elongation factor 2 (EF2). *Microb. Cell* 1 (6): 203–205. https://doi.org/10.15698/mic2014.06.151.
Schilling, S., Wasternack, C., and Demuth, H.U. (2008). Glutaminyl cyclases from animals and plants: a case of functionally convergent protein evolution. *Biol. Chem.* 389 (8): 983–991. https://doi.org/10.1515/BC.2008.111.
Schlott, A.C., Holder, A.A., and Tate, E.W. (2018). *N*-myristoylation as a drug target in malaria: exploring the role of *N*-myristoyltransferase substrates in the inhibitor mode of action. *ACS Infect. Dis.* 4 (4): 449–457. https://doi.org/10.1021/acsinfecdis.7b00203.
Sengupta, R. and Holmgren, A. (2013). Thioredoxin and thioredoxin reductase in relation to reversible *S*-nitrosylation. *Antioxid. Redox Signal.* 18 (3): 259–269. https://doi.org/10.1089/ars.2012.4716.
Seth, D. and Stamler, J.S. (2011). The SNO-proteome: causation and classifications. *Curr. Opin. Chem. Biol.* 15 (1): 129–136. https://doi.org/10.1016/j.cbpa.2010.10.012.
Shah, P.T., Martin, R., Yan, Y. et al. (2016). Carbonylation modification regulates Na/K-ATPase signaling and salt sensitivity: a review and a hypothesis. *Front. Physiol.* 7: 256. https://doi.org/10.3389/fphys.2016.00256.
Shiba, T., Mizote, H., Kaneko, T. et al. (1971). Hypusine, a new amino acid occurring in bovine brain. Isolation and structural determination. *Biochim. Biophys. Acta.* 244 (3): 523–531.
Song, Y. and Brady, S.T. (2015). Post-translational modifications of tubulin: pathways to functional diversity of microtubules. *Trends Cell Biol.* 25 (3): 125–136. https://doi.org/10.1016/j.tcb.2014.10.004.
Stahl, S., da Silva Mateus Seidl, A.R., Ducret, A. et al. (2015). Loss of diphthamide pre-activates NF-κB and death receptor pathways and renders MCF7 cells hypersensitive to tumor necrosis factor. *Proc. Natl. Acad. Sci. U.S.A.* 112 (34): 10732–10737. https://doi.org/10.1073/pnas.1512863112.
Su, X., Lin, Z., and Lin, H. (2013). The biosynthesis and biological function of diphthamide. *Crit. Rev. Biochem. Mol. Biol.* 48 (6): 515–521. https://doi.org/10.3109/10409238.2013.831023.
Thazhath, R., Liu, C., and Gaertig, J. (2002). Polyglycylation domain of beta-tubulin maintains axonemal architecture and affects cytokinesis in *Tetrahymena*. *Nat. Cell Biol.* 4 (3): 256–259.
Thompson, P.R. and Fast, W. (2006). Histone citrullination by protein arginine deiminase: is arginine methylation a green light or a roadblock? *ACS Chem. Biol.* 1 (7): 433–441.
Vinh, J., Loyaux, D., Redeker, V., and Rossier, J. (1997). Sequencing branched peptides with CID/PSD MALDI-TOF in the low-picomole range: application to the structural study of the posttranslational polyglycylation of tubulin. *Anal. Chem.* 69 (19): 3979–3985.

Wallin, R. and Hutson, S.M. (2004). Warfarin and the vitamin K-dependent gamma-carboxylation system. *Trends Mol. Med.* 10 (7): 299–302.

Wisniewski, J.R., Zougman, A., and Mann, M. (2008). Nepsilon-formylation of lysine is a widespread post-translational modification of nuclear proteins occurring at residues involved in regulation of chromatin function. *Nucleic Acids Res.* 36 (2): 570–577.

Woody, N.C. and Pupene, M.B. (1973). Excretion of hypusine by children and by patients with familial hyperlysinemia. *Pediatr. Res.* 7 (12): 994–995.

Woolery, A.R., Luong, P., Broberg, C.A., and Orth, K. (2010). AMPylation: something old is new again. *Front. Microbiol.* 1: 113. https://doi.org/10.3389/fmicb.2010.00113.

Wu, G., Miller, R.A., Connolly, B. et al. (2014). Pyroglutamate-modified amyloid-β protein demonstrates similar properties in an Alzheimer's disease familial mutant knock-in mouse and Alzheimer's disease brain. *Neurodegener. Dis.* 14 (2): 53–66. https://doi.org/10.1159/000353634.

Xia, L., Hai, B., Gao, Y. et al. (2000). Polyglycylation of tubulin is essential and affects cell motility and division in *Tetrahymena thermophila*. *J. Cell Biol.* 149 (5): 1097–1106.

Yu, L., Vizel, A., Huff, M.B. et al. (2006). Investigation of N-terminal glutamate cyclization of recombinant monoclonal antibody in formulation development. *J. Pharm. Biomed. Anal.* 42 (4): 455–463.

Zhang, J., Ye, Z.W., Singh, S. et al. (2018). An evolving understanding of the *S*-glutathionylation cycle in pathways of redox regulation. *Free Radic. Biol. Med.* 120: 204–216. https://doi.org/10.1016/j.freeradbiomed.2018.03.038.

Zhu, A., Sun, H., Raymond, R.M. Jr. et al. (2007). Fatal hemorrhage in mice lacking gamma-glutamyl carboxylase. *Blood* 109 (12): 5270–5275.

# Appendix A

In this book, Chapters 1–21 provides an overview of the co-translational modifications (CTMs) and post-translational modifications (PTMs) of proteins including recombinant therapeutic proteins (RTPs). The list of RTPs currently approved by US FDA includes binding proteins, activating proteins, enzymes, monoclonal antibodies (mAbs), antibody fusion proteins, antibody fragments, etc. In many RTPs, the CTMs and PTMs plays an important role in defining their mechanism of actions (MOAs), mechanism of toxicities (MOTs), immunogenicity, etc. In some RTPs, the CTMs and PTMs may be required for their biological functions. For example, some RTPs are being used as enzyme replacement therapies that contain phosphorylated glycans that are important for their biological activities. Further, many of the $IgG_1$k-based mAbs exhibit antibody-mediated effector functions in which Fc glycosylation is required. In some RTPs, many of the CTMs and PTMs may pose safety risks. For example, proteolysis, deamidation, oxidation may negatively impact the biological functions and hence may raise safety issues. In addition, the nature of CTMs and PTMs of RTPs may vary with the expression systems that might impact the MOAs, MOTs, immunogenicity, pharmacokinetic properties, etc. Table A.1 provides a list of RTPs approved by US FDA as of March 2018 along with their expression system, molecular nature, and the possible common PTMs. The data presented in Table A.1 was generated using www.FDA.gov and prescribing information from the listed products as sources.

*Co- and Post-Translational Modifications of Therapeutic Antibodies and Proteins*, First Edition. T. Shantha Raju.

**Table A.1** List of approved biological products in the United States and their possible common PTMs.

| Proprietary name | Proper name | Approval date (M/D/Y) by US FDA | Expression system | Molecular nature | Possible common CTMs and/or PTMs |
|---|---|---|---|---|---|
| Actemra | Tocilizumab | 01/08/10 | Mammalian (CHO cells) | Humanized $IgG_1 \kappa$ antibody against human IL-6 receptor | Glycosylation, glycation, deamidation, oxidation, C-terminal clipping, etc. |
| Actemra | Tocilizumab | 10/21/13 | Mammalian (CHO cells) | Humanized $IgG_1 \kappa$ antibody against human IL-6 receptor | Glycosylation, glycation, deamidation, oxidation, C-terminal clipping, etc. |
| Actimmune | Interferon γ-1b | 02/25/99 | Bacteria (*Escherichia coli*) | Human interferon γ-1b | Glycation, deamidation, oxidation, etc. |
| Activase | Alteplase, cathflo activase | 11/13/87 | Mammalian | Recombinant human tissue type plasminogen activator (rht-PA) | Glycosylation, glycation, deamidation, oxidation, etc. |
| Adcetris | Brentuximab vedotin | 08/19/11 | Mammalian | Chimeric $IgG_1$ antibody drug conjugate | Glycosylation, glycation, deamidation, oxidation, C-terminal Lys clipping, free thiols, etc. |
| Aimovig | Erenumab-aooe | 05/17/18 | Mammalian (CHO cells) | Human $IgG_2$ antibody against calcitonin gene-related peptide receptor | Glycosylation, glycation, deamidation, oxidation, C-terminal clipping, etc. |
| Aldurazyme | Laronidase | 04/30/03 | Mammalian (CHO cells) | Human L-iduronidase | Glycosylation, glycation, deamidation, oxidation, etc. |
| Alferon N Injection | Interferon alfa-n3 | 10/10/89 | NA | Human interferon α-N3 | Glycation, deamidation, oxidation, etc. |

| | | | | | |
|---|---|---|---|---|---|
| Amjevita | Adalimumab-atto | 09/23/16 | Mammalian | Human IgG biosimilar of humira | Glycosylation, glycation, deamidation, oxidation, etc. |
| Anthim | Obiltoxaximab | 03/18/16 | Mammalian | Chimeric $IgG_1 \kappa$ antibody against PA component of *Bacillus anthracis* toxin | Glycosylation, glycation, deamidation, oxidation, C-terminal clipping, etc. |
| Aranesp | Darbepoetin alfa | 09/17/01 | Mammalian (CHO cells) | Protein engineered version of human erythropoietin | Glycosylation, glycation, deamidation, oxidation, etc. |
| Arcalyst | Rilonacept | 02/27/08 | Mammalian (CHO cells) | Human $IgG_1$ Fc fusion protein consisting of the ligand-binding domains of the extracellular portions of the human interleukin-1 receptor component (IL-1RI) and IL-1 receptor accessory protein (IL-1RAcP) linked in-line to the Fc portion of human $IgG_1$ | Glycosylation, glycation, deamidation, oxidation, C-terminal clipping, etc. |
| Arzerra | Ofatumumab | 10/26/09 | Mammalian (mouse myeloma, NS0) | Human $IgG_1 \kappa$ antibody against CD20 antigen | Glycosylation, glycation, deamidation, oxidation, C-terminal clipping, etc. |
| Avastin | Bevacizumab | 02/26/04 | Mammalian (CHO cells) | Humanized $IgG_1$ against VEGF | Glycosylation, glycation, deamidation, oxidation, C-terminal Lys clipping, etc. |
| Avonex | Interferon β-1a | 05/17/96 | Mammalian (CHO cells) | Human interferon β-1a | Glycosylation, glycation, deamidation, oxidation, etc. |

*(Continued)*

**Table A.1** (Continued)

| Proprietary name | Proper name | Approval date (M/D/Y) by US FDA | Expression system | Molecular nature | Possible common CTMs and/or PTMs |
|---|---|---|---|---|---|
| Bavencio | Avelumab | 03/23/17 | Mammalian | Human $IgG_1\lambda$ | Glycosylation, glycation, deamidation, oxidation, C-terminal Lys clipping, etc. |
| Benlysta | Belimumab | 03/09/11 | Mammalian (NS0 cells) | Human $IgG_1\lambda$ against BLyS | Glycosylation, glycation, deamidation, oxidation, C-terminal Lys clipping, etc. |
| Besponsa | Inotuzumab ozogamicin | 08/17/17 | Mammalian (CHO cells) | Humanized $IgG_4$ antibody against CD22 conjugated to *N*-acetyl γ calicheamicin | Glycosylation, glycation, deamidation, oxidation, C-terminal clipping, etc. |
| Betaseron | Interferon β-1b | 07/23/93 | Mammalian | Avonex biosimilar | Glycosylation, glycation, deamidation, oxidation, etc. |
| Blincyto | Blinatumomab | 12/03/14 | Mammalian (CHO cells) | Bispecific IgG against CD3 and CD19 | Glycosylation, glycation, deamidation, oxidation, C-terminal Lys clipping, etc. |
| Botox | Onabotulinumtoxina | 12/29/89 | Bacteria (*Clostridium botulinum*) | Botulinum toxin type A | Glycation, deamidation, oxidation, etc. |
| Brineura | Cerliponase alfa | 04/27/17 | Mammalian (CHO cells) | Human tripeptidyl peptidase | Glycosylation, glycation, deamidation, oxidation, etc. |
| Campath, Lemtrada | Alemtuzumab | 05/07/01 | Mammalian | Chimeric IgG | Deamidation, glycation, glycosylation, oxidation, etc. |
| Cimzia | Certolizumab pegol | 04/22/08 | Bacteria (*E. coli*) | Humanized antibody Fab' fragment conjugated to 40 kDa PEG | Glycation, deamidation, oxidation, etc. |

| | | | | | |
|---|---|---|---|---|---|
| Cinqair | Reslizumab | 03/23/16 | Mammalian (mouse myeloma, NS0) | Humanized $IgG_4\kappa$ antibody against human interleukin-5 | Glycosylation, glycation, deamidation, oxidation, C-terminal clipping, etc. |
| Cosentyx | Secukinumab | 01/21/15 | Mammalian (CHO cells) | Human $IgG_1\kappa$ against IL-17A antigen | Glycosylation, glycation, deamidation, oxidation, C-terminal clipping, etc. |
| Crysvita | Burosumab-twza | 04/17/18 | Mammalian (CHO cells) | Human $IgG_1$ against FGF23 | Glycosylation, glycation, deamidation, oxidation, C-terminal Lys clipping, etc. |
| Cyltezo | Adalimumab-adbm | 08/25/17 | Mammalian | Human IgG Biosimilar | Glycosylation, glycation, deamidation, oxidation, etc. |
| Cyramza | Ramucirumab | 04/21/14 | Mammalian (mouse myeloma, NS0) | Human $IgG_1$ against VEGFR2 | Glycosylation, glycation, deamidation, oxidation, C-terminal clipping, etc. |
| Darzalex | Daratumumab | 11/16/15 | Mammalian (CHO cells) | Human $IgG_1$ antibody against CD38 | Glycosylation, glycation, deamidation, oxidation, C-terminal clipping, etc. |
| Dupixent | Dupilumab | 03/28/17 | Mammalian (CHO cells) | Human $IgG_4$ antibody against IL-4Rα | Glycosylation, glycation, deamidation, oxidation, C-terminal clipping, etc. |
| Dysport | Abobotulinumtoxin A | 04/29/09 | Bacteria | Neurotoxin | Deamidation, oxidation, glycation, etc. |
| Elaprase | Idursulfase | 07/24/06 | Mammalian (human cell line) | Human iduronate-2-sulfatase | Glycosylation, glycation, deamidation, oxidation, etc. |
| Elitek | Rasburicase | 07/12/02 | Yeast (*Saccharomyces cerevisiae*) | Urate oxidase enzyme | Glycation, deamidation, oxidation, etc. |
| Elspar | Asparaginase | 01/10/78 | Bacteria | Asparagine amido-hydrolase | Deamidation, glycation, oxidation, etc. |

(Continued)

**Table A.1** (Continued)

| Proprietary name | Proper name | Approval date (M/D/Y) by US FDA | Expression system | Molecular nature | Possible common CTMs and/or PTMs |
|---|---|---|---|---|---|
| Empliciti | Elotuzumab | 11/30/15 | Mammalian (NS0) | Human $IgG_1$ against SLAMF7 | Glycosylation, glycation, deamidation, oxidation, C-terminal clipping, etc. |
| Enbrel | Etanercept | 11/02/98 | Mammalian (CHO cells) | Human $IgG_1$ Fc fusion protein and is a TNF blocker | Glycosylation, glycation, deamidation, oxidation, C-terminal clipping, etc. |
| Entyvio | Vedolizumab | 05/20/14 | Mammalian (CHO cells) | Humanized $IgG_1$ antibody against α4β7 integrin | Glycosylation, glycation, deamidation, oxidation, C-terminal clipping, etc. |
| Epogen/Procrit | Epoetin alfa | 06/01/89 | Mammalian (CHO cells) | Human erythropoietin | Glycosylation, glycation, deamidation, oxidation, etc. |
| Erbitux | Cetuximab | 02/12/04 | Mammalian (murine myeloma cells) | Human/mouse chimeric $IgG_1$ antibody against EGFR | Glycosylation, glycation, deamidation, oxidation, C-terminal Lys clipping, etc. |
| Erelzi | Etanercept-szzs | 08/30/16 | Mammalian (CHO cells) | Biosimilar to Enbrel | Glycosylation, glycation, deamidation, oxidation, C-terminal clipping, etc. |
| Erwinaze | Asparaginase erwinia chrysanthemi | 11/18/11 | Bacteria | Asparaginase from *Erwinia chrysanthemi* | Deamidation, glycation, oxidation, etc. |
| Extavia | Interferon β-1b | 08/14/09 | Mammalian | Avonex biosimilar | Glycosylation, glycation, deamidation, oxidation, etc. |
| Eylea | Aflibercept | 11/18/11 | Mammalian | Fc fusion protein | Deamidation, glycation, glycosylation, oxidation, etc. |
| Fabrazyme | Agalsidase beta | 04/24/03 | Mammalian | α-Galactosidase enzyme | Deamidation, glycation, glycosylation, oxidation, etc. |

| | | | | | |
|---|---|---|---|---|---|
| Fasenra | Benralizumab | 11/14/17 | Mammalian (CHO cells) | Humanized $IgG_1$ against IL-5Rα | Glycosylation, glycation, deamidation, oxidation, C-terminal Lys clipping, etc. |
| Fulphila | Pegfilgrastim-jmdb | 06/04/18 | Bacteria (*E. coli*) | Pegfilgrastim biosimilar | Glycation, deamidation, oxidation, etc. |
| Gazyva | Obinutuzumab | 11/01/13 | Mammalian (CHO cells) | Humanized $IgG_1\kappa$ antibody against human CD20 antigen | Glycosylation, glycation, deamidation, oxidation, C-terminal clipping, etc. |
| Granix | Tbo-filgrastim | 08/29/12 | Bacteria (*E. coli*) | Methionyl human G-CSF | Glycation, deamidation, oxidation, etc. |
| Hemlibra | Emicizumab-kxwh | 11/16/17 | Mammalian (CHO cells) | Bispecific humanized $IgG_4$ against factor IXa and factor X | Glycosylation, glycation, deamidation, oxidation, C-terminal clipping, etc. |
| Herceptin | Trastuzumab | 09/25/98 | Mammalian (CHO cells) | Humanized $IgG_1\kappa$ antibody against ECD of HER2 | Glycosylation, glycation, deamidation, oxidation, C-terminal clipping, etc. |
| Humira | Adalimumab | 12/31/02 | Mammalian | Human IgG | Glycosylation, glycation, deamidation, oxidation, etc. |
| Ilaris | Canakinumab | 06/17/09 | Mammalian | Human $IgG_1$ against IL-1β | Glycosylation, glycation, deamidation, oxidation, C-terminal Lys clipping, etc. |
| Ilumya | Tildrakizumab-asmn | 03/20/18 | Mammalian (CHO cells) | Humanized $IgG_1\kappa$ antibody against P19 subunit of IL-23 | Glycosylation, glycation, deamidation, oxidation, C-terminal clipping, etc. |
| Imfinzi | Durvalumab | 05/01/17 | Mammalian (CHO cells) | Human $IgG_1$ antibody against PD-L1 | Glycosylation, glycation, deamidation, oxidation, C-terminal clipping, etc. |
| Inflectra | Infliximab-dyyb | 04/05/16 | Mammalian | Remicade biosimilar | Glycosylation, glycation, deamidation, oxidation, C-terminal clipping, etc. |

*(Continued)*

**Table A.1** (Continued)

| Proprietary name | Proper name | Approval date (M/D/Y) by US FDA | Expression system | Molecular nature | Possible common CTMs and/or PTMs |
|---|---|---|---|---|---|
| Intron A | Interferon alfa-2b | 06/04/86 | Bacteria (*E. coli*) | Human Interferon α 2b | Glycation, deamidation, oxidation, etc. |
| Ixifi | Infliximab-qbtx | 12/13/17 | Mammalian | Remicade biosimilar | Glycosylation, glycation, deamidation, oxidation, C-terminal clipping, etc. |
| Jetrea | Ocriplasmin | 10/17/12 | Yeast (*Pichia pastoris*) | Truncated human plasmin | Glycation, deamidation, oxidation, etc. |
| Kadcyla | Ado-trastuzumab emtansine | 02/22/13 | Mammalian | Antibody drug conjugate | Deamidation, glycation, glycosylation, oxidation, etc. |
| Kalbitor | Ecallantide | 12/01/09 | Yeast (*P. pastoris*) | Human plasma kallikrein inhibitor | Glycation, deamidation, oxidation, etc. |
| Kanuma | Sebelipase alfa | 12/08/15 | Avian (chicken eggs white) | Human lysosomal acid lipase (LAL) | Glycation, deamidation, oxidation, etc |
| Kepivance | Palifermin | 12/15/04 | Bacteria (*E. coli*) | Truncated human keratinocyte growth factor (KGF) | Glycation, deamidation, oxidation, etc. |
| Kevzara | Sarilumab | 05/22/17 | Mammalian (CHO cells) | Human $IgG_1 \kappa$ against human IL-6 receptor | Glycosylation, glycation, deamidation, oxidation, C-terminal clipping, etc. |
| Keytruda | Pembrolizumab | 09/04/14 | Mammalian (CHO cells) | Humanized $IgG_4$ antibody that blocks the interactions between PD-1 and its ligands, PD-L1 and PD-L2 | Glycosylation, glycation, deamidation, oxidation, C-terminal clipping, etc. |
| Kineret | Anakinra | 11/14/01 | Bacteria | Modified interleukin-1 receptor antagonist | Deamidation, oxidation, etc. |

| | | | | | |
|---|---|---|---|---|---|
| Krystexxa | Pegloticase | 09/14/10 | Bacteria (modified *E. coli*) | Mammalian urate oxidase conjugated to PEG | Glycation, deamidation, oxidation, etc. |
| Lartruvo | Olaratumab | 10/19/16 | Mammalian (mouse myeloma, NS0) | Human $IgG_1$ κ antibody against human platelet-derived growth factor receptor α (PDGFR-α) | Glycosylation, glycation, deamidation, oxidation, C-terminal clipping, etc. |
| Leukine | Sargramostim | 03/05/91 | Yeast (*S. cerevisiae*) | Human granulocyte macrophage colony stimulating factor (GM-CSF) | Glycosylation, glycation, deamidation, oxidation, etc. |
| Lucentis | Ranibizumab | 06/30/06 | Bacteria (*E. coli*) | Humanized $IgG_1$ κ antibody fragment against VEGF-A | Glycation, deamidation, oxidation, etc. |
| Lumizyme | Alglucosidase alfa | 05/24/10 | Mammalian | α-Glucosidase | Deamidation, glycation, glycosylation, oxidation, phosphorylation, etc. |
| Mepsevii | Vestronidase alfa-vjbk | 11/15/17 | Mammalian (CHO cells) | Human lysosomal β-glucuronidase | Glycosylation, glycation, deamidation, oxidation, etc. |
| Mircera | Methoxy polyethylene glycol-epoetin beta | 11/14/07 | Mammalian (CHO cells) | Epoetin β is conjugated to methoxy polyethylene glycol (mPEG) | Glycosylation, glycation, deamidation, oxidation, etc. |
| Mvasi | Bevacizumab-awwb | 09/14/17 | Mammalian (CHO cells) | Biosimilar of Avastin | Glycosylation, glycation, deamidation, oxidation, C-terminal Lys clipping, etc. |
| Myalept | Metreleptin | 02/24/14 | Bacteria (*E. coli*) | Methionyl-human leptin | Glycation, deamidation, oxidation, etc. |
| Mylotarg | Gemtuzumab ozogamicin | 09/01/17 | Mammalian (mouse myeloma NS0 cells) | Humanized $IgG_4$ linked to *N*-acetyl γ calcheamicin through primary amine groups | Glycosylation, glycation, deamidation, oxidation, C-terminal clipping, etc. |

*(Continued)*

**Table A.1** (Continued)

| Proprietary name | Proper name | Approval date (M/D/Y) by US FDA | Expression system | Molecular nature | Possible common CTMs and/or PTMs |
|---|---|---|---|---|---|
| Myobloc | Rimabotulinumtoxinb | 12/08/00 | Bacteria (*C. botulinum*) | Neurotoxin | Glycation, deamidation, oxidation, etc. |
| Myozyme | Alglucosidase alfa | 04/28/06 | Mammalian | α-Glucosidase | Deamidation, glycation, glycosylation, oxidation, phosphorylation, etc. |
| Naglazyme | Galsulfase | 05/31/05 | Mammalian (CHO cells) | Glycosaminoglycan *N*-acetylgalactosamine 4-sulfatase | Glycosylation, glycation, deamidation, oxidation, C-α formylglycine, etc. |
| Natpara | Parathyroid hormone | 01/23/15 | Bacteria (*E. coli*) | Parathyroid hormone | Glycation, deamidation, oxidation, etc. |
| Neulasta | Pegfilgrastim | 01/31/02 | Bacteria (*E. coli*) | Methionyl human G-CSF conjugated to monomethoxy PEG | Glycation, deamidation, oxidation, etc. |
| Neumega | Oprelvekin | 11/25/97 | Bacteria (*E. coli*) | Human IL-11 | Glycation, deamidation, oxidation, etc. |
| Neupogen | Filgrastim | 02/20/91 | Bacteria (*E. coli*) | Human granulocyte colony-stimulating factor (hG-CSF) | Glycation, deamidation, oxidation, etc. |
| Nivestym | Filgrastim-aafi | 07/20/18 | Bacteria (*E. coli*) | Biosimilar to Neupogen | Glycation, deamidation, oxidation, etc. |
| Nplate | Romiplostim | 08/22/08 | Bacteria (*E. coli*) | Fc-peptide fusion protein | Glycation, deamidation, oxidation, etc. |
| Nucala | Mepolizumab | 11/04/15 | Mammalian (CHO cells) | Humanized $IgG_1$ antibody against IL-5 | Glycosylation, glycation, deamidation, oxidation, C-terminal clipping, etc. |

| | | | | | |
|---|---|---|---|---|---|
| Nulojix | Belatacept | 06/15/11 | Mammalian | CTLA4-$IgG_1$ Fc fusion protein | Glycosylation, glycation, deamidation, oxidation, C-terminal Lys clipping, etc. |
| Ocrevus | Ocrelizumab | 03/28/17 | Mammalian (CHO cells) | Humanized $IgG_1\kappa$ against human CD20 antigen | Glycosylation, glycation, deamidation, oxidation, C-terminal clipping, etc. |
| Ogivri | Trastuzumab-dkst | 12/01/17 | Mammalian (CHO cells) | Herceptin biosimilar | Glycosylation, glycation, deamidation, oxidation, C-terminal clipping, etc. |
| Oncaspar | Pegaspargase | 02/01/94 | Bacteria (*E. coli*) | L-Asparaginase conjugated to mPEG | Glycation, deamidation, oxidation, etc. |
| Ontak | Denileukin diftitox | 02/05/99 | Bacteria (*E. coli*) | Diptheria toxin A and B | Glycation, deamidation, oxidation, etc. |
| Opdivo | Nivolumab | 12/22/14 | Mammalian (CHO cells) | Human $IgG_4\kappa$ antibody to block the interactions between PD1 and its ligands PD-L1 and PD-L2 | Glycosylation, glycation, deamidation, oxidation, C-terminal clipping, etc. |
| Orencia | Abatacept | 12/23/05 | Mammalian | IgG fusion protein | Glycosylation, glycation, deamidation, oxidation, etc. |
| Palynziq | Pegvaliase-pqpz | 05/24/18 | Bacteria (*E. coli*) | Phenylalanine ammonia lyase conjugated to *N*-hydroxysuccinimide (NHS) methoxy PEG | Deamidation, glycation, oxidation, etc. |
| Pegasys | Peginterferon alfa-2a | 10/16/02 | Bacteria (*E. coli*) | Human α-2a interferon conjugated to bis-monomethoxy PEG | Glycation, deamidation, oxidation, etc. |

*(Continued)*

**Table A.1** (Continued)

| Proprietary name | Proper name | Approval date (M/D/Y) by US FDA | Expression system | Molecular nature | Possible common CTMs and/or PTMs |
|---|---|---|---|---|---|
| Pegasys Copegus Combination Pack | Peginterferon alfa-2a copackaged with ribavirin | 06/04/04 | Bacteria (*E. coli*) | Combination product of Peginterferon α-2a and Ribavirin | Glycation, deamidation, oxidation, etc. |
| PegIntron, Sylatron | Peginterferon alfa-2b | 01/19/01 | Bacteria (*E. coli*) | Human α interferon is conjugated to monomethoxy PEG | Glycation, deamidation, oxidation, etc. |
| Perjeta | Pertuzumab | 06/08/12 | Mammalian (CHO cells) | Humanized $IgG_1$ antibody against the ECD of HER2 | Glycosylation, glycation, deamidation, oxidation, C-terminal clipping, etc. |
| Plegridy | Peginterferon β-1a | 08/15/14 | Mammalian (CHO cells) | Human interferon β-1a is conjugated to modified PEG | Glycosylation, glycation, deamidation, oxidation, etc. |
| Portrazza | Necitumumab | 11/24/15 | Mammalian (mouse myeloma, NS0) | Human $IgG_1 κ$ antibody against EGFR | Glycosylation, glycation, deamidation, oxidation, C-terminal clipping, etc. |
| Praluent | Alirocumab | 07/24/15 | Mammalian | Human $IgG_1$ | Deamidation, glycation, glycosylation, oxidation, etc. |
| Praxbind | Idarucizumab | 10/16/15 | Mammalian (CHO cells) | Humanized $IgG_1$ Fab fragment that binds and neutralizes thrombin inhibitor dabigatran | Glycation, deamidation, oxidation, etc. |
| Proleukin | Aldesleukin | 05/05/92 | Bacteria | Engineered interleukin-2 | Deamidation, glycation, oxidation, etc. |

| | | | | | |
|---|---|---|---|---|---|
| Prolia, Xgeva | Denosumab | 06/01/10 | Mammalian (CHO cells) | Human $IgG_2$ that binds receptor activator of nuclear factor kappa-B ligand (RANKL) | Glycosylation, glycation, deamidation, oxidation, C-terminal clipping, etc. |
| ProstaScint | Capromab pendetide | 10/28/96 | Mammalian | Murine $IgG_1$ against prostate specific membrane antigen (PSMA) | Glycosylation, glycation, deamidation, oxidation, C-terminal Lys clipping, etc. |
| Pulmozyme | Dornase alfa | 12/30/93 | Mammalian (CHO cells) | Human DNAse | Glycosylation, glycation, deamidation, oxidation, etc. |
| Raxibacumab | Raxibacumab | 12/14/12 | Mammalian (murine cells) | Human $IgG_1\lambda$ antibody against PA component of *Bacillus anthracis* toxin | Glycosylation, glycation, deamidation, oxidation, C-terminal clipping, etc. |
| Rebif | Interferon β-1a | 03/07/02 | Mammalian | Avonex biosimilar | Glycosylation, glycation, deamidation, oxidation, etc. |
| Regranex | Becaplermin | 12/16/97 | Yeast | Recombinant human platelet-derived growth factor | Glycosylation, glycation, deamidation, oxidation, etc. |
| Remicade | Infliximab | 08/24/98 | Mammalian (mouse myeloma, SP2/0) | Human mouse chimeric $IgG_1$ antibody against TNF-α | Glycosylation, glycation, deamidation, oxidation, C-terminal clipping, etc. |
| Renflexis | Infliximab-abda | 04/21/17 | Mammalian | Remicade biosimilar | Glycosylation, glycation, deamidation, oxidation, C-terminal clipping, etc. |
| **ReoPro** | **Abciximab** | 12/22/94 | Mammalian | Chimeric Fab | Deamidation, oxidation, glycation, etc. |
| Repatha | Evolocumab | 08/27/15 | Mammalian (CHO cells) | Human $IgG_2$ against human proprotein convertase subtilisin-kexin 9 (PCSK9) | Glycosylation, glycation, deamidation, oxidation, C-terminal clipping, etc. |

*(Continued)*

**Table A.1** (Continued)

| Proprietary name | Proper name | Approval date (M/D/Y) by US FDA | Expression system | Molecular nature | Possible common CTMs and/or PTMs |
|---|---|---|---|---|---|
| Retacrit | Epoetin alfa-epbx | 05/15/18 | Mammalian (CHO cells) | Biosimilar to Epogen | Glycosylation, glycation, deamidation, oxidation, C-terminal clipping, etc. |
| Retavase | Reteplase | 10/30/96 | Bacteria (*E. coli*) | Modified form of human tissue type plasminogen activator | Glycation, deamidation, oxidation, etc. |
| Rituxan | Rituximab | 11/26/97 | Mammalian (CHO cells) | Chimeric $IgG_1 \kappa$ antibody against CD20 antigen | Glycosylation, glycation, deamidation, oxidation, C-terminal clipping, etc. |
| Rituxan Hycela | Rituximab and hyaluronidase human | 06/22/17 | Mammalian (CHO cells) | Combination of Rituxan and hyaluronidase | Glycosylation, glycation, deamidation, oxidation, C-terminal clipping, etc. |
| Santyl | Collagenase | 06/04/65 | Bacteria (*C. histolyticum*) | Collagenase to digest collagen in necrotic tissues | Glycation, deamidation, oxidation, etc. |
| Siliq | Brodalumab | 02/15/17 | Mammalian (CHO cells) | Human $IgG_2$ antibody against IL-17 RA | Glycosylation, glycation, deamidation, oxidation, C-terminal Lys clipping, etc. |
| Simponi | Golimumab | 04/24/09 | Mammalian (mouse myeloma, SP2/0) | Human $IgG_1$ against TNF-α | Glycosylation, glycation, deamidation, oxidation, C-terminal clipping, etc. |
| Simponi Aria | Golimumab | 07/18/13 | Mammalian (mouse myeloma, SP2/0) | Human $IgG_1$ against TNF-α | Glycosylation, glycation, deamidation, oxidation, C-terminal clipping, etc. |
| Simulect | Basiliximab | 05/12/98 | Mammalian | Chimeric $IgG_1$ | Glycosylation, glycation, deamidation, oxidation, C-terminal Lys clipping, etc. |

| Soliris | Eculizumab | 03/16/07 | Mammalian (mouse myeloma) | Humanized $IgG_{2/4}$ antibody against human complement | Glycosylation, glycation, deamidation, oxidation, C-terminal clipping, etc. |
|---|---|---|---|---|---|
| Stelara | Ustekinumab | 09/25/09 | Mammalian (mouse myeloma cells) | Human $IgG_1\kappa$ antibody against p40 subunit of IL-12 and IL-23 cytokines | Glycosylation, glycation, deamidation, oxidation, C-terminal clipping, etc. |
| Stelara | Ustekinumab | 09/23/16 | Mammalian (mouse myeloma cells) | Human $IgG_1\kappa$ antibody against p40 subunit of IL-12 and IL-23 cytokines | Glycosylation, glycation, deamidation, oxidation, C-terminal clipping, etc. |
| Strensiq | Asfotase alfa | 10/23/15 | Mammalian | Human $IgG_1$ Fc fusion protein | Glycosylation, glycation, deamidation, oxidation, etc. |
| Sylvant | Siltuximab | 04/23/14 | Mammalian (CHO cells) | Chimeric monoclonal antibody against IL-6 antigen | Glycosylation, glycation, deamidation, oxidation, C-terminal clipping, etc. |
| Synagis | Palivizumab | 06/19/98 | Mammalian | Humanized $IgG_1\kappa$ against F protein of respiratory syncytial virus (RSV) | Glycosylation, glycation, deamidation, oxidation, C-terminal clipping, etc. |
| Taltz | Ixekizumab | 03/22/16 | Mammalian | Humanized $IgG_4$ antibody against IL-17A | Glycosylation, glycation, deamidation, oxidation, C-terminal clipping, etc. |
| Tanzeum | Albiglutide | 04/15/14 | Modified yeast | GLP-1 peptide conjugated to albumin | Deamidation, glycation, oxidation, etc. |
| Tecentriq | Atezolizumab | 05/18/16 | Mammalian | Nonglycosylated humanized $IgG_1$ | Deamidation, glycation, oxidation, etc. |
| TNKase | Tenecteplase | 06/02/00 | Mammalian (CHO cells) | Protein engineered version of human t-PA | Glycosylation, glycation, deamidation, oxidation, etc. |
| Tremfya | Guselkumab | 07/13/17 | Mammalian | Human $IgG_1\lambda$ against human IL-23 | Glycosylation, glycation, deamidation, oxidation, C-terminal clipping, etc. |

(Continued)

**Table A.1** (Continued)

| Proprietary name | Proper name | Approval date (M/D/Y) by US FDA | Expression system | Molecular nature | Possible common CTMs and/or PTMs |
|---|---|---|---|---|---|
| Trogarzo | Ibalizumab-uiyk | 03/06/18 | Mammalian (mouse myeloma, NS0) | Humanized $IgG_4$ against CD4 domain-2 | Glycosylation, glycation, deamidation, oxidation, C-terminal clipping, etc. |
| Trulicity | Dulaglutide | 09/18/14 | Mammalian | GLP $IgG_4$ Fc fusion protein | Glycosylation, glycation, deamidation, oxidation, C-terminal clipping, etc. |
| Tysabri | Natalizumab | 11/23/04 | Mammalian (mouse myeloma) | Humanized $IgG_4$ against α-4-integrin | Glycosylation, glycation, deamidation, oxidation, C-terminal clipping, etc. |
| Unituxin | Dinutuximab | 03/10/15 | Mammalian (murine myeloma SP2/0 cells) | Chimeric $IgG_1$ antibody against disialoganglioside (GD2) | Glycosylation, glycation, deamidation, oxidation, C-terminal clipping, etc. |
| Vectibix | Panitumumab | 09/27/06 | Mammalian (CHO cells) | Human $IgG_2κ$ antibody against human EGFR | Glycosylation, glycation, deamidation, oxidation, C-terminal clipping, etc. |
| Vimizim | Elosulfase alfa | 02/14/14 | Mammalian (CHO cells) | Human *N*-acetylgalactosamine-6-sulfatase | Glycosylation, glycation, deamidation, oxidation, etc. |
| Voraxaze | Glucarpidase | 01/17/12 | Bacteria (*E. coli*) | Carboxypeptidase | Glycation, deamidation, oxidation, etc. |
| Xeomin | IncobotulinumtoxinA | 07/30/10 | Bacteria (*C. botulinum*) | Botulinum toxin A inhibits the release of neurotransmitter acetylcholine | Glycation, deamidation, oxidation, etc. |

| | | | | | |
|---|---|---|---|---|---|
| Xiaflex | Collagenase clostridium histolyticum | 02/02/10 | Bacteria (*C. histolyticum*) | Mixture of collagenases to treat Peyronie's disease | Glycation, deamidation, oxidation, etc. |
| Xolair | Omalizumab | 06/20/03 | Mammalian (CHO cells) | Humanized $IgG_1 \kappa$ antibody against human IgE antibodies | Glycosylation, glycation, deamidation, oxidation, C-terminal clipping, etc. |
| Yervoy | Ipilimumab | 03/25/11 | Mammalian (CHO cells) | Human $IgG_1 \kappa$ antibody against CTLA4 | Glycosylation, glycation, deamidation, oxidation, C-terminal clipping, etc. |
| Zaltrap | Ziv-aflibercept | 08/03/12 | Mammalian (CHO cells) | Human $IgG_1$ Fc fusion protein containing the extracellular domains of human VEGF receptors 1 and 2 fused human $IgG_1$ Fc | Glycosylation, glycation, deamidation, oxidation, C-terminal clipping, etc. |
| Zarxio | Filgrastim-sndz | 03/06/15 | Bacteria (*E. coli*) | Biosimilar to Neupogen | Glycation, deamidation, oxidation, etc. |
| Zenapax | Daclizumab | 12/10/97 | Mammalian | Humanized IgG with $\gamma$ heavy chains and $\kappa$ light chains | Glycosylation, glycation, deamidation, oxidation, C-terminal clipping, etc. |
| Zevalin | Ibritumomab tiuxetan | 02/19/02 | Mammalian | Immunoconjugate of murine $IgG_1$ against CD20 and a thiourea linker chelator of Yttrium-90 | Glycosylation, glycation, deamidation, oxidation, C-terminal clipping, etc. |
| Zinbryta | Daclizumab | 05/27/16 | Mammalian | Humanized IgG with $\gamma$ heavy chains and $\kappa$ light chains | Glycosylation, glycation, deamidation, oxidation, C-terminal clipping, etc. |
| Zinplava | Bezlotoxumab | 10/21/16 | Mammalian | Human $IgG_1$ against C. difficile toxin B | Glycosylation, glycation, deamidation, oxidation, C-terminal Lys clipping, etc. |

# Index



*Co- and Post-Translational Modifications of Therapeutic Antibodies and Proteins*, First Edition. T. Shantha Raju.